Springer Series on Bio- and Neurosystems

Volume 11

Series Editor

Nikola Kasabov, Knowledge Engineering and Discovery Research Institute, Auckland University of Technology, Penrose, New Zealand

The Springer Series on Bio- and Neurosystems publishes fundamental principles and state-of-the-art research at the intersection of biology, neuroscience, information processing and the engineering sciences. The series covers general informatics methods and techniques, together with their use to answer biological or medical questions. Of interest are both basics and new developments on traditional methods such as machine learning, artificial neural networks, statistical methods, nonlinear dynamics, information processing methods, and image and signal processing. New findings in biology and neuroscience obtained through informatics and engineering methods, topics in systems biology, medicine, neuroscience and ecology, as well as engineering applications such as robotic rehabilitation, health information technologies, and many more, are also examined. The main target group includes informaticians and engineers interested in biology, neuroscience and medicine, as well as biologists and neuroscientists using computational and engineering tools. Volumes published in the series include monographs, edited volumes, and selected conference proceedings. Books purposely devoted to supporting education at the graduate and post-graduate levels in bio- and neuroinformatics, computational biology and neuroscience, systems biology, systems neuroscience and other related areas are of particular interest.

The books of the series are submitted for indexing to Web of Science.

More information about this series at http://www.springer.com/series/15821

Nergis Tomen · J. Michael Herrmann ·
Udo Ernst
Editors

The Functional Role of Critical Dynamics in Neural Systems

Springer

Editors
Nergis Tomen
Institute for Theoretical Physics
University of Bremen
Bremen, Germany

Udo Ernst
Institute for Theoretical Physics
University of Bremen
Bremen, Germany

J. Michael Herrmann
Institute of Perception
Action and Behaviour
University of Edinburgh
Edinburgh, UK

ISSN 2520-8535 ISSN 2520-8543 (electronic)
Springer Series on Bio- and Neurosystems
ISBN 978-3-030-20964-3 ISBN 978-3-030-20965-0 (eBook)
https://doi.org/10.1007/978-3-030-20965-0

This Springer imprint is published by the registered company Springer Nature Switzerland AG
The registered company address is: Gewerbestrasse 11, 6330 Cham, Switzerland

Foreword

Brain criticality emerged at the turn of the century as a promising theoretical framework to address the plethora of scale-invariant phenomena observed in neuroscience. From neuronal avalanches measured in vitro across millimeters of brain slices to long-range time correlations measured non-invasively across the entire human brain, theoreticians and experimentalists alike have been investigating the mechanisms which allow these phenomena to emerge and the impacts they have on brain function. At the interdisciplinary frontier where the physics of complex systems meets the daunting minutiae governing brain dynamics, the literature on brain criticality has seen an effervescent growth in the last decade. Despite the enthusiasm in the field and the richness of the results that have since emerged, coherent collections with the current state of the art have been scarce.

This volume helps to fill that gap, focussing on novel approaches for establishing concrete links between brain function and criticality. Examples include modelling/theoretical work with biophysically plausible schemes for harnessing critical states in information processing; experimental studies linking neural signatures of criticality to key aspects of brain function, such as learning, memory, and perception; and expanding the focus of 'classic' research on criticality to complementary concepts.

The authors came together in the wonderful conference "Dynamical Network States, Criticality and Cortical Function", which was organized by Udo Ernst and Nergis Tomen at Delmenhorst in March of 2017. I am sure it will be very helpful to researchers in the field as well as students and newcomers.

Recife, Brazil
March 2019

Mauro Copelli
Federal University of Pernambuco

Contents

Introduction

Background

Self-organised criticality has been described as the way nature works (Bak 1996), and even systems that emerge due to human activity, such as economies or traffic jams, show similar behaviour. A general definition of this phenomenon is still elusive, but a common property of critical systems is that they are the result of a complex interaction of many active components which are not tightly controlled by a central agency.

Criticality has been studied for a long time in the context of phase transitions in physical systems where it relates to sudden qualitative changes as a function of a parameter. A slight decrease in the temperature turns liquid water into ice if the temperature passes a critical point where ice and liquid water are present at the same time. Near this point, the interaction of the two phases can produce complex structures. The behaviour of a system near a phase transition is rather special and may display interesting properties which decay as one moves away from the critical point. This special behaviour might be characterised by the presence of power laws in the distributions of certain quantitative measures of the system. For example, the power-law distribution of ranges of interaction at the transition point represents the transition from short-range to extended correlations.

The fact that a similar characterisation applies to a variety of different phenomena as a typical, rather than an exceptional behaviour, does appear surprising (Frigg 2003; Mora and Bialek 2011). While, in biology, it can be attributed to the evolutionary drive towards the functional benefits that are implied by criticality (Shew and Plenz 2013), in systems in geo-, astro-, plasma or quantum physics, this is not easily explained by putative complex control mechanisms. Theoretical approaches to explain the typicality of criticality include the maximal entropy production principle (Dewar 2003), which is a generalisation of the maximal entropy principle from equilibrium statistical mechanics, or the Tweedie exponential dispersion model (Kendal 2015), which also relates to random matrix theory for ensembles that obey an invariance (Kendal and Jørgensen 2011). Alternative explanations are also mentioned in Watkins et al. (2016).

The idea of self-organisation first occurred in Vico's New Science (Vico 1984) and was popularised in the context of living systems by Ashby (1991), who also pointed to the logical problem inherent in the notion: the distinction between the dynamics of the states of the system and the adaptivity of mechanism that is necessary to reach criticality is not always unambiguous, which contributes to the difficulty of a general definition of the concept. In other words, the concept of self-organised criticality is but a name for the phenomenon of critical dynamics without fine tuning, unless specific mechanism is revealed that leads to this effect.

Specifically, if a control parameter, such as temperature or average connectivity, is identified, then a self-organizing critical system should approach the critical value of this parameter without being given explicit directions. This means the critical parameter value represents the attractor of the system. The lack of necessity for fine tuning of the parameters is often referred to in order to distinguish SOC from phase transitions at a specific critical point, but it is also one of the controversial aspects. A number of mechanisms and explanations have been suggested as to how such systems emerge, following the seminal paper of Bak et al. (1987). However, the fact that other specifications of the system, which are not obvious from the system dynamics (e.g. a "hidden" control parameter), need to be precise for SOC to exist remains one of the main criticisms of the concept.

Facets of Criticality

Properties of Criticality

Although criticality is often presented as a common feature of a variety of different systems, it appears as a multifaceted concept which is characterised by a set of properties (see Dickman et al. 2000; Muñoz 2018), such as the presence of power laws in event distributions or spatial or temporal correlation functions. This implies the absence of a typical length or time scale, i.e. the system is scale-invariant, at least as scales larger than the individual elements and smaller than the total system size are concerned. Often, scale invariance is accompanied by the occurrence of fractals, and indeed, critical dynamics represent an important mechanism for the generation of fractal structures.

In addition, critical systems are on the brink of losing stability, which means that the average time to return to the attractor diverges at the critical point, a phenomenon known as critical slowing down. This marginal stability is also the reason for the sensitive response to external perturbations. Critical systems show very high susceptibility. This can be produced by the amplification of small events which implies the possibility of the formation of avalanches. Small inputs can thus be detected by means of strong responses of the system, which is seen as one of the reasons why neural systems are critical. Due to the scale invariance, larger inputs also lead to corresponding effects and avoid saturation, which leads to a high dynamical range, providing another advantage for a perceptual system.

Characterisation of Criticality

There exist a number of ways to characterise criticality in a more general sense. They are different in the way that they set the precedence of the properties or mechanisms which imply criticality.

So far we have asked, why systems have power-law event distributions. This approach is usually referred to as dynamical criticality and relates to the idea of the edge of instability, namely the marginal case of a stable system that will show large fluctuations under noise. The condition for a dynamical system to be marginally stable is the presence of zero eigenvalues (or unit eigenvalues in the case of a map). Whether or not a high-dimensional marginally stable system displays power laws depends on a number of factors, and thus, zero eigenvalues are only a necessary condition. A linear system, in particular, will typically not show power-law behaviour.

In nonlinear systems, there are many ways to couple dimensions, some of which have received special attention, including the dynamics on small-world or scale-free graphs. Such networks were numerically shown to produce a behaviour that is more similar to criticality in finite size compared to random or fully connected networks, such that it is tempting to talk about a form of structural criticality. Although this form of criticality is only possible if the system is also dynamically critical, the notion is not unreasonable as many classes of complex networks have a larger fraction of near-zero eigenvalues than random networks.

It is often possible to relate the supposed underlying mechanisms, for example, in a neural network model that exhibits critical behaviour, to a mathematical paradigm such as the Tweedie exponential dispersion model (Kendal and Jørgensen 2011), critical branching processes (Otter 1949; Kimmel and Axelrod 2015; Alava and Lauritsen 2009) or directed percolation (Stauffer and Aharony 1991; Vázquez and Costa 1999). The paradigms, e.g. branching processes and percolation, are related and lead to the same predictions in the infinite case (Levina 2008), and all are compatible with the idea of dynamical criticality.

A different characterisation is a statistical criticality. As dynamical criticality manifests itself in the power-law event distribution, the aim in statistical approaches is a direct match of the event distribution to statistical models, without considering the time evolution or dynamical dependencies between the events. A more in-depth comparison of statistical and dynamical criticality can be found in Chapter "Statistical Models of Neural Activity, Criticality, and Zipf's Law".

Conditions for Criticality

Typically, the external perturbations which drive a critical system are slow. In addition, the slow driving force which eventually causes an avalanche may follow a different physical principle than the behaviour of the avalanche itself. For example, an earthquake is a series of sudden rock fractures and stick–slip friction of layers of the earth's crust, while the driving force originates in the deeper regions of the planet and manifests itself in the slow continental drift. In sandpiles, however, the driving input to the system and the toppling of the sand grains during the avalanche are the same physical phenomenon. What is common to both systems is that, in either case, the driving force is assumed not to interfere with the fast processes of the avalanche. This is true if the driving force is very slow, so slow that even in a sandpile of substantial size, a previous large scale avalanche will have concluded before the next grain of sand from the driving force arrives. This separation of two-time scales is formally necessary for self-organised criticality. If the external input would "intelligently" wait until the avalanche dynamic has ceased, then we would not be able to speak of self-organisation.

The larger the system, the smaller the driving force must be in order to exclude any overlap with the avalanche dynamics. In theory, when the limit case of an infinite system is considered, the strength of the driving force—be it as a temporal average or as the strength of a continuous signal—consequentially needs to vanish. This is a reasonable assumption, for example, for earthquake dynamics, where the accumulation of stress can build up over centuries.

Moreover, the accumulation of the driving inputs and how it affects avalanche dynamics are related to a conservation law. If a sand grain is added to a sandpile at a certain location, it will stay there unless it is affected by an avalanche. Conservation in this case is simply the local constancy of the number of grains.

In terms of conservation in the whole system, one possibility would be to construct finite size models where sand grains are lost through toppling out of the boundaries of the system, but are replaced by the driving input. Theoretically, one may also consider an infinite sandpile where avalanches depart towards the infinitely far bottom, and the fraction of sand grains which arrive within a finite time is negligible compared to the infinite volume of the theoretical pile. The opposite case of studying the behaviour of a finite pile in infinite time may work as well. For example, in Eurich et al. (2002) the asymptotic form of the event distribution was derived for a finite size model, which can then be evaluated in the limit of infinite system size which, however, requires the input to become arbitrarily small as mentioned above.

Conservation on the scale of the full system is thus a questionable concept: in a finite system, the driving force and the avalanche-related fluctuations will lead to departures from conservativity, whereas in the infinite case small departures may go unnoticed. A more useful concept is local conservativeness which either refers to the redistribution of activity during the avalanche, where sand grains leaving a site must arrive somewhere else or to the time between avalanches or sites which are unaffected by the current avalanche, where the sand grains simply stay in place.

Criticality and the Brain

When addressing criticality in the brain, sooner or later one has to link the concept to function. Over time, brains have evolved to generate meaningful behaviour for ensuring survival in complex dynamic environments. If particular neural populations, or the whole brain, exhibit signatures of criticality, then such states can be inferred to likely correlate with certain advantages, either for processing sensory information, for adapting computation to task requirements or for generating appropriate behavioural responses. However, it turns out that before being able to assess the link between criticality and function one has to cope with a range of other challenges since the brain is neither a spin glass nor an idealised sandpile.

First of all, it is obvious that the brain is not a homogeneous system constructed from identical units with simple dynamics. Instead, the brain has a modular structure, comprising a multitude of specialised areas, linked by connections with a rich and diverse topological structure, and consisting of billions of highly nonlinear units with miscellaneous and complex microscopic dynamics. Moreover, neural systems are composed of highly adaptive components which themselves contain processes for structural self-organisation. These include the processes governing the geometry of the dendritic tree, the arrangement of ion channels, the maintenance of organelles and structures within the cell or the interactions with non-neural neighbouring cells. All of these processes interact with the signalling with other neurons, such that it is a bold assumption to consider those elements as representing a static function.

To the scientific community, it thus came somewhat as a surprise that neural systems are nevertheless capable of displaying critical dynamics (Beggs and Plenz 2003), which was previously only studied in computer simulations and mathematical analyses of rather "hypothetical" neural networks (e.g. Eurich et al. 2002). One can speculate that this initial empirical observation of criticality was facilitated by the fact that the observations took place in vitro and the corresponding networks were organotypic cultures and acute slices of the rat cortex. This implies relatively homogeneous and isolated networks, receiving neither sensory input nor recurrent signals from other brain areas, in which activity emerges spontaneously and on a time scale much slower than the resulting avalanches.

Nevertheless, a plethora of follow-up studies provided further and ample evidence for criticality in brain dynamics, focusing on a range of different signatures in a variety of different species observed in both active and passive brain states. The appealing fact about criticality in the brain, as opposed to, for example, geological or astrophysical systems, is that experimental conditions are controllable to a good extent, can be carried out in situ and have enriched the discussion of the applicability of the concept in brain science from an early stage. It is, however, still difficult to describe the system in questions beyond drastically simplified theoretical models, which consequently do not reach strong explanatory power.

Signatures and Conditions of Neural Criticality

Studying the macroscopic phenomenon of criticality is largely independent of the specific underlying model. Complex neural models are rarely used (Wojcik et al. 2007) in studies of neural avalanches, so that any findings may still be artefacts of the idealisations. While other features of biological neural networks may support criticality, the more interesting theoretical results on neuronal avalanches, which have been obtained for simple model neurons, may be lost with inclusion of further biocomplexity. Thus being confronted with the full complexity of the brain, one often has to adapt, or otherwise newly develop, suitable methods for identifying signatures of criticality before addressing its benefits for computation. The contributions in this book provide a range of interesting ideas and solutions.

A particularly difficult problem is the lack of a clear separation of time scales: if a sensory stimulus drives a neural system, input signals arrive with the same rate at which collective events are taking place, making it effectively impossible to identify individual avalanches and obtain precise avalanche statistics. In this case, one had to instead focus on alternative measures and properties of criticality.

One idea is to look at the eigenvalues of the systems' slow dynamics—if the brain is in a critical state, one expects the system to be marginally stable, with some eigenvalues having a small positive real part, thus inducing a small local instability, which is kept in check by nonlinear contributions, keeping the system globally stable Chapter "Linear Stability of Spontaneously Active Local Cortical Circuits: Is There Criticality on Long Time Scales?". This concept is intrinsically linked to "critical slowing down", which leads to the emergence of correlations on long time scales. Interestingly, Chapter "From Neurons to Networks: Critical Slowing Down Governs Information Processing Across Vigilance States" demonstrates that this phenomenon appears on a systems level (measured with EEG recordings), as well as on a single neuron level. Under certain conditions long-range correlations will be accompanied by nested oscillations, such as high-amplitude low-frequency β -oscillations and low-amplitude high-frequency γ -oscillations, Chapter "Avalanche Dynamics and Correlations in Neural Systems" or possibly involving more frequency bands as observed in somatosensory neocortex slices Chapter "Linear Stability of Spontaneously Active Local Cortical Circuits: Is There Criticality on Long Time Scales?". A novel concept for linking criticality to oscillatory brain activity emerges from the idea of decomposing functional interactions into a so-called connectome harmonics, i.e. oscillatory eigenmodes which are engaged and disengaged in dependence on current needs, and exhibit scale-free power spectra over spatial scales Chapter "Playing at the Edge of Criticality: Expanded Whole-Brain Repertoire of Connectome-Harmonics" when the brain dynamics is critical.

An aspect which makes brain research nontrivial, also in other contexts, is observability: even with advanced methods, experimentalists can only observe part of a region of interest in the brain. Methods such as fMRI are also subject to those limitations since they trade measurements from the whole system for a much coarser spatial and temporal resolution. This limitation is known as the "subsampling" problem. Luckily, a subsampling-invariant estimator has recently been developed for quantifying the distance of a system from criticality Chapter "Assessing Criticality in Experiments".

A further challenge is to reliably determine under which conditions neural systems become critical. Typically, the emergence of critical dynamics in a neural system involves more than just one system parameter, for example, Chapter "Complexity of Network Connectivity Promotes Self-organized Criticality in Cortical Ensembles" a combination of structural aspects (scale-free and small-world features) and dynamical aspects (balance of excitation and inhibition). In Chapter "The Role of Criticality in Flexible Visual Information Processing", a similar interplay between a particular ratio of excitation and inhibition in combination with a connection structure matching the statistics of the sensory input is required. However, if it is indeed necessary that so many different factors have to be precisely controlled in order to poise a system at a particularly useful functional state, how is this achieved by the brain? This important question has been addressed by many studies. In this book, we present examples investigating growth processes in the developing brain Chapter "Homeostatic Structural Plasticity Can Build Critical Networks" and studying activity-dependent plasticity in neural circuits Chapter "Fading Memory, Plasticity, and Criticality in Recurrent Networks". In the latter example networks evolve into a state where the typical signatures of criticality only show up under certain conditions, but not when the network is actively processing a stimulus.

Taken together, such examples taken from the chapters of this book demonstrate that relating criticality in neural systems to brain function is in many cases intrinsically linked to identifying conclusive signatures or necessary conditions for criticality in strongly driven, inhomogeneous systems in the first place.

Linking Criticality to Brain Function

Theoretical work has developed elegant ideas about how critical states might support function (see, e.g. Shew and Plenz 2013). However, if we were to have a closer and rather critical look at these concepts, many unsolved questions become apparent. For instance, let us consider the notion that criticality maximises the dynamic range of a neural system (Kinouchi and Copelli 2006), bringing about optimal sensitivity to sensory stimuli (Chialvo 2006): such a property could be used, for example, to represent a set of stimuli to be discriminated by inducing maximally different neural responses. But how should those stimuli be encoded to optimally use the full dynamic range? How will such networks scale as new stimulus representations are added? Can noise in the system and complex interactions between different areas lead to avalanches propagating everywhere, such that in the end, all neurons display maximum activation and discrimination between stimuli is not possible anymore? And even if an appropriate and noise-robust encoding can be found, what would be the corresponding read-out mechanism for the information contained in the emerging dynamics? This is just one example of the wide spectrum of related questions the contributions in this book are addressing.

One of the central ideas surrounding the function of criticality, or in a more general sense having a system operate at the transition point between two different dynamical regimes, is functional flexibility: small changes in the system configuration can lead to dramatic effects in the network dynamics and allow smooth and fast transitions between network states Chapter "Complexity of Network Connectivity Promotes Self-organized Criticality in Cortical Ensembles". One example is switching between anticipatory behaviour and active sensing in rodents, caused by switching synchronisation between superficial and deep cortical layers

(Kodama et al. 2018), which is a functional role consistent with observations of signatures of critical dynamics Chapter "Linear Stability of Spontaneously Active Local Cortical Circuits: Is There Criticality on Long Time Scales?".

Another idea which has been explored is that attention serves to move neural systems towards or away from a critical point, which can greatly enhance neural processing depending on the behavioural task, with only small changes to network configuration Chapter "The Role of Criticality in Flexible Visual Information Processing". Corresponding network dynamics often involve the emergence of nested oscillations, for example in the beta–gamma range Chapter "Avalanche Dynamics and Correlations in Neural Systems", which has been related, by complementary experimental work (Bosman et al. 2012), to selective information processing and signal routing in the visual system.

Memory is another prominent example where switching between two modes is pivotal, namely between faithful replay of learned patterns or sequences (recall) and a faithful representation of the stimulus (storage) Chapter "Critical Behavior and Memory Function in a Model of Spiking Neurons with a Reservoir of Spatio-Temporal Patterns". Since both functions use the same neural substrate, the capability to rapidly switch between the different modes is both theoretically interesting and of functional importance. A complementary example also linking criticality to memory postulates an optimum for a "latching" dynamics, i.e. maximizing the ability of a network to perform an unstructured stochastic hopping between attractors while still exhibiting an unambiguous overlap with the stored pattern (discriminability) Chapter "The Challenge of Taming a Latching Network Near Criticality".

Apart from flexibility and optimisation between competing functional requirements, information integration over space and time is another focal point in linking criticality to function. In the critical state, avalanches of spikes propagate quickly over large neuronal ensembles. This ability can serve to transmit information from the senses quickly and reliably towards the motor components, in order to control the body in agreement with the affordances of the environment (Bertschinger and Natschlager 2004; Kinouchi and Copelli 2006). It could also be used to link distributed information into coherent representations, which are easy to read out by neural coincidence detectors Chapter "The Role of Criticality in Flexible Visual Information Processing". A "meaningful" sensory stimulus therefore serves as a template which can activate a matching, critical subnetwork in the cortex thus engaging a linking process which generates stimulus-specific, characteristic activation patterns Chapter "Optimal Fisher Decoding of Neural Activity Near Criticality".

With respect to the temporal dimension, critical networks exhibit long temporal correlations Chapter "Linear Stability of Spontaneously Active Local Cortical Circuits: Is There Criticality on Long Time Scales?" related to the property of critical slowing down, enabling the brain to successfully integrate information over time. Chapter "From Neurons to Networks: Critical Slowing Down Governs Information Processing Across Vigilance States" demonstrates that this ability breaks down under fatigue, but recovers slowly by the corresponding networks reassuming the critical state during sleep. Naturally, optimally linking (sensory) information over space and time is equally beneficial when it comes to learning or storing spatio-temporal patterns Chapter "Fading Memory, Plasticity, and Criticality in Recurrent Networks", Chapter "Critical Behavior and Memory Function in a Model of Spiking Neurons with a Reservoir of Spatio-Temporal Patterns".

From theoretical work outside of neural systems, self-organised criticality appears as the normal or ideal behaviour of a system. Under realistic conditions or in practical use, however, the system may be found to deviate to either side and represent a compromise between the benefits of criticality and other system characteristics. Neural systems are subject to additional constraints or limitations, such as finiteness, risk-avoidance or feedback. We may therefore find both sub- and supercritical systems that can still serve as examples of self-organised criticality when studied theoretically.

In fact, it has been suggested that the brain operates in a driven, slightly subcritical state (Priesemann et al. 2014) Chapter "Assessing Criticality in Experiments". In particular in the presence of parametric fluctuations, a subcritical system will be less prone to enter the risky supercritical phase (Priesemann et al. 2014; Tomen et al. 2014). Criticality in metaheuristic optimisation is also optimal only for temporally and spatially unbounded systems. In real problems, a slightly subcritical parameter setting produces better results (Erskine et al. 2017). If the optimal function is only engaged on demand, for example, when attention is directed towards a sensory stimulus, it may be reasonable for the system to stand by in a subcritical state until a control input brings it closer to criticality (Tomen et al. 2014) Chapter "The Role of Criticality in Flexible Visual Information Processing".

In addition to temporal modulations, the distance of a network state from the critical point may also be spatially non-uniform. Functionally speaking, the brain is continuously bombarded with a large number of sensory signals, of which only a subset might contain behaviourally relevant information. If criticality emerges as a consequence of part of a sensory input matching an internal representation Chapter "The Role of Criticality in Flexible Visual Information Processing", only the dynamics in parts of the network engaged in processing the corresponding, task-relevant signals might become critical, while the remaining, larger part of the network processing irrelevant information remains in a subcritical state.

In other contexts, deviations towards supercriticality might be useful: a neural system for motor control, for example, will need to enter the supercritical regime in order to provide a substantial driving of the actuators. In this case, critical dynamics represent the readiness to execute a movement which can then be actualised by an excursion into the supercritical regime. Negative feedback from the actuator can be used to terminate the excursion and to reestablish the critical regime. More generally, if the system has the function to produce a particular structure, then the wealth of structural forms that is present at the critical point may serve as a reservoir Chapter "The Challenge of Taming a Latching Network Near Criticality", but the eventual formation of the structure requires an increased permanence of the structure which is present in the supercritical state. For example, in a critical neural network that serves the function of memory (Uhlig et al. 2013), the cue for a memory pattern raises the dynamics into the supercritical regime, where an activation of a firing pattern is possible that exceeds the likelihood of the pattern implied by the activity distribution of the uncued network.

Taken together, the examples collected in this book demonstrate both, the existing challenges but also a range of innovative and fresh ideas on how to cope with them when relating the concept of criticality to brain function. For our readers, it offers the unique chance to get a comprehensive overview of methods, concepts and examples of critical phenomena in neural systems and their relation to function, computation, and behaviour. We hope readers will enjoy the contents of this book as much as we enjoyed the variety of contributions to our workshop in March 2017!

Thanks: We would like to express our special thanks to our office manager Agnes Janßen, who not only organised a smoothly running workshop together with the staff at the Hanse-Wissenschaftskolleg in Delmenhorst (Germany), but also helped us greatly in putting together this book. Furthermore, we would like to extend our heartfelt gratitude to everyone who provided us with well thought out reviews of the chapter manuscripts, who were recruited both from amongst the participants of the workshop and contributors to this book and from members of the scientific community interested in criticality and network dynamics in the brain.

Acknowledgements: The editors Nergis Tomen and Udo Ernst were supported by the BMBF (Bernstein Award Udo Ernst, Grant no. 01GQ1106).

List of Reviewers in Alphabetical Order

Lucilla de Arcangelis
Kimberlyn Bailey
Federica Capparelli
Gustavo Deco
Udo Ernst
Roberto F. Galán
Thilo Gross
J. Michael Herrmann
Klaus Linkenkaer-Hansen
Paolo Massobrio
Christian Meisel
Raoul-Martin Memmesheimer
Arjen van Ooyen
Klaus Pawelzik
Dietmar Plenz
Viola Priesemann
David Rotermund
Silvia Scarpetta
Jean-Philippe Thivierge
Alessandro Treves
Jochen Triesch
Jens Wilting
Florentin Wörgötter

Udo Ernst
Nergis Tomen
J. Michael Herrmann

References

Alava, M.J., Lauritsen, K.B.: Branching processes. In: Meyers, R.A. (ed.) *Encyclopedia of Complexity and Systems Science*, pp. 644–657. Springer New York, New York (2009)

Ashby, W.R.: Principles of the self-organizing system. In: Klir, G.J. (ed.) *Facets of Systems Science, International Federation for Systems Research International Series on Systems Science and Engineering*, pp. 521–536. Springer US, Boston, MA (1991)

Bak, P.: How Nature Works: The Science of Self-organized Criticality, Copernicus, Copernicus (1996)

Bak, P., Tang, C., Wiesenfeld, K.: Self-organized criticality: An explanation of the 1/f noise. Phys. Rev. Lett. **59**(4), 381–384 (1987)

Beggs, J.M., Plenz, D.: Neuronal avalanches in neocortical circuits. J. Neurosci.: Official J. Soc. Neurosci. **23**(35),11167–11177 (2003)

Bertschinger, N., Natschläger, T.: Real-time computation at the edge of chaos in recurrent neural networks. Neural Comput. **16**(7), 1413–1436 (2004)

Bosman, C.A., Schoffelen, J.-M., Brunet, N., Oostenveld, R., Bastos, A.M., Womelsdorf, T., Rubehn, B., Stieglitz, T., De Weerd, P., Fries, P.: Attentional stimulus selection through selective synchronization between monkey visual areas. Neuron **75**(5), 875–888 (2012)

Chialvo, D.R.: Psychophysics: Are our senses critical? Nat. Phys. **2**(5), 301–302 (2006)

Dewar, R.: Information theory explanation of the fluctuation theorem, maximum entropy production and self-organized criticality in non-equilibrium stationary states. J. Phys. A: Math. Gen. **36**(3), 631–641 (2003)

Dickman, R., Muñoz, M.A., Vespignani, A., Zapperi, S.: Paths to self-organized criticality. Braz. J. Phy. **30**(1), 27–41 (2000)

Erskine, A., Joyce, T., Herrmann, J.M.: Stochastic stability of particle swarm optimisation. Swarm Intell. **11**, 295–315 (2017)

Eurich, C.W., Herrmann, J.M., Ernst, U.A.: Finite-size effects of avalanche dynamics. Phys. Rev. E, Stat., Nonlinear, Soft Matter Phys. **66**(6 Pt 2), 066137 (2002)

Frigg, R.: Self-organised criticality—what is is and what it isn't. Stud. Hist. Philos. Sci. Part A. **34** (3), 613–632 (2003)

Kendal, W.S.: Self-organized criticality attributed to a central limit-like convergence effect. Phys. Stat. Mech. Appl. **421**, 141–150 (2015)

Kendal, W.S., Jørgensen, B.: Tweedie convergence: A mathematical basis for Taylor's power law, 1/f noise, and multifractality. Phys. Rev. E. **84**(6), 066120 (2011)

Kimmel, M., Axelrod, D.: Interdisciplinary applied mathematics. In: Branching Processes in Biology, 2nd edn. Springer-Verlag, New York (2015)

Kinouchi, O., Copelli, M.: Optimal dynamical range of excitable networks at criticality. Nat. Phys. **2**(5), 348–351 (2006)

Kodama, N.X., Feng, T., Ullett, J.J., Chiel, H.J., Sivakumar, S.S., Galán, R.F.: Anti-correlated cortical networks arise from spontaneous neuronal dynamics at slow timescales. Sci. Rep. **8**, 66 (2018)

Levina, A.: A mathematical approach to self-organized criticality in neural networks. PhD thesis, Univ. Göttingen. (2008)

Mora, T., Bialek, W.: Are biological systems poised at criticality? J. Stat. Phy. **144**(2), 268–302 (2011)

Muñoz, M.A.: Colloquium: Criticality and dynamical scaling in living systems. Rev. Mod. Phys. **90**(3), 031001 (2018)

Otter, R.: The multiplicative process. Ann. Math. Stat. **20**(2), 206–224 (1949)

Priesemann, V., Wibral, M., Valderrama, M., Pröpper, R., Le Van Quyen, M., Geisel, T., Triesch, J., Nikolić, D., Munk, M.H.J.: Spike avalanches in vivo suggest a driven, slightly subcritical brain state. Front. Syst. Neurosci. **8** (2014)

Shew, W.L., Plenz, D.: The functional benefits of criticality in the cortex. *Neuroscientist: Rev. J. Bringing Neurobiol. Neurol. Psychiatry*. **19**(1), 88–100 (2013)
Stauffer, D., Aharony, A.: Introduction To Percolation Theory, 2nd edn. Taylor and Francis (1991)
Tomen, N., Rotermund, D., Ersnst, U.: Marginally subcritical dynamics explain enhanced stimulus discriminability under attention. Front. Syst. Neurosci. **8**, 151 (2014)
Uhlig, M., Levina, A., Geisel, T., Herrmann, M.: Critical dynamics in associative memory networks. Front. Comput. Neurosci., **7** (2013)
Vázquez, A., Costa, O.S.: Self-organized criticality and directed percolation. J. Phys. A: Math. Gen. **32**(14), 2633–2644 (1999)
Vico, G.: *The New Science of Giambattista Vico: Unabridged Translation of the Third Edition*. Cornell University Press, Ithaca, unabridged ed edition (1984)
Watkins, N.W., Pruessner, G., Chapman, S C., Crosby, N.B., Jensen, H.J.: 25 Years of self-organized criticality: concepts and controversies. Space Sci. Rev. **198**(1), 3–44 (2016)
Wojcik, G.M., Kaminski, W.A., Matejanka, P.: Self-organised criticality in a model of the rat somatosensory cortex. In: Malyshkin, V.(ed.) Parallel Computing Technologies, Lecture Notes in Computer Science, pp. 468–476. Springer Berlin, Heidelberg (2007)

Avalanche Dynamics and Correlations in Neural Systems

Fabrizio Lombardi, Hans J. Herrmann and Lucilla de Arcangelis

Abstract The existence of power law distributions is only a first requirement in the validation of the critical behavior of a system. Long-range spatio-temporal correlations are fundamental for the spontaneous neuronal activity to be the expression of a system acting close to a critical point. This chapter focuses on temporal correlations and avalanche dynamics in the spontaneous activity of cortex slice cultures and in the resting fMRI BOLD signal. Long-range correlations are investigated by means of the scaling of power spectra and of Detrended Fluctuations Analysis. The existence of $1/f$ decay in the power spectrum, as well as of power-law scaling in the root mean square fluctuation function for the appropriate balance of excitation and inhibition suggests that long-range temporal correlations are distinctive of "healthy brains". The corresponding temporal organization of neuronal avalanches can be dissected by analyzing the distribution of inter-event times between successive events. In rat cortex slice cultures this distribution exhibits a non-monotonic behavior, not usually found in other natural processes. Numerical simulations provide evidences that this behavior is a consequence of the alternation between states of high and low activity, leading to a dynamic balance between excitation and inhibition that tunes the system at criticality. In this scenario, inter-times show a peculiar relation with avalanche sizes, resulting in a hierarchical structure of avalanche sequences. Large avalanches correspond to low-frequency oscillations, and trigger cascades of smaller ones that are part of higher frequency rhythms. The self-regulated balance of excitation and inhibition observed in cultures is confirmed at larger scales, i.e. on

---

F. Lombardi
Keck Laboratory for Network Physiology, Department of Physics, Boston University, Boston, MA, USA
e-mail: fabri@bu.edu

H. J. Herrmann
Institute of Computational Physics for Engineering Materials, IfB, ETH Zürich, Zürich, Switzerland
e-mail: hans@ifb.baug.ethz.ch

L. de Arcangelis (✉)
Department of Engineering, University of Campania "Luigi Vanvitelli", INFN sez. Napoli Gr. Coll., Salerno, Aversa (CE), Italy
e-mail: lucilla.dearcangelis@unicampania.it

N. Tomen et al. (eds.), *The Functional Role of Critical Dynamics in Neural Systems*, Springer Series on Bio- and Neurosystems 11,
https://doi.org/10.1007/978-3-030-20965-0_1

fMRI data from resting brain activity, and appears to be closely related to critical features of avalanche activity, which could play an important role in learning and other functional performance of neuronal systems.

1 Introduction

Critical systems are characterized by divergent correlations in space and time. As a consequence such systems lack a characteristic length scale, and the statistics of events is governed by power laws. At the beginning of this century, a novel experimental approach has revealed that spontaneous cortical activity is organized in bursts made of neuronal avalanches [1]. An avalanche is defined as an ensemble of neurons that fire close-in-time, namely at a temporal distance smaller than a given time interval, and is usually characterized by its size and duration. The particular definition of size and duration depends on the specific signal that is used to investigate spontaneous network activity. In general, those definitions provide a measure for the population of firing neurons and the time interval covered by their almost-synchronous firings.

Neuronal avalanches have been first identified in the organotypic cultures from coronal slices of rat somatosensory cortex [1], where they are stable for many hours [2]. The size and duration of neuronal avalanches follow power law distributions with very robust exponents, which is a typical feature of a system acting in a critical state, where large fluctuations are present and the response does not have a characteristic size. The same critical behavior has been observed also in vivo on rat cortical layers during early post-natal development [3], and on the cortex of awake adult rhesus monkeys [4], using microelectrode array recordings, as well as on dissociated neurons from rat hippocampus and cortex [5, 6] or leech ganglia [5]. Recently, avalanche dynamics has been also identified in the resting state of the human brain by means of non-invasive techniques such as magneto-encephalography (MEG) [7].

All these experimental results consistently exhibit power law size distributions decaying with the exponent -1.5, and duration distributions that follow a power law with exponent -2. These exponents are consistent with the universality class of the mean field branching process [8] and are therefore independent of the dimensionality of the system. This property, verified experimentally, is unusual in critical phenomena and it has been shown numerically to depend on the combined action of plastic adaptation and refractory time [9]: Starting from a fully connected network, refractory time selects a preferential direction in the synaptic connection between two neurons, while the repeated adaptation of less used connections leads to the progressive depletion of loops in the network. As a result, the network becomes "tree-like" providing the mean field universality class.

Brain activity having features typical of systems at a critical point represents a crucial ingredient for learning. Indeed, a neuronal network model reproducing quantitatively the experimentally observed critical state of the brain, is able to learn and remember logical rules including the exclusive OR, which has posed difficulties to several previous attempts [10]. Learning occurs via plastic adaptation of synaptic strengths and exhibits universal features: The learning performance and the average time required to provide the right answer are controlled by the strength of plastic adaptation, in a way independent of the specific task assigned to the system. Interestingly, even complex rules can be learned provided that the plastic adaptation is sufficiently slow. The implemented learning dynamics is a cooperative mechanism where all neurons contribute to select the right answer. In fact, because the system acts in a critical state, the response to a given input can be highly flexible, adapting more easily to different rules. The analysis of the dependence of the performance of the system on the average connectivity confirms that learning is a truly collective process, where a high number of neurons may be involved and the system learns more efficiently if more branched paths are possible [10]. The investigation of the learning performance in neuronal networks provides also insights into the role of inhibitory synapses. Performing more tasks in parallel is a typical feature of real brains characterized by the coexistence of excitatory and inhibitory synapses, whose percentage in mammals is measured to have a typical value of 20–30%. By investigating parallel learning of more Boolean rules in neuronal networks [11], it has been evidenced that that multi-task learning results from the alternation of learning and forgetting the individual rules. Interestingly, a fraction of 30% inhibitory synapses optimizes the overall performance, since it guarantees, at the same time, the network excitability necessary to express the response and the variability required to confine the employment of resources.

Slow relaxation is a fundamental feature of systems acting at the critical point. Hence, temporal correlations are relevant over long time scales, and give rise to bursts of events. This phenomenology is distinctive of many natural phenomena, where not only power law distributions arise, but also a complex temporal organization of events is observed. Examples are earthquakes and solar flares, which share a number of statistical laws evidencing the presence of temporal correlations between successive events [12]. Several statistical tools are currently available for detecting the presence of such correlations and investigate their temporal range. In the case of brain activity, these methods can be applied either to the raw neuronal signal, defined as a continuous variable, or else to the sequence of neuronal avalanches, considered as a point-process in time. An intriguing question is indeed how to reconcile well-known properties of neuronal signals, such as oscillations with different characteristic frequencies, with neuronal avalanches lacking characteristic spatial and temporal scales [4, 13, 14].

Such techniques, applied to numerical and experimental data from systems at different scales, constitute the necessary ingredient towards a definite assessment of criticality. In parallel, numerical models are of crucial importance to identify the dynamic mechanisms controlling correlations, and to understand the role of criticality in brain functions. In this chapter, we will present a summary of different statistical methods and their application for the investigation of avalanche dynamics in neural networks, with a particular focus on temporal correlations. Experimental results will be supported by numerical studies with the objective of providing a coherent understanding of the temporal features of neuronal activity.

2 Avalanche Activity and Power Spectra

A first indication of long-range temporal correlations is provided by the observation of $1/f$ noise. Indeed, by the Wiener–Kintchine theorem [15], this implies that the time correlation function decays to zero in an infinitely long time interval. Analyses of experimental recordings of spontaneous neural activity, from EEG, to MEG and LFPs, generally show a power law decay in the power spectral density (PSD), $S(f) \propto f^{-\beta}$, with superimposed peaks at the characteristic frequencies of dominant brain rhythms. However, estimated values of the exponent β vary over a rather broad interval, and appear to depend on brain areas and patient conditions. In particular, for healthy subjects β takes values within the interval $[0.8, 1.5]$ [16–20], whereas it is much larger in epileptic patients [21].

To understand some of the basic mechanisms affecting the PSD scaling of neural signals, in this section we investigate the relationship between network inhibition, i.e. the percentage of inhibitory synapses, and the scaling exponent β in a neuronal network of integrate and fire neurons driven by slow external stimulation that simulates spontaneous activity. We consider neurons on a scale-free network and consistently show that inhibitory neurons determine the scaling behavior of the PSD: For a neuronal network of only excitatory neurons, the PSD follows a power law with an exponent $\beta \simeq 2$. By introducing inhibitory neurons, β decreases, and exhibits values in the interval $[1, 1.4]$ for a percentage of inhibitory synapses between 20 and 30%, in agreement with experimental findings [16–19].

2.1 *Neuronal Model and Avalanche Activity*

We simulate the intrinsic activity of a neuronal network by means of an integrate and fire model [22–24] inspired in self-organized criticality (SOC) [25]. In this model, each neuron i is characterized by a membrane potential v_i and fires if and only if v_i is equal or above a firing threshold v_c. To trigger activity, namely to bring a neuron at or above the firing threshold, we apply a small stimulation to a randomly chosen neuron. Then, whenever at time t the potential of neuron i fulfils the condition $v_i \geq v_c$, the neuron fires, changing the potential of all connected neurons.

The N neurons are located at the nodes of a network which can have a very general topology. In the following, we present results obtained for neurons randomly distributed in a square and connected by a scale-free network of directed synapses. More precisely, to each neuron i we assign an out-going connectivity degree, $k_{out_i} \in [2, 100]$, according to the degree distribution $P(k_{out}) \propto k_{out}^{-2}$ of the functional network measured in [26], and two neurons are then connected with a distance-dependent probability, $P(r) \propto e^{-r/r_0}$, where r is their Euclidean distance [27] and r_0 a typical edge length. To each synaptic connection we assign an initial random strength $g_{ij} \in [0.15, 0.3]$ and to each neuron an excitatory or inhibitory character by fixing a percentage p_{in} of inhibitory synapses. Outgoing synapses are excitatory if their presynaptic neuron is excitatory, inhibitory otherwise. Since synapses are directed, $g_{ij} \neq g_{ji}$, in general out-degree and in-degree of a neuron do not coincide. Therefore once the network of out-connections is established, we identify the resulting degree of in-connections, k_{in_j}, for each neuron j, namely we identify the number of synapses directed to j.

The change in the membrane potential of the postsynaptic neuron j due to the firing of neuron i is proportional to the relative synaptic strength $g_{ji}/\sum_l g_{li}$

$$v_j(t+1) = v_j(t) \pm \frac{v_i \cdot k_{out_i}}{k_{in_j}} \frac{g_{ji}}{\sum_{l=1}^{k_{out_i}} g_{li}}. \tag{1}$$

Here the normalization by the synaptic strengths ensures that during the propagation of very large avalanches the membrane potential assumes finite values, while the plus or minus sign is for excitatory or inhibitory synapses, respectively. After firing, the membrane potential of the neuron is set to $v_i = 0$ and the neuron remains in a refractory state for $t_{ref} = 1$ time step, during which it is unable to receive or transmit any signal. Each neuron in the network is an integrate and fire unit, therefore it will change its potential by summing the successive stimulations from presynaptic firing neurons according to Eq. 1. With respect to traditional neuronal models implementing partial differential equations for the temporal dependence of the membrane potential, this model is a cellular automaton where time is a discrete variable. The time unit is the time delay between the triggering of the action potential in the presynaptic neuron and the change of the membrane potential in the postsynaptic neuron, which corresponds to few ms in real neuronal system. Within this temporal window our model is unable to provide the state of the neuron, however it is numerically very efficient thus allowing simulations of very large systems.

When a neuron i fires, its out-going connections induce a potential variation in the k_{out_i} postsynaptic neurons. The strength g_{ji} of these active synapses is increased proportionally to the membrane potential variation $|\delta v_j|$ that occurred at the postsynaptic neuron j

$$g_{ji} = g_{ji} + |\delta v_j|/v_c. \tag{2}$$

Conversely, the strength of all inactive synapses during an avalanche is reduced by the average strength increase per bond

$$\Delta g = \sum_{ij} \delta g_{ji}/N_B, \tag{3}$$

where N_B is the number of bonds. We set a minimum and a maximum value for the synaptic strength g_{ij}, $g_{min} = 0.0001$ and $g_{max} = 1.0$. Whenever $g_{ij} < g_{min}$, the synapse g_{ij} is pruned, i.e. permanently removed.

These rules constitute a Hebbian-like scheme for synaptic plastic adaptation. The network memorizes the most used synaptic paths by increasing their strengths, whereas less used synapses eventually atrophy. They implement a sort of long-term synaptic plasticity, a homeostatic mechanism that dynamically balances synaptic strengthening and weakening in the network. Short-term plasticity is not taken into account in the present version of the model. Plastic adaptation is implemented during an initial sequence of external stimuli and shapes the network of synaptic strengths. Its extent can be viewed as a measure for the experience of the trained network. Synaptic strengths are initially uniformly distributed in the interval $[0.15, 0.3]$. The distribution of g_{ij} resulting from plastic adaptation is shown in the inset of Fig. 1b, and closely resembles the distribution of synaptic strengths measured experimentally [28]. Then avalanche activity is measured for fixed synaptic connections by applying a new sequence of random stimulations. Figure 1a and b show the network activity as function of time for different percentage of inhibitory neurons.

An avalanche is defined as a cascade of successively firing neurons after an external stimulation and can involve a variable number of neurons. The avalanche size is defined as the number of firing neurons s, or, alternatively, as the sum $s_{\Delta V}$ of all positive potential variations (depolarizations) δv_i^+ that occurred in the network, namely $s_{\Delta V} = \sum_i \delta v_i^+$. Avalanches are also characterized by their duration T, which is defined as the number of iterations taken by the activity propagation. The distributions of avalanche sizes, $P(s)$, and durations, $P(T)$, obtained by this model, are in agreement with experimental data [1, 7], namely they exhibit a power law behavior with exponents $\alpha \simeq -1.5$ and $\tau \simeq -2.0$, respectively, followed by an exponential cutoff (Figs. 1c and 2b). As shown in the lower inset of Fig. 1c, a scaling relation exists between s and $s_{\Delta V}$, namely $s_{\Delta V}(s) \sim s^{\gamma_s}$ with $\gamma_s \simeq 1$, implying that $P(s_{\Delta V})$ follows the same scaling behavior as the distribution of avalanche sizes measured in terms of firing neurons.

The scaling behavior of such distributions is independent of the network topology and of the percentage p_{in} of inhibitory synapses. On the other hand, p_{in} significantly affects the exponential cutoff, as shown in Fig. 1c: By increasing the percentage of inhibitory synapses the exponential cutoff gradually moves towards smaller avalanche sizes s and the scaling regime shrinks, evidencing the universal scaling behavior $P(s) \sim s^{-\alpha} f(s/p_{in}^{-\theta})$. The scaling function confirms that the size of the largest possible avalanche decreases with the percentage of inhibitory synapses, whereas the critical exponent α shows a stable value 1.5. Similar behavior is observed for the duration distributions [24].

2.2 *Power Spectra*

The analysis of the size and duration distributions evidences the crucial role played by inhibition in avalanche dynamics [24]. A very first step in the investigation of temporal correlations in neuronal signals is the analysis of the power spectrum [24]. In real neuronal network, one generally studies the electric signal resulting from the firing of a more or less localized population of neurons, such as local field potentials (LFPs) [1]. In a close analogy with LFPs, for each network configuration we construct a temporal signal, $V(t)$, as the sum of all potential variations occurring at each time

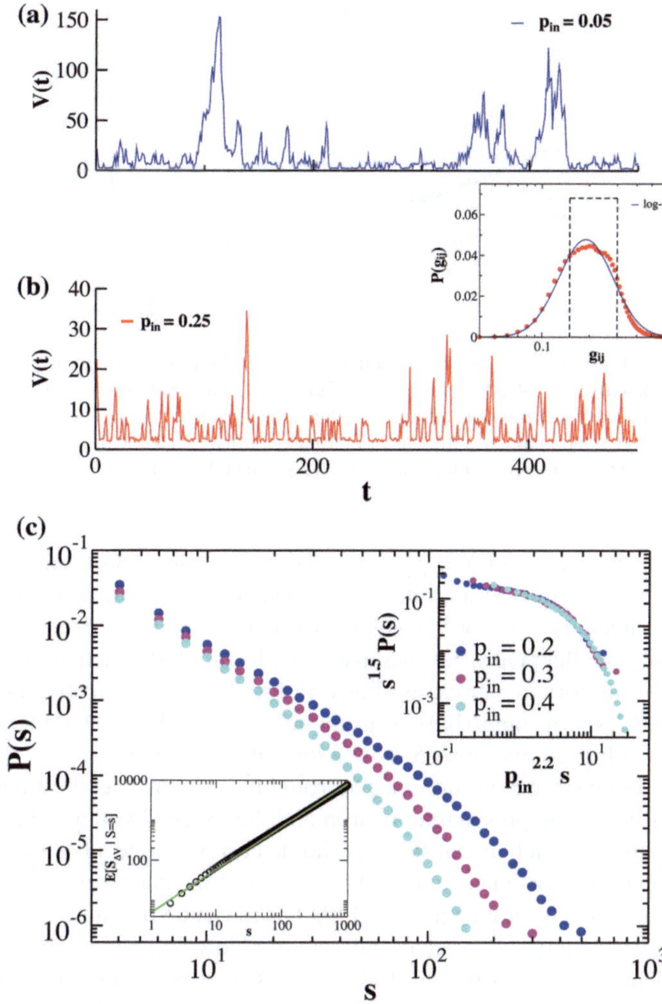

Fig. 1 Network activity and avalanche size distributions for different fraction p_{in} of inhibitory synapses. The intensity of activity $V(t)$ is the sum of all potential variations occurring in the network at each time step, namely $V(t) = \sum_i \delta v_i(t)$. Network activity for $p_{in} = 0.05$ (**a**) and $p_{in} = 0.25$ (**b**). Inset of **b**: Distribution $P(g_{ij})$ of synaptic strengths at the end of the initial period of plastic adaptation. The dashed lines indicate the initial uniform distribution; **c** Avalanche size distributions on a scale-free network for different values of p_{in}. Upper inset: Plotting $s^\alpha P(s)$ versus $p_{in}^\theta s$, with $\alpha = 1.5$ and $\theta = 2.2$, data collapse onto a universal scaling function; Lower inset: Relation between the number of firing neurons s and the sum $s_{\Delta V}$ of all positive potential variations (depolarizations) δv_i^+ occurred in the network, namely $s_{\Delta V} = \sum_i \delta v_i^+$

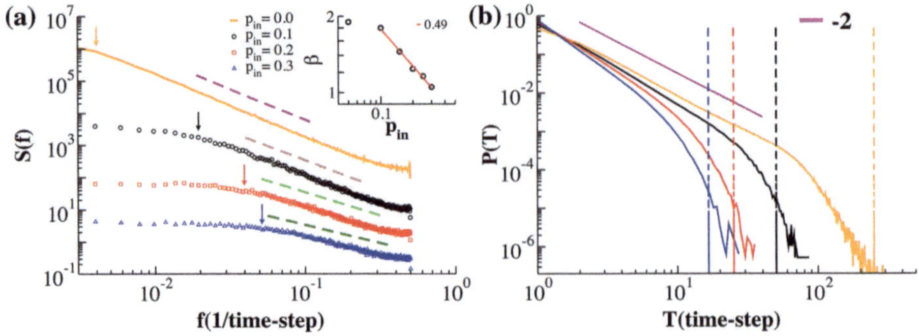

Fig. 2 Power spectral density of avalanche activity (**a**) and duration distribution $P(T)$ (**b**) for a scale-free network with $N = 64{,}000$ neurons and different p_{in}. The PSD follows a power law whose exponent β depends on p_{in} (Inset) and approaches the value $\beta = 1$ for $p_{in} = 0.3$. The cutoff at low frequencies (arrows), which indicates the transition to white noise, scales with p_{in} and corresponds to avalanche durations in the exponential cutoff of $P(T)$ (dashed lines)

step of network activity, i.e $V(t) = \sum_i \delta v_i(t)$. We then evaluate the Fourier transform of this signal, whose squared amplitude averaged over all configurations provides the PSD of the network activity. The signal $V(t)$ significantly depends on p_{in}, and exhibits not only smaller peaks, but also sparser high amplitude fluctuations for larger percentage of inhibitory synapses, as shown in Fig. 1a. The dissipative property of inhibitory synapses turns out to have a strong influence on the PSD of the avalanche activity, $S(f)$. In Fig. 2a, we show $S(f)$ for several values of p_{in}.

In the simulations the time unit can be roughly estimated as the time interval between the firing of the presynaptic neuron and the induced voltage change in the postsynaptic neuron, which in real systems should correspond to a few milliseconds. With this rough correspondence, our PSD frequency range goes from approximately 1 Hz to about 100 Hz. We notice that the PSD has the same qualitative behavior for each value of p_{in}, namely a $f^{-\beta}$ decay and a cutoff at low frequencies that, as the cutoff in the avalanche size distribution shifts towards smaller s values, moves towards higher frequencies with increasing p_{in}. The low-frequency cutoff indicates the transition to white noise, which characterizes an uncorrelated process, and corresponds to avalanche durations in the exponential cutoff of the distribution $P(T)$ of avalanche durations. The exponent β decreases for increasing values of the fraction of inhibitory synapses. In particular, for a purely excitatory network we find that β is close to 2, an exponent associated to the PSD of brown noise and larger than values found in experimental studies of spontaneous cortical activity in healthy subjects [16–19], but close to values measured for epileptic patients [21]. However, when p_{in} becomes closer to the fraction of inhibitory synapses characteristic of real neuronal systems, i.e. about 0.3, we find that β is in the interval [1, 1.4], the range of experimentally measured β values [16–19]. More specifically, for $0.1 \leq p_{in} \leq 0.3$ the exponent β decays as $p_{in}^{-\delta}$, where $\delta \simeq 0.49$, and tends to one as $p_{in} \rightarrow 0.3$ (Fig. 2a). The range of the scaling regime varies as a function of p_{in}, and, according to our rough estimate of numerical time units, goes from a few to nearly 100 Hz for $0.1 \leq p_{in} \leq 0.3$ (Fig. 2a). The low-frequency cutoff (arrows in Fig. 2a) is indeed at about 4Hz, 8Hz, and 10Hz, for $p_{in} = 0.1, 0.2$, and $p_{in} = 0.3$, respectively. As shown in [24], for a fixed value of p_{in} the low-frequency cutoff scales with the systems size, and decreases for increasing sizes.

The $1/f$ scaling regime observed in the resting state brain activity approximately spanned the interval [1, 10] Hz in [19], while more recent studies evidenced that this regime extends to higher frequencies [16, 18], and shows superimposed peaks corresponding to dominant brain rhythms [29]. In particular, Novikov et al. showed that the $1/f$ behavior extends up to 50 Hz in the MEG of resting human brain, a range that considerably overlap with our numerical results. Moreover, it is interesting to notice that experiments on dissociated networks of rat hippocampal neurons and leech ganglia [5] exhibit a very similar behavior as a function of network inhibition: The power spectra for both systems exhibit a $1/f$ power law decay, however if inhibition is hindered by introducing picrotoxin (PTX) or bicuculline in the physiological bath, the power-law decay becomes closer to brown noise, i.e. $1/f^{-2}$, as obtained in simulations of purely excitatory networks.

The scaling exponent β of the PSD is related to the critical exponent $\frac{\tau-1}{\alpha-1}$, which connects avalanche sizes and durations [25], $s(T) \sim T^{\frac{\tau-1}{\alpha-1}}$. It has been shown that, for purely excitatory models with $\alpha < 2$, $\beta = \frac{\tau-1}{\alpha-1}$ [30]. Our model very closely follows this analytical prediction: Indeed for a purely excitatory network we find $\beta \simeq 2$ and $\frac{\tau-1}{\alpha-1} \simeq 2$. On the other hand, to our knowledge no analytical derivation of the relation between β and $\frac{\tau-1}{\alpha-1}$ is available for systems with excitatory and inhibitory interactions. Deriving such a relationship for SOC-like models with inhibitory interactions is a general problem of great interest and implies the introduction of anti-ferromagnetic interactions in the model in [30]. The $1/f$ behavior of power spectra provides a first important evidence for the existence of long-range correlations in systems that are self-tuned into a critical state, and appears to be connected to the right percentage of inhibition in the system. In particular, de los Rios et al. [31] have shown that a dissipative term in the dynamics of the original sand pile model gives rise to avalanche activity whose PSD decays as $1/f$. However, in their model the avalanche sizes are not distributed according to a power law. In our neuronal model instead, inhibition gives rise to $1/f$ power spectra and avalanche size distributions still follow a power law behavior with a universal scaling function.

3 Inter-event Time Distributions

For a wide variety of natural stochastic phenomena the presence and extent of temporal correlations is analyzed in terms of the distribution of inter-event times. This is the distribution of time interval durations between successive events. The advantage of this analysis is that the distribution of inter-event times has a simple exponential decay for a pure Poisson process, whereas it exhibits a power law regime over the temporal range where correlations are relevant. In particular, it has been shown analytically that the Omori temporal decay of the earthquake rate after the occurrence of a large mainshock gives rise to a power law decay in the inter-event time distribution with an exponent related to the Omori exponent [32]. More generally for several processes, such as earthquakes, solar flares,or acoustic emissions due to rock fracture, this distribution is monotonic and behaves as a Gamma distribution, exhibiting an initial power law decay followed by an exponential cutoff. This behavior is robust and universal with respect to different event catalogs and the lower threshold imposed to the event size in the temporal sequence [12, 33]. Following this approach, we identified neuronal avalanches in rat cortex slice cultures and measured the distribution $P(\Delta t)$ of time intervals separating consecutive avalanches [34]. Interestingly, the inter-avalanche time distribution exhibits a novel, non-monotonic behavior in different culture samples, with common features: An initial power-law regime that is characterized by exponent values between -2 and -2.3, a local minimum located at 200 ms $< \Delta t_{min} <$ 1 s, and a more or less pronounced maximum at $\Delta t \simeq 1-2$ s. This complex behavior is not usually observed in other natural phenomena and suggests that temporal correlations are not only relevant in avalanche occurrence, but reflect complex underlying dynamical mechanisms.

3.1 Up-States and Down-States

Numerical simulations evidence that a traditional integrate and fire neuronal model is unable to reproduce the non-monotonic behavior of the inter-avalanche event distributions found experimentally. To achieve this goal we hypothesize that non-monotonicity arise from the transition between two substantially different network states, and introduce in the model (2.1) the concept of up and down-states.

Spontaneous activity exhibits a complex alternation between bursty periods, called up-states, where neuronal avalanches are detected, and quiet periods, named down-states. These are characterized by a general disfacilitation in the system, i.e. absence of synaptic activity, causing long-lasting returns to resting potentials in a large population of neurons [35]. Action potentials are rare during down-states, however small amplitude depolarizing potentials originating from spontaneous synaptic release, may occur. The non-linear amplification of small amplitude signals leads to the generation of larger depolarizing events, bringing the system back into the up-state. Moreover, together with this network features, experimental observations indicate that neurons are characterized by two preferred values of the membrane potential: A very negative one, below resting potential, in the down-state, and a more depolarized one in the up-state. Since the neuron up-state is just a few millivolts below the action potential threshold, during the up-state neurons respond faster to synaptic stimulations, giving rise to close-in.time avalanches. The down-state instead is characterized by long periods of quiescence during which the network recovers from biomolecular mechanisms that hamper activity, as the exhaustion of available synaptic vesicles, the increase of the nucleoside adenosine inhibiting glutamate release, or the blockade of receptor channels by the presence of external magnesium [35–38].

We implement up and down-states in the original model by monitoring the avalanche activity [34]. We measure the size of each avalanche in terms of depolarizations δv_i of all active neurons, $s_{\Delta v}$ (see Sect. 2.1). As soon as an avalanche is larger than a threshold value, $s_{\Delta v}^{min}$, the system transitions into a down-state and neurons active in the last avalanche become hyperpolarized proportionally to their previous activity, namely we reset

$$v_i = v_i - h\delta v_i \tag{4}$$

where $h > 0$. This equation implies that each neuron is hyperpolarized proportionally to its previous activity, i.e. its potential is the lower, the higher its potential variation in the previous avalanche. This rule introduces a short-range memory at the level of a single neuron and models a number of possible mechanisms for local inhibition [36–38]. Conversely, if an avalanche has a size smaller than $s_{\Delta v}^{min}$, the system remains or transitions into an up-state. In the up-state, the potentials v of all neurons firing in an avalanche are not set equal to zero resting potential, but to the depolarized value

$$v_i = v_c(1 - s_{\Delta v}/s_{\Delta v}^{min}) \tag{5}$$

The neuron membrane potential in the up-state then depends on the response of the whole network and remains close to the firing threshold. The threshold $s_{\Delta v}^{min}$ controls the extension of the up-state and therefore the level of excitability of the system, whereas h controls the level of hyperpolarization of neurons in the down-state and therefore its duration. Interestingly, the last two equations each depend on a single parameter, h and $s_{\Delta v}^{min}$, which introduce a memory effect at the level of single neuron activity and of the entire system, respectively. These two parameters can be tuned separately in numerical simulations to obtain the best fit with experimental data. The implementation of up and down-states in numerical simulations provides a temporal activity made of bursts of avalanches (up-states) followed by long periods of quiescence (down states). In Fig. 3 the duration distributions of up and down-states indeed show that down-states can last much longer that up-states, in agreement with experimental data on rat visual cortex [39] and simulations of integrate and fire neuron networks [40].

The inter-avalanche time distribution for the temporal sequence of up and down-states is shown in Fig. 4, together with experimental results for critical and super-critical samples of rat cortex slices. Super-critical samples are generally obtained by suppressing inhibition, or enhancing excitation. They are characterized by an excess of large avalanches and a smaller α power law exponent in the avalanche size distribution as compared to the normal, critical samples (Fig. 4d). The non-monotonic behavior of experimental data can be reproduced with very good agreement by tuning separately the two parameters h and $s_{\Delta v}^{min}$. The close-in time avalanches occurring in the up-state are responsible for the initial power law regime with an exponent close to -2 (Fig. 4a), as for experimental data (Fig. 4b), evidencing that temporal

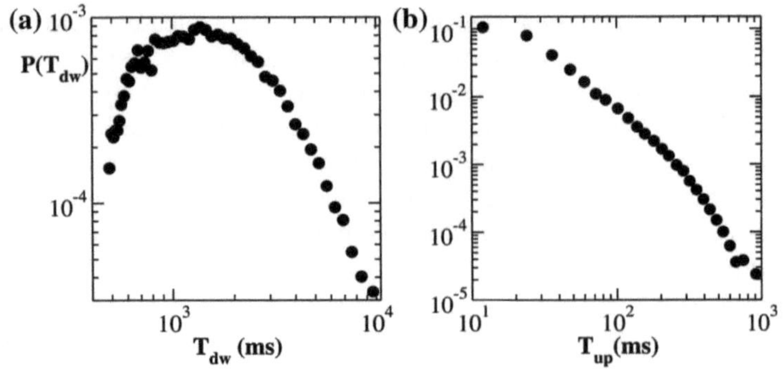

Fig. 3 Distribution of durations of down-states (**a**) and up-states (**b**) averaged over 100 configurations of networks with $N = 64{,}000$ neurons with $p_{in} = 0.1$

correlations are extended over a range at least comparable to the up-state duration. Conversely, the long-lasting down-states originate the bell-shaped behavior at large Δt centered at a value that depends on h and $s_{\Delta v}^{min}$. The superposition of these two different regimes then provides the local minimum and therefore the non monotonic behavior of the distribution.

The overall behavior of the inter-event time distribution, and thus the temporal organization of avalanches, is controlled by a single parameter, the ratio $R = h/s_{\Delta v}^{min}$, which expresses the balance between excitation and inhibition, dynamically realized by the alternation of up and down-states. By increasing the level of excitability, the system becomes super-critical, and the distribution of avalanche sizes closely resembles the one for disinhibited cultures (Fig. 4c). At the same time, we observe major changes in the inter-event time distribution (Fig. 4a). In particular, the power law regime becomes shorter, its exponent changes, and the local minimum gets more pronounced. Similar changes are observed in cultures treated with PTX (Fig. 4b) [41].

The model described here, consistently connects avalanche statistics with dynamical properties of neural systems that are captured by the inter-event time distribution. Very similar values of R have to be implemented to reproduce the experimental distributions for different critical samples. This means that the up and down-state features consistently change to produce a specific temporal organization of avalanches at criticality. In the next Section we will discuss the intimate connection between inter-event time and avalanche size.

3.2 *Avalanches and Oscillations*

According to similar studies performed for other dynamical processes, we analyze the experimental inter-event time distribution by constructing temporal sequences of events containing only avalanches whose size is larger than a given threshold s_c. Usually this selection provides a sparser sequence, where avalanches are larger and

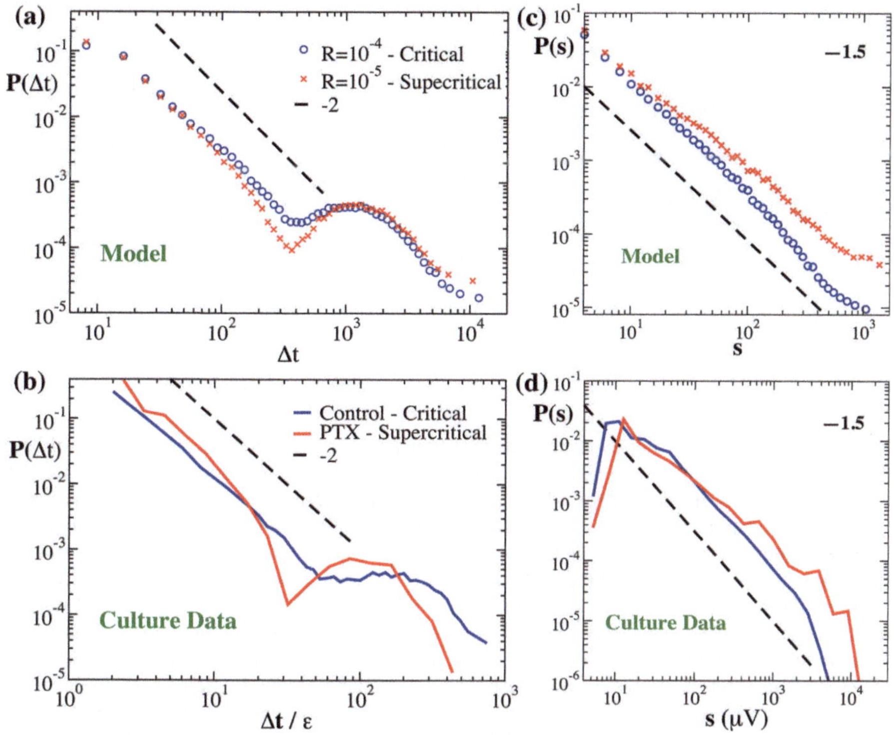

Fig. 4 Numerical and experimental distribution of inter-event time and avalanche size in critical and supercritical state. **a** Numerical inter-event time distributions averaged over 100 scale-free network configurations with $N = 16{,}000$ neurons for values of $s_{\Delta v}^{min}$ and h providing the best agreement with experimental data in normal (critical) condition with $R \simeq 10^{-4}$ (blue circles). For smaller values of R activity becomes supercritical (red crosses) and the distribution becomes more similar to data from disinhibited cultures. **b** Experimental inter-event time distribution for a slice of rat cortex in normal condition (blue line) and disinhibited by PTX (red line). **c** Numerical avalanche size distribution in a critical condition (blue circles), $R = 10^{-4}$, and in a super-critical condition, $R = 10^{-5}$ (red squares). **d** Experimental avalanche size distribution measured for the culture in normal, critical condition (blue line) and for the culture treated by PTX, i.e. in disinhibited, super-critical condition (red line)

more distant in time. In many cases, the scaling behavior of the inter-event time distributions obtained using such a procedure for different thresholds turns out to be universal if the inter-event time Δt is rescaled by the average rate τ_0, namely all distribution collapse onto a universal function f [33],

$$P(\Delta t; s_c) = \tau_0 f(\tau_0 \Delta t) \tag{6}$$

where f is a Gamma function. This result evidences the presence of a unique time scale in the process, namely the average inter-event time or the inverse average rate. The same analysis performed on experimental data for seven samples of rat cortical

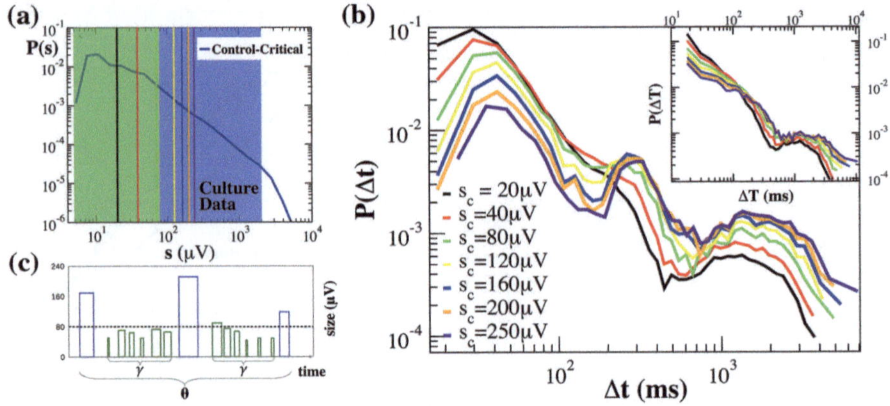

Fig. 5 **a** Experimental inter-event time distribution $P(\Delta t; s_c)$ are evaluated by setting different thresholds s_c for the minimum avalanche size. The color code of the vertical bars corresponds to the legend of panel **b**. Avalanches whose size falls in the blue region tend to occur at frequency θ or smaller, whereas avalanches with size in the green region tend to occur at higher frequency. **b** Experimental inter-event time distributions for different values of the threshold s_c on avalanche size. Already for $s_c = 80\,\mu$V, the distribution clearly exhibits an additional peak. Beside the one at large time scales, $\Delta t \simeq 1000 - 2000$ ms, which is related to the characteristic time of up-state recurrence, the peak at about 300 ms corresponds to the period of θ oscillations, and the peak around 40 ms to the β/γ rhythms. Inset: The same distributions evaluated for the reshuffled avalanche time series and for different values of the threshold s_c on avalanche size. **c** Hierarchical organization of avalanches with different sizes (bar heights) corresponding to temporal organization of nested $\theta - \beta/\gamma$ oscillations. Large avalanches (blue bars) occur with θ frequency and trigger smaller avalanches related to faster γ oscillations (green bars). Here bar widths indicate durations. Spacing between blue bars corresponds to a θ period. Spacing between the starting points of green bars corresponds to the γ period. Sizes s of avalanches related to θ cycles tend to fall within the blue region of the size distribution $P(s)$ plotted in **a**, whereas the sizes corresponding to nested γ oscillations fall within the green region

slices [14] evidences a more complex behavior, as shown in Fig. 5 for the critical sample discussed in Sect. 3. By increasing the minimum size threshold from 20 to 250 μV, the distributions, as expected, show a decrease in the probability to observe small Δt and a corresponding increase at large Δt. More strikingly, the functional form of the probability distribution is clearly non-universal, since for increasing s_c values pronounced peaks emerge at Δts corresponding to the characteristic periods of θ and β/γ rhythms, $\Delta t_\theta \simeq 300$ ms and $\Delta t_\gamma \simeq 40$ ms, respectively (5b). In particular, we notice that the probability $P(\Delta t_\theta; s_c)$ is nearly independent of s_c (Fig. 5b), which means that the ratio $N(\Delta t_\theta; s_c)/N(s_c) \simeq const$, and thus the number $N_\theta(s_c)$ of avalanches related to θ oscillations decreases proportionally to $N(s_c)$ for increasing values of s_c. On the other hand, the probability $P(\Delta t; s_c)$ increases with s_c for $\Delta t > \Delta t_\theta$ and decreases for $\Delta t < \Delta t_\theta$, implying that the number of avalanches separated by longer (shorter) Δt, decreases slower (faster) than $N(s_c)$. Hence, long inter-event times tend to separate large avalanches, whereas shorter inter-event times tend to separate smaller avalanches.

The same analysis has been applied to a sequence in which avalanche sizes are reshuffled by keeping their occurrence time fixed. In this case the peaks observed in Fig. 5b disappear (inset of Fig. 5b). Since reshuffling sizes does not change the inter-event time distribution but destroys the underlying relationship between avalanche sizes and inter-event times, we must conclude that the peaks emerging in the distributions are a consequence of the correlations between sizes and Δts. While short quiet times and fast β/γ oscillations tend to be associated with smaller avalanches, slower oscillations are in general related to larger avalanches. Indeed, varying the threshold s_c in a range of values within the power law regime of the size distribution $P(s)$, typically between 30 and 400 μV (Fig. 5a), the probability $P(\Delta t_\theta; s_c)$ of Δt associated with θ or slower oscillations (Fig. 5b), remains nearly unchanged. In particular, the coexistence of a θ peak with a faster decrease of the probability of Δt_γ suggests an hierarchical structure in the avalanche sequence, reminiscent of the temporal organization of nested $\theta - \beta/\gamma$ oscillations [3] (Fig. 5c): Large avalanches occur with θ frequency and trigger smaller ones in faster β/γ cycles (Fig.5c). In a previous study [13] it has been numerically shown that critical-state dynamics of avalanches and oscillations jointly emerge in a neuronal network model when excitation and inhibition is balanced. The present analysis of experimental data enlightens that oscillations in neuronal activity are the outcome of the temporal organization of avalanches with different size and provides a first empirical evidence for the coexistence of avalanches and oscillations in the critical regime of neuronal activity.

On the basis of these observations, it is clear that no universal function as in Eq. 6 can be obtained for the cortex slice data. Indeed, one can see that only the second regime of the distribution, the bell-shaped bump originated by the down-states, collapses onto a unique function when inter-event times are appropriately rescaled [14, 42]. Concerning the up-states, the appearance of peaks indicates that avalanche occurrence is not controlled by a unique time scale, and that the temporal structure is more complex.

4 Detrended Fluctuation Analysis

In the previous Section we have seen that a predominant characteristic of the intrinsic activity in cortex slice cultures is the continuous alternation of two distinct network states, one with prominent correlations in neural firing, the up-state, and another one with sparse and weakly correlated activity, the down-state. The analysis of inter-event time distributions shows that consecutive avalanches are correlated over a time-scale of about 1 s, and raises the question whether avalanches separated by a longer temporal distance, namely by a down-state, are significantly correlated. In other words, we aim to understand more closely the role of down-states in the context of avalanche dynamics: Are they a sort of memory resetting period? Do they keep memory of past activity and thus correlate consecutive up-states [41]?

In order to address this questions, we first consider the spontaneous activity signal recorded in cortex slice cultures, $V(t) = \sum_{i=1}^{n_e} \delta v_i(t)$ [41], i.e. the sum of all potential variations recorded at time t on the n_e electrodes placed on the cortical slice, and apply the detrended fluctuation analysis (DFA) to quantify its long-range power-law correlations. The DFA [43] involves the following steps: (i) Calculate the integrated signal $I(t) = \sum_{t'=1}^{N_{max}} (V(t') - < V >)$, where $< V >$ is the mean activity and N_{max} is the length of the signal; (ii) Divide $I(t)$ into boxes of equal length n and fit $I(t)$ with a polynomial $I_n(t)$ of order 1, which represents the trend in that box; (iii) Detrend $I(t)$ by subtracting the local trend, $I_n(t)$, in each box and calculate the root-mean-square (r.m.s.) fluctuation $F(n) = \sqrt{1/N_{max} \sum_{t=1}^{N_{max}} [I(t) - I_n(t)]^2}$; (iv) Repeat this calculation over a broad range of box sizes and obtain a functional relation between $F(n)$ and n. For a power-law correlated time series, the average r.m.s. fluctuation function $F(n)$ and the box size n are connected by a power-law relation, that is $F(n) \sim n^H$. The exponent H is a parameter which quantifies the long-range power-law correlation properties of the signal. Values of $H < 0.5$ indicate the presence of anti-correlations in the time series, $H = 0.5$ absence of correlations (white noise), and $H > 0.5$ indicates the presence of positive correlations in the time series.

For the spontaneous activity in cortex slice cultures, the fluctuation function tends to follow either of the representative behaviors that we present in Fig. 6. The first one (Fig. 6a) corresponds to the culture whose inter-event time distribution is discussed in Sect. 3, and exhibits a peculiar structure characterized by a power-law with an exponent $H \simeq 0.8$ for $n < T_0$, a crossover to a short flat region at $n \simeq T_0$, and again a power-law regime with $H \simeq 0.8$ up to $n = T_1$. For $n > T_1$, the fluctuation function becomes completely flat. We begin the analysis of this complex behavior by discussing the plateau for $n > T_1$. According to previous studies on the effects of specific trends on the scaling behavior of the r.m.s. fluctuation function [44, 45], this plateau is due to the peculiar periodic structure of the signal that can be clearly recognized in the integrated signal $I(t)$ (Fig. 6a, lower inset) and in the average network activity as a function of time (Fig. 6b, lower panel). Furthermore, the auto-correlation (Fig. 6a, upper inset) shows a clear peak around $T_1 = 50$ s, that approximately corresponds to the length of the period observed in $I(t)$ (Fig. 6a, lower inset).

Inside this fundamental cycle one can easily identify a further periodicity at $T_0 \simeq 1$ s (Fig. 6a, upper inset), that corresponds to the characteristic duration of down-states (Fig. 4) [14, 34]. Following Refs. [44, 45], this periodic trend could be responsible for the crossover observed in the fluctuation function at $n \simeq T_0$. Indeed, this kind of crossover between two regimes with approximately the same exponent H, has been associated with either periodic trends [44] or non-stationarities [45] in an otherwise long-range correlated signal, such as long segments with zero amplitude. In the signal we consider here, zero amplitude segments correspond to down-state durations and have indeed a characteristic length of about T_0. As pointed out in Refs. [44, 45], the correlation features of the 'clean' signal dominate for $n << T_0$ and $n >> T_0$. Hence, this behavior confirms the presence of strong correlations in the up-states, i.e. $n < T_0$, and, at the same time, reveals that temporal correlations extend over a much longer range, spanning the entire period T_1. Interestingly, the investigation of the local behavior of the signal and its evolution over time (Fig. 6b), shows that the exponent H fluctuates between 0.7 and 0.8 and closely follows the periodic behavior of the network activity, being larger when the network activity is more intense (Fig. 6b).

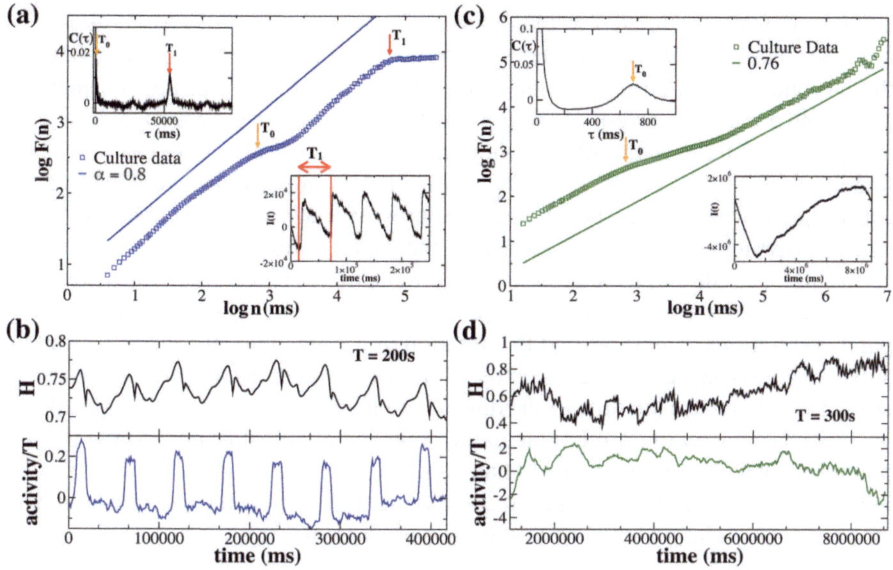

Fig. 6 Detrended fluctuation analysis of spontaneous activity for two representative samples of cortex slice cultures. The logarithm $logF(n)$ of the r.m.s. fluctuation function is plotted versus the logarithm $logn$ of the window size n. **a** $F(n)$ for the network activity $V(t)$ corresponding to the experimental inter-avalanche time distribution (blue curve) shown in Fig. 4. T_0 indicates the characteristic duration of down-states, while T_1 refers to a longer periodicity that may appear in the activity of cultures [14]. Upper inset: Auto-correlation of the network activity $V(t)$; Lower inset: Integrated activity $I(t)$. Vertical red lines delimit the period T_1. **b** Upper panel: DFA performed using a sliding window $T = 200$ s with a step of 1 s. H is estimated in each window with a least square fit. Lower panel: Average network activity in consecutive sliding windows T. **c** $F(n)$ for the activity of a culture that only exhibits the periodicity related to the alternation of up and down-states [34]. T_0 indicates the characteristic duration of down-states. Upper inset: Auto-correlation of the network activity $V(t)$; Lower inset: Integrated activity $I(t)$. **d** Upper panel: DFA performed using a sliding window $T = 300$ s with a step of 1 s. H is estimated in each window with a least square fit. Lower panel: Average network activity in the sliding window T

In Fig. 6c, we show a second representative behavior observed in the r.m.s. fluctuating function of culture network activity. In this case there is only one peak in the auto-correlation at $\tau = T_0 \simeq 600$ ms (Fig. 6c, upper inset), which corresponds to the characteristic down-state duration of the sample, and $F(n)$ exhibits a first crossover at $n \simeq T_0$ from a region with $H \simeq 0.75$ to one with $H \simeq 0.4$, followed by a second crossover to the power law regime with $H \simeq 0.75$ (Fig. 6c). As for the first representative case (Fig. 6a), we observe that the short and long timescale behavior of the fluctuation function are very similar, which suggests that the network activity features long-range temporal correlations transcending up-state durations. Similarly to the scenario in Fig. 6a, here the short and long-range power law scaling of $F(n)$ are separated by a crossover that starts around $n \simeq T_0$. However, the interpretation of this extended crossover to a region with $H \simeq 0.4$ is more delicate. On the one hand, it seems to be related to the down-state duration T_0, as discussed above. On the other hand, the analysis of the local behavior of the signal (Fig. 6d), evidences the presence of anti-correlated segments with $H \simeq 0.4$ (Fig. 6d, upper panel). To better understand the origin of the crossover at $n \simeq T_0$, we have repeated the DFA in segments of the signal with and without anti-correlations. In both cases, we have found a very similar behavior for $10^3 < n < 10^4$ as in the $F(n)$ averaged over the entire signal, which suggests that again the crossover at $n \simeq T_0$ has to be associated with the rhythmic alternation between up and down-states.

In summary, although the presence of periodic trends and non-stationarities gives rise to crossovers and non-constant scaling exponents, the DFA indicates that temporal correlations extend beyond the duration of a single up-state and transcend the down-state time scale. As suggested by numerical simulations, the crossover located around the characteristic time of the down-state may be indicative of a change in the nature and intensity of correlations in the down-state. To better understand this last point and provide a more refined characterization of the temporal structure of the intrinsic activity, in the following Section we introduce a method to systematically investigate the relation between avalanches and inter-times.

5 Conditional Probability Analysis

The DFA represents a preliminary approach in the query for temporal correlations in neuronal signals. As we have seen above, DFA can provide information about the nature of the correlations and help in identifying distinct dynamical mechanisms that operate on specific timescales. However, this approach does not allow to investigate the temporal organization of the process, namely the correlations between the avalanche sizes and their occurrence times. In order to address this question, we apply a powerful statistical method to the temporal sequence of neuronal avalanches, where avalanches are characterized by their size s_i, their starting and ending time, t_i^i and t_i^f, respectively.

The method, developed for the analysis of seismic catalogs [46], is based on the analysis of conditional probabilities on real and surrogate data. Specifically, here we evaluate the probability $P(s_{i+1}/s_i > \lambda | \Delta t < t_0)$ to observe two consecutive avalanches with size ratio $s_{i+1}/s_i > \lambda$ under the condition that their temporal separation $\Delta t = t_{i+1}^i - t_i^f < t_0$, where λ and t_0 are parameters. For each couple of parameter values, we evaluate the same probability in temporal sequences where avalanche sizes are reshuffled in time by keeping their occurrence times fixed. Since in reshuffled sequences avalanche sizes are uncorrelated by construction, the conditional probability P evaluated for several realizations of reshuffled sequences follows a Gaussian distribution, whose average value P^* and standard deviation σ depend on the parameters λ and t_0 [41]. Then, we compare $P(s_{i+1}/s_i > \lambda | \Delta t < t_0)$ with P^* by considering the difference $\delta P = P(s_{i+1}/s_i > \lambda | \Delta t < t_0) - P^*$ for each couple of parameters. If $|\delta P > 2\sigma|$, we conclude that non-zero correlations exist between the size of avalanches separated by $\Delta t < t_0$, and distinguish two cases: $\delta P > 0$ and $\delta P < 0$. In the first case, it is more likely to observe two consecutive avalanches distant in time $\Delta t < t_0$ with a size ratio $s_{i+1}/s_i > \lambda$ in the original avalanche sequence rather than in a reshuffled sequence. This indicates that consecutive avalanche sizes are positively correlated. Conversely, we say that consecutive avalanches are anti-correlated if $\delta P < 0$, meaning that it is more likely to observe two consecutive avalanches distant in time $\Delta t < t_0$ with a size ratio $s_{i+1}/s_i > \lambda$ in a random sequence rather than in the original one.

In Fig. 7a we plot δP averaged over seven samples of cortex slices as function of λ for different values of the temporal distance t_0. Data show that the intensity of correlations depends on t_0 since curves clearly separate and the amplitude decreases if farther in time avalanches are included in the probability evaluation. The maximum of the conditional probability indicates the most probable size ratio for different t_0. Interestingly the same analysis performed for other stochastic processes, as earthquakes and solar flares, provides a maximum position not changing as t_0 increases. Conversely, we observe that, if two consecutive avalanches have a temporal distance smaller than 100 ms, the maximum is located at $\lambda < 1$, indicating that the second avalanche tends to be smaller than the first one. By increasing the temporal distance, the maximum drifts towards the region $\lambda > 1$, implying that for consecutive far-in-time avalanches, the second one tends to be larger.

This complex temporal organization can be understood in the context of the alternation of up and down-states pointed out in Sect. 3. In the up-state avalanches are close in time and their sizes are correlated. Due to the longstanding activity, avalanche sizes tend to gradually decrease. Conversely, far-in-time avalanches occur after a down-state, during which the system is able to recover resources and then trigger larger avalanches.

Crucially, for disinhibited cultures (PTX) this temporal organization is strongly altered [41], and individual cultures may exhibit either single or multiple peaks in

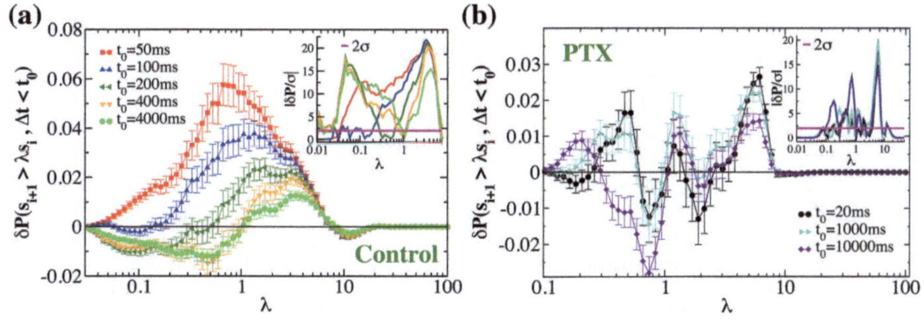

Fig. 7 **a** The quantity $\delta P(s_{i+1} > \lambda s_i, \Delta t_i < t_0)$ as a function of λ for different values of t_0 in normal conditions. The bar on each data point is 2σ. Each curve represents an average over all experimental samples. Inset: The ratio $|\delta P|/\sigma$ as a function of λ for different values of t_0. The magenta horizontal line indicates the value 2σ. **b** The same quantities for disinhibited cultures treated with PTX

δP. As a consequence, the group average shown in Fig. 7b exhibits multiple peaks located at different values of λ. Importantly, all analyzed cultures exhibit a maximum for $\lambda > 1$ that is nearly independent of t_0, evidencing that in the disinhibited condition, the avalanche process is extremely unbalanced, and an avalanche can be considerably larger than the previous one, independently of their time separation. These observations further suggest that the temporal structure of neuronal avalanches is the signature of a healthy system, and is controlled by fundamental biomolecular mechanisms.

The same analysis can be performed on neuronal signals from systems at larger scale, for instance to study correlations between large events in the brain blood oxygenated level dependent (BOLD) signal measured by fMRI on healthy patients [47]. The activity $B(\mathbf{r}_i, t)$ is monitored at each voxel i as function of time. Data are recorded in time every $\delta t = 2.5$ s, therefore time is measured in units of δt. The study focuses only on extreme activity events and therefore analyzes voxels for which $B(\mathbf{r}_i, t)$ is larger than a given threshold $B_c = 18{,}000$, value that selects the largest 10% of the entire activity range. The below-threshold values are substituted by zero. Rather than the activity itself, the interesting quantity is the activity variation at each voxel, $s_i(t) = B(\mathbf{r}_i, t + \delta t) - B(\mathbf{r}_i, t)$, which can have positive and negative values. Following the same procedure outlined before, $\delta P = P(\Delta s < s_0 | \Delta t < t_0) - P^*$ measures the probability to observe the difference between successive variations occurring at any couple of voxels l and m, $\Delta s = s_m(t') - s_l(t) < s_0$, at a temporal distance $\Delta t = t' - t < t_0$, compared to a reshuffled, uncorrelated sequence. To provide a more detailed information, data can be analysed by separating the four different cases of successive variations having the same/opposite sign. This study evidences [47] that successive variations with the same sign (both positive or both negative) are anti-correlated over a short time scale, i.e. less than few seconds, and they are uncorrelated over longer time scales. Conversely, consecutive variations with opposite sign appear to be strongly correlated over the time scale of few seconds (Fig. 8). The overall analysis suggests that successive activity enhancements or depressions

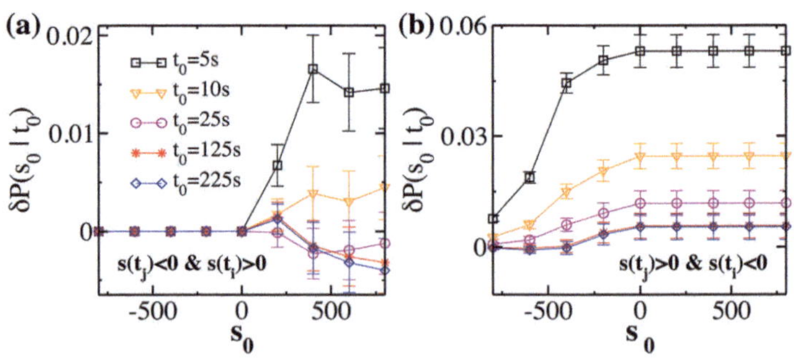

Fig. 8 The quantity $\delta P(s_0, t_0)$ as a function of s_0 for $t_0 = 5, 25, 125, 175, 225$ s. For each t_0 and s_0 the error bar is the standard deviation $\sigma(s_0, t_0)$. Different panels consider different combinations of successive variations with opposite sign

are very unlikely if close in time, and completely uncorrelated over longer temporal scales. Conversely, decreasing activity in some voxels triggers activity enhancements in other voxels after a short time delay, and vice versa. Variations of different signs show, indeed, a strong evidence of correlations suggesting that the system activity realizes a sort of homeostatic balance compensating local enhancements and depressions.

6 Conclusions

In this review we have presented a number of statistical analyses of experimental data aimed at understanding avalanche dynamics in neural networks through the study of temporal correlations. We focused on spontaneous local field potential activity recorded in cortex slice cultures, and on the resting fMRI BOLD signal. In order to achieve a deeper comprehension of the microscopic mechanisms leading to the temporal organization observed at the macroscopic scale, we compare experimental results to numerical simulations of a network of integrate and fire neurons. A first achievement of numerical simulation was the identification of the crucial role of inhibition in the scaling behavior of the EEG power spectra. We have shown that the PSD follows a power-law whose exponent scales with the percentage of inhibitory synapses, and tends to $1/f$ noise for a percentage of about 30%, a value usually estimated for mammal brains. Conversely, fully excitatory networks exhibit $1/f^2$ PSD, i.e. brown noise, which suggests that long range correlations can be obtained only if a certain level of inhibition is present in the network [24]. These evidences indicated that the frequency spectrum of resting brain activity may be controlled by the ratio between excitation and inhibition, and could be used as a tool to measure this ratio experimentally, as more recently shown in [48], and thus investigate pathological conditions. Indeed the analysis of the PSD in epileptic patients provides a scaling exponent close to 2, while in healthy subjects it generally falls in the interval [0.8,1.4].

We further investigated the role of balanced excitation and inhibition focusing on the sequence of avalanches [14, 34] in normal and disinhibited condition. Within this approach, avalanche activity is considered as a point process and attention is shifted to the time elapsed between successive avalanches, i.e. the inter-event time. The analysis of inter-event times is widely used to investigate temporal correlations, which are often associated with the presence of a power-law regime in the inter-event time distribution. For instance, in seismology such distribution behaves as a Gamma distribution, where the initial power-law decay is the expression of correlations in earthquake occurrence due to aftershock sequences triggered by large mainshocks. In spontaneous activity of rat cortex slice cultures, we consistently observe a power-law regime that indicates significant temporal correlations between consecutive avalanches over a temporal range of several hundreds of milliseconds. However, the general functional behavior is more complex than a Gamma distribution, since the inter-event time distribution exhibits non-monotonicity and a characteristic inter-event time of about one second. Numerical simulations are able to provide insights into this peculiar behavior, and show that the underlying avalanche dynamics results from the alternation of up and down-states, corresponding to long bursts and quiescent periods, respectively. The crucial observation is that data from different samples are reproduced by numerical simulations if the ratio of the two parameters controlling those two distinct network states is tuned to the same, "optimal" value. This ratio controls the dynamic balance of excitation and inhibition that is realized by the alternation of high and low activity periods, and goes beyond the structural inhibition, i.e. the percentage of inhibitory synapses. The combined influence of dynamical and structural inhibition on the PSD of the network activity will be subject of upcoming studies. In this review we have shown that unbalanced excitation disrupt the temporal organization of avalanches and alters the functional form of the inter-avalanche time distribution, at the same time moving the system away from criticality. These conclusions are supported by a remarkable agreement with ours and other experimental results [49]. Indeed, the comparison of critical and disinhibited, supercritical cultures evidences the same changes observed in numerical simulations [50].

The analysis of inter-event time distributions has also shown that avalanche occurrence preserves the temporal features of θ and β/γ oscillations [14], and has uncovered a hierarchical structure where large avalanches occurring with θ frequency trigger cascades of smaller avalanches corresponding to the higher frequency oscillations, reminiscent of the temporal organization of nested $\theta - \beta/\gamma$ oscillations [51]. Remarkably, our analysis showed that characteristic brain rhythm time scales do not imply characteristic avalanche sizes, and indicates that different rhythms interact and organize in time as avalanches with scale-free size. The connection between nested oscillations and neuronal avalanches has been also pointed out in [3]. Investigation of spontaneous neuronal activity in the rat cortex layer 2/3 has shown that bursts develop a temporal organization of higher frequency oscillations, β and γ, nested into θ oscillations, while the spatio-temporal organization of LFPs is characterized by the scaling behavior of neuronal avalanches.

The self-regulated balance of excitation and inhibition is widely considered as a fundamental property of 'healthy' neural systems. Pharmacological alterations of this balance can drive those systems to a pathological conditions and thus away from criticality, as seen above. Such kind of perturbation not only influences the avalanche statistics, as previously reported [49], but determines a general reorganization of avalanches in time, as we discussed here. In order to address this point, we used a technique that compares conditional probabilities evaluated in real and reshuffled time series, particularly suitable for understanding the relationship between bursty and quiescent periods [12, 41]. In a recent study we have shown that suppressing inhibition disrupts the relationship between bursts and quiescence that characterizes cortical networks at criticality. In particular, while in the critical state longer recovery periods are needed for the system to generate large avalanches, in the disinhibited condition large avalanches also arise after short periods of quiescence [41]. As a consequence, in critical cultures an avalanche tends to be smaller than the preceding one in the up-states, i.e. for Δt shorter than about 200 ms, and viceversa when the time separation is larger, namely when two avalanches are separated by a down-states. In the case of network disinhibition, this organization is strongly altered and an avalanche can be considerably larger than the previous one, independently of their time separation.

Overall, the study of inter-event time and their relation with avalanche sizes reveals basic features of avalanche dynamics: Temporal correlations are certainly relevant in the up-states, and the analysis of conditional probabilities indicates that also avalanches belonging to consecutive up-states, namely separated by a down-state, are correlated, although the nature and strength of their correlation may be different. The detrended fluctuations analysis confirms these conclusions and indicates that network activity is long-range correlated. The r.m.s. fluctuation function is generally characterized by the same scaling exponent at short and long timescale, $H \simeq 0.8$, and these two power law regimes are separated by a crossover to a region with $H \simeq 0.5$. This crossover is always located around the characteristic time of the down-state and may be a consequence of a change in the nature and intensity of correlations in this network state [44, 45].

Finally, we have shown that balanced network activity can be also recognized at larger scale, in the fMRI BOLD signal [47]. Variations in the signal at each voxel exhibit clear correlations in time: A local increase in activity is followed by a close in time decrease in activity at other voxels and vice-versa. Therefore, at different scales, for different neuronal systems and signal kinds, results consistently indicate that healthy neuronal systems self-regulate the dynamic balance of excitation and inhibition, which is the expression of relevant temporal correlations in network activity. Balanced networks exhibiting critical avalanche dynamics are able to accomplish tasks, as learning Boolean rules and multi-task learning, characterized by properties observed in real systems [10, 11]. Criticality could therefore play an important role in the optimal response to external stimuli, in information processing, and in the functional performances of neuronal systems in general.

References

1. Beggs, J.M., Plenz, D.: J. Neurosci. **23**, 11167 (2003)
2. Beggs, J.M., Plenz, D.: J. Neurosci. **24**, 5216 (2004)
3. Gireesh, E.D., Plenz, D.: Proc. Natl. Acad. Sci. USA **105**, 7576–7581 (2008)
4. Petermann, T., Thiagarajan, T.C., Lebedev, M.A., Nicolelis, M.A.L., Chialvo, D.R., Plenz, D.: Proc. Natl. Acad. Sci. USA **106**, 15921–15926 (2009)
5. Mazzoni, A., Broccard, F.D., Garcia-Perez, E., Bonifazi, P., Ruaro, M.E., Torre, V.: PLoS One **2**(5), e439, 1–12 (2007). https://doi.org/10.1371/journal.pone.0000439
6. Pasquale, V., Massobrio, P., Bologna, L.L., Chiappalone, M., Martinoia, S.: Neuroscience **153**, 1354–1369 (2008)
7. Shriki, O., Alstott, J., Carver, F., Holroyd, T., Hanson, R.N.A., Smith, M.L., Coppola, R., Bullmore, E., Plenz, D.: J. Neurosci. **33**, 7079–7090 (2013)
8. Zapperi, S., Lauritsen, K.B., Stanley, E.: Phys. Rev. Lett. **75**, 4071 (1995)
9. Michiels van Kessenich, L., de Arcangelis, L., Herrmann, H.J.: Sci. Rep. **6**, 32071 (2016)
10. de Arcangelis, L., Herrmann, H.J.: Proc. Natl. Acad. Sci. USA **107**, 39773981 (2009)
11. Capano, V., Herrmann, H.J., de Arcangelis, L.: Sci. Rep. **5**, 9895 (2015)
12. de Arcangelis, L., Godano, C., Grasso, J.R., Lippiello, E.: Phys. Rep. **628**, 1–91 (2016)

13. Poil, S., Hardstone, R., Mansvelder, H.D., Linkenkaer-Hansen, K.: J. Neurosci. **32**, 98179823 (2012)
14. Lombardi, F., Herrmann, H.J., Plenz, D., de Arcangelis, L.: Front. Syst. Neurosci. **8**, 204 (2014)
15. Pathria, R.K.: Statistical Mechanics. Pergamon Press, Oxford (1972)
16. Novikov, E., Novikov, A., Shannahoff-Khalsa, D., Schwartz, B., Wright, J.: Phys. Rev. E **56**, R2387 (1997)
17. Bedard, C., Kröger, H., Destexhe, A.: Phys. Rev. Lett. **97**, 118102 (2006)
18. Dehghani, N., Bedard, C., Cash, S.S., Halgren, E., Destexhe, A.: J. Comput. Neurosci. **21**, 405–421 (2010)
19. Pritchard, W.: Int. J. Neurosci. **66**, 119–129 (1992)
20. Zarahn, E., Aguirre, G.K., Esposito, M.D.: Neuroimage **5**, 179 (1997)
21. He, B.J., Zempel, J.M., Snyder, A.Z., Raichle, M.E.: Neuron **66**, 353–369 (2014)
22. de Arcangelis, L., Perrone-Capano, C., Herrmann, H.J.: Phys. Rev. Lett. **96**, 028107 (2006)
23. Pellegrini, G.L., de Arcangelis, L., Herrmann, H.J., Perrone-Capano, C.: Phys. Rev. E **76**, 016107 (2007)
24. Lombardi, F., Herrmann, H.J., de Arcangelis, L.: Chaos **27**, 047402 (2017)
25. Jensen, H.J.: Self-organized criticality: emergent complex behavior in physical and biological systems. Cambridge University Press, Cambridge (1998)
26. Eguiluz, V.M., Chialvo, D.R., Cecchi, G.A., Baliki, M., Apkarian, A.V.: Phys. Rev. Lett. **94**, 018102 (2005)
27. Roerig, B., Chen, B.: Cereb. Cortex **12**, 187–198 (2002)
28. Song, S., Sjstrm, P.J., Reigl, M., Nelson, S., Chklovskii, D.B.: PLoS Biol. **3**, e68 (2005)
29. Buzsaki, G., Draguhn, A.: Science **304**, 1926–1929 (2004)
30. Kuntz, M., Sethna, J.: Phys. Rev. B **62**, 17 (2000)
31. Rios, P.D.L., Zhang, Y.C.: Phys. Rev. Lett. **82**, 472 (1999)
32. Utsu, T.: International Handbook of Earthquake and Engineering Seismology, vol. 81A, p. 719. Elsevier, Amsterdam (2002)
33. Corral, A.: Phys. Rev. Lett. **92**, 108501 (2004)
34. Lombardi, F., Herrmann, H.J., Perrone-Capano, C., Plenz, D., de Arcangelis, L.: Phys. Rev. Lett. **108**, 228703 (2012)
35. Wilson, C.: Scholarpedia **3**, 1410 (2008)
36. Timofeev, I., Grenier, F., Bazhenov, M., Sejnowski, T.J., Steriade, M.: Cereb. Cortex **10**, 1185–1199 (2000)
37. Timofeev, I., Grenier, F., Steriade, M.: Proc. Natl. Acad. Sci. USA **98**, 1924–1929 (2001)
38. Thompson, S.M., Haas, H.L., Ghwiler, B.H.: J. Physiol. **451**, 347–363 (1992)
39. Cossart, R., Aronov, D., Yuste, R.: Nature **423**, 283–288 (2003)
40. Mendoza, M., Kaydul, A., de Arcangelis, L., Andrade, Jr., J.S., Herrmann, H.J.: Nat. Commun. **5**, 5035 (2014)
41. Lombardi, F., Herrmann, H.J., Plenz, D., de Arcangelis, L.: Sci. Rep. **24690** (2016)
42. Lombardi, F., de Arcangelis, L.: Eur. Phys. J. Spec. Top. **223**, 2119–2130 (2014)
43. Peng, C.K., Buldyrev, S.V., Havlin, S., Simons, M., Stanley, H.E., Goldberger, A.L.: Phys. Rev. E **49**, 1685 (1994)
44. Hu, K., Ivanov, P.C., Chen, Z., Carpena, P., Stanley, H.E.: Phys. Rev. E **64**, 011114 (2001)
45. Chen, Z., Ivanov, P.C., Hu, K., Stanley, H.E.: Phys. Rev. E **65**, 041107 (2002)
46. Lippiello, E., de Arcangelis, L., Godano, C.: Phys. Rev. Lett. **100**, 038501 (2008)
47. Lombardi, F., Chialvo, D.R., Herrmann, H.J., de Arcangelis, L.: Chaos Solitons Fractals **55**, 102 (2013)
48. Gao, Richard, Peterson, Erik J., Voytek, Bradley: Neuroimage **158**, 70–78 (2017)
49. Shew, W., Yang, H., Petermann, T., Roy, R., Plenz, D.: J. Neurosci. **29**, 15595–15600 (2009)
50. Scarpetta, S., de Candia, A.: Front. Syst. Neurosci. **8**, 88 (2014)
51. He, B.J., Zempel, J.M., Snyder, A.Z., Raichle, M.E.: Neuron **66**, 353–369 (2010)

Dr. Fabrizio Lombardi studied physics at the University of Naples "Federico II" and received his Ph.D. in physics from ETH Zurich. He is presently working as a research scientist at Boston University. His research interests fall into the realm of theoretical and computational statistical physics, with a particular inclination towards non-equilibrium dynamics and information processing in biological and physiological systems. Currently, Dr. Lombardi focuses on bursting dynamics and criticality in brain activity across the sleep-wake cycle, with the objective of uncovering the relation between emergent scale invariance, network collective behavior, and brain functions. In parallel, he is developing a holistic approach to human physiology that aims to associate distinct conditions with networks of interactions inferred from synchronous recordings of several organs across the human body, and predict their evolution in response to perturbations (e.g. organ failure, medical treatments).

Hans Jürgen Herrmann theoretical physicist and professor at ETH Zürich Ph.D. in Cologne (1981) in statistical physics and then after one year post-doc in the US member of CNRS at the Service de Physique Théorique in Saclay (France) being today Directeur de Recherche 1ère Class in section 02; from 1990 till 1994 head of the many-body group at HLRZ of KFA Jülich (Germany) for four years, then from 1994 till 2000, chaired professor and director of the PMMH of ESPCI, Paris for six years; from 1996 to 2006, full professor and director of the Institute of Computer Physics at the University of Stuttgart (Germany); since April 2006 full Professor at the Institute of Building Materials at ETH Zürich (Switzerland); Guggenheim Fellow (1986), member of the Brazilian Academy of Science, Max-Planck prize recipient (2002) and Gentner-Kastler prize (2004); ERC Advanced (2012); Aneesur Rahman Prize (2018) managing editor of International Journal of Modern Physics C and of Granular Matter and member of several editorial boards and committees including the Research Commission of ETH; author of over 700 publications and co-editor of 13 books.
His education is based on theoretical solid state physics (master) and statistical physics of critical phenomena (Ph.D.). After his Ph.D. he worked among others on gelation and irreversible growth and built a special purpose computer to calculate the conductivity of percolation clusters. He has studied the fracture of heterogeneous materials since 1986 and since 1992, he has investigated the properties of granular media. Highlights in this research were the construction of space-filling bearings and the establishment of the equations of motion of dunes. His present research subjects include watersheds, the formation of river deltas, quicksand, relativistic fluids, the failure of fibrous and polymeric composites and complex networks.

Lucila de Arcangelis received the Ph.D. in Physics from the BostonUniversity. She was visiting scientist at the University of Cologne and the CEA in Saclay, she obtained in 1990 a CNRS position at the ESPCI in Paris and in 1993 a Faculty position in Italy. Her research interests span from percolation, fractals, cellular automata to spin glass, models for fracture and gelation. Recently, she has focused on statistical properties of earthquake occurrence and neuronal activity on complex networks.

Playing at the Edge of Criticality: Expanded Whole-Brain Repertoire of Connectome-Harmonics

Selen Atasoy, Gustavo Deco and Morten L. Kringelbach

Abstract In order for us to survive, our behaviour has to be perched somewhere between stability and flexibility, or between exploitation and exploration of available resources. This requires the underlying spatiotemporal brain dynamics to be delicately balanced between order and disorder, drawing upon a large repertoire of available brain states. Beyond survival, in order to thrive the brain has to be sufficiently flexible to be able to seek novel trajectories and expand the dynamical repertoire. Here we propose that a key ingredient could be play, the active exploration of novelty beyond exploiting existing potentially scarce resources. Using a novel analysis method called 'connectome harmonics' we not only demonstrate that brain activity resides close to criticality—at the delicate balance between order (stability) and disorder (flexibility)—but this whole-brain criticality is also intrinsically linked to oscillatory brain dynamics. We show that compared to wakefulness, other conscious states are related to different connectome-harmonic repertoires and differ in their proximity to criticality, where the critical regime may enhance the ability to flexibly seek new brain states. In particular, we propose that these brain dynamics may underlie the creative process found in play and improvisation, and as such may shed new light on discovering how the brain optimizes the balance between exploitation and exploration needed for behavioural flexibility.

S. Atasoy (✉) · M. L. Kringelbach
Department of Psychiatry, University of Oxford, Oxford, UK
e-mail: selenatasoy@gmail.com

Center for Music in the Brain, Aarhus University, Aarhus, Denmark

G. Deco
Center of Brain and Cognition, Universitat Pompeu Fabra, Barcelona, Spain

ICREA, Institució Catalana de Recerca i Estudis Avancats (ICREA), Barcelona, Spain

Department of Neuropsychology, Max Planck Institute for Human Cognitive and Brain Sciences, Leipzig, Germany

School of Psychological Sciences, Monash University, Melbourne, Australia

N. Tomen et al. (eds.), *The Functional Role of Critical Dynamics in Neural Systems*, Springer Series on Bio- and Neurosystems 11,
https://doi.org/10.1007/978-3-030-20965-0_2

1 Introduction

What are the conditions in the brain that allow for the optimal behavioural flexibility including creativity; for the creation of valuable novelty in behaviour and thought? Here we propose that *play*, i.e. the engagement in activities with freedom to explore and predict the outcomes without a specific, enforced final goal, may provide one of the best routes to allow creativity to flourish. Play can provide stability and flexibility, which are necessary not only for creativity but also for the emergence of complexity and complex behaviour in dynamical systems. Play imposes the necessary structure and stability while also allowing freedom to explore and is found in all mammalian species, although it does seem to be optimally expressed in humans; not just in childhood but also later in scientific and artistic expression such as music. Similarly, at criticality, complex dynamical systems also try to attain the optimal balance between two opposing forces, between organization and flexibility, between order and chaos, between integration and segregation and between regularity and randomness.

The critical regime, where two opposing processes are balanced [1], enables the essential dualism necessary for complex dynamics; i.e. certain amount of stability (order) for coherent functioning and certain degree of disorder to enable flexibility [2]. Notably, criticality has also been proposed to underlie the resting state dynamics of healthy human brain. Due to its various functional advantages; e.g. larger capacity for information encoding [3] and faster information processing [3, 4], criticality stands out as an important necessary candidate mechanism in the dynamics of spontaneous activity and cognitive function of the brain.

Generally, the key characteristic of criticality manifests itself in certain statistical properties [2]. A fundamental prerequisite for critical dynamics is that the distribution of observables and their fluctuations follow power-laws indicating a scale-free organization [2, 5]. Remarkably, supporting the hypothesis that brain dynamics reside at the edge of criticality, experimental studies have revealed this key characteristic of critical dynamics—the power-law distributions—in large scale brain networks in functional magnetic resonance imaging (fMRI) [6, 7], electroencephalography (EEG) [8–10], magnetoencephalography (MEG) [6, 8, 10, 11] and intracranial depth recordings in humans [12] as well as in numerical simulations of computational models of brain dynamics [13, 14], mostly with small deviations from criticality to the subcritical (ordered) regime. Although diverse generative mechanisms can give rise to power-law distributions [15–17], the observation of power-laws in multiple observables with consistent exponents at various scales of neural activity ranging from neuronal avalanches (bursts of activity) measured in LFP to large-scale brain activity as well as in numerical simulations of various computational neural models [13, 14] provide strong evidence for the hypothesis that brain dynamics reside at the edge of criticality.

Despite this growing body of experimental and theoretical evidence, however, the relation between the critical dynamics in the brain and one of the most fundamental characteristics of human brain activity—spontaneous cortical oscillations—remains poorly understood. Here, we review findings of a novel analysis on functional neuroimaging data using harmonic brain states, which not only reveals signatures of criticality in brain activity but also illustrates the crucial link between whole-brain criticality and neural oscillations. To this end, we take advantage of a newly introduced technique, i.e. 'connectome harmonics' [18], to decompose brain activity into the activation of a set of brain states [19]. Remarkably, when expressed in terms of these harmonic brain states (connectome harmonics), the power-frequency distribution of the dynamical repertoire observed in functional magnetic resonance imaging (fMRI) data follows power-laws for all power-frequency relations. These findings firstly provide confirming evidence for the hypothesis that whole-brain criticality underlies brain activity in resting state. Secondly, the connectome harmonic decomposition reveals that brain activity in an altered state of consciousness, the psychedelic state, exhibits a closer fit to power-laws implying that these pharmacological alterations may tune brain dynamics closer towards criticality and in fact the proximity of brain dynamics to criticality may be crucial for understanding the neural signatures of different mental states. More importantly, the implicit link of these harmonic brain states; i.e. spatial patterns of synchronous neural activity on the cortex, to the frequency of temporal oscillations, establishes the missing link between whole-brain criticality and oscillatory brain dynamics.

2 Oscillations, Synchrony and Harmonics in Brain Activity

A fundamental aspect of the mammalian brain activity persistent across various anatomical scales is coherent *oscillations*. The cortical oscillations of the mammalian brain range approximately from 0.05 Hz to all 500 Hz [20], which are notably *coherent* among spatially distributed cortical regions. For instance, the discovery of strong temporal correlations within widely distributed cortical regions in spontaneous slow (<0.1 Hz) fluctuations of the blood oxygen level dependent (BOLD) signal measured with fMRI has revealed that human brain at rest exhibits a large-scale spatiotemporal organization into distinct networks. These networks, termed resting state networks (RSNs), show synchronous fluctuations of neural activity [21–23]. Remarkably, the topography of these functional connectivity patterns closely resembles the functional networks of the human brain identified by various sensory, motor, and cognitive paradigms [22] suggesting the important functional significance of these oscillatory networks.

How do the spatially distributed parts of the human brain exhibit synchronized activity over time and is there a fundamental principle that predicts which anatomical regions will exhibit such synchrony? In other words, can one predict the particular shape of these oscillatory resting state networks? Recently, it has been demonstrated that the spatial patterns of resting state networks were in fact predicted by the harmonic waves emerging on the structural connectivity of the human brain; i.e. the human connectome [18].

Harmonic patterns are ubiquitous in nature emerging as building blocks of pattern formation in various physical and biological phenomena: standing wave patterns emerging in sound-induced vibrations of a guitar string or a metallic plate (first demonstrated as complex sand patterns by Chladni [24]), patterns of ion motion emerging from electro-magnetic interactions [25, 26], electron wave function of a free particle given by time-independent Schrödinger equation [27, 28] and even patterns emerging in complex dynamical systems such as the reaction-diffusion models introduced by Turing [29], which can explain various instances of biological pattern formation [30]. To gain a more intuitive understanding of the emergence of these harmonic patterns, consider the simple and famous example of vibrating metal plates. Every vibrating system, such as a musical instrument or even a simple metal plate, has certain preferred frequencies, called *natural frequencies*. When the system is excited at one of its natural frequencies, it vibrates, forming standing wave patterns. As firstly illustrated by Ernst Chladni in the 18th century, even a simple metal plate can form these standing wave patterns [24], where each point on the metal plate oscillates up and down with the same frequency (Fig. 1).

Notable, in this simple example of resonance or synchronization phenomena, the shape of the standing wave (i.e. pattern of synchronous oscillations) is determined by only two factors; (1) the activated natural frequency and (2) the shape of the metal plate. When a different natural frequency is activated, the forming standing wave pattern changes automatically. Furthermore, the emerging standing wave patterns naturally adapt their shape to the boundary conditions posed by the geometry of the metal plate. Each of these harmonic patterns—the standing wave patterns—provides a mode of vibration, in which the whole system; i.e. the metal plate, is fully synchronized. Hence, the harmonics can be seen as the fully synchronous modes of oscillations.

Mathematically, these harmonic patterns are given by the eigenfunctions of the Laplace operator, which lies at the heart of theories of heat, light, sound, electricity, magnetism, gravitation and fluid mechanics [32]. The Laplace operator is defined as the combination of two differential operators; i.e. the divergence of the gradient, which in Cartesian coordinates corresponds to the sum of second partial derivatives. The spectral decomposition of the Laplacian yields a set of eigenfunction-eigenvalue pairs (Eq. 1):

$$\Delta \psi_k = \lambda_k \psi_k, \quad \text{with} \quad 0 < \lambda_1 < \lambda_2 < \cdots . \tag{1}$$

The eigenfunctions ψ_k of this spectral decomposition are the solutions of the time-independent (standing) wave equation (known as Helmholtz equation) yielding the spatial patterns of synchronous oscillations in standing waves and the eigenvalues relate to the frequency of oscillation. Interestingly, when applied to a one-dimensional domain with circular boundary conditions, the Laplace eigenfunctions correspond to the sine and cosine functions with different frequencies, i.e. to the well-known Fourier basis (Fig. 1). Furthermore, the spherical harmonics, which have been commonly used in shape and geometry processing and recently been proposed to underlie human brain activity [31], also correspond to the eigenfunctions of the Laplace operator on a sphere (Fig. 1). Notably, application of the graph theoretical counterpart of the Laplace operator, the graph Laplacian, to the human connectome has also enabled the computation of these harmonic patterns on the particular structural connectivity

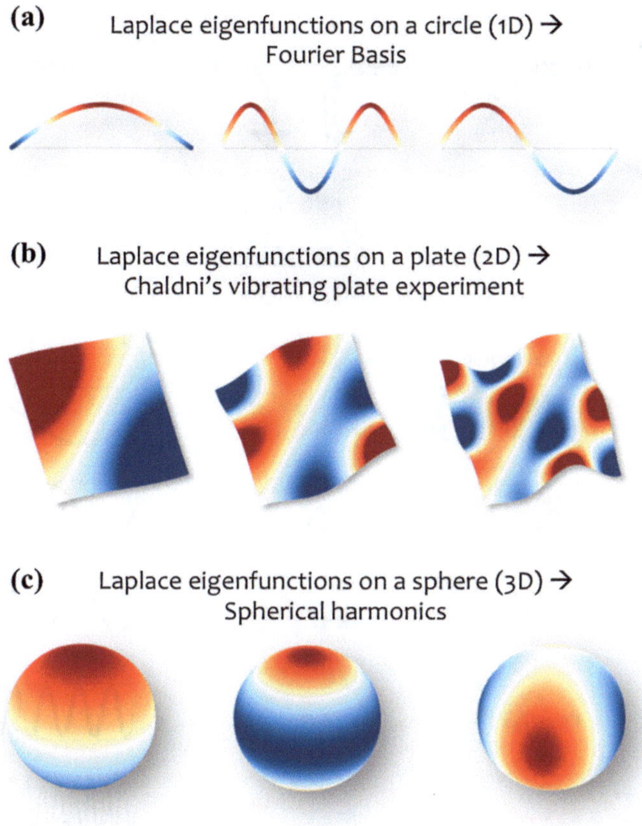

Fig. 1 Harmonic patterns. **a** Harmonics, i.e. eigenfunctions of the Laplace operator applied to a 1D domain yield the sine and cosine functions with different frequencies, which provide the well-known Fourier basis. **b** Standing wave patterns emerging on vibrating metal plates, firstly illustrated by Chladni [24], are also described by the eigenfunctions of the Laplacian. **c** The eigenfunctions of the Laplace operator applied to a sphere reveal the spherical harmonics, which have been intensively used in shape and geometry processing. Furthermore, close analogs of spherical harmonics, i.e. harmonics of the cortical surface, have recently been also proposed to underlie human brain activity [31]

of the human brain [18] (Fig. 2). Importantly, due to this implicit relation, connectome harmonics provide a connectome-specific extension of the Fourier basis [19]. The same way that any continuous signal can be reconstructed from the Fourier basis, any pattern of cortical activity pattern, e.g. measured in fMRI or MEG, can also be decomposed and reconstructed from the set of connectome harmonics [19]. Hence, the connectome harmonics provide a new function basis, a harmonic language, to describe brain activity as a combination of fully synchronous patterns of activity, where each harmonic corresponds to a unique frequency of temporal oscillation and a unique spatial wavelength [19].

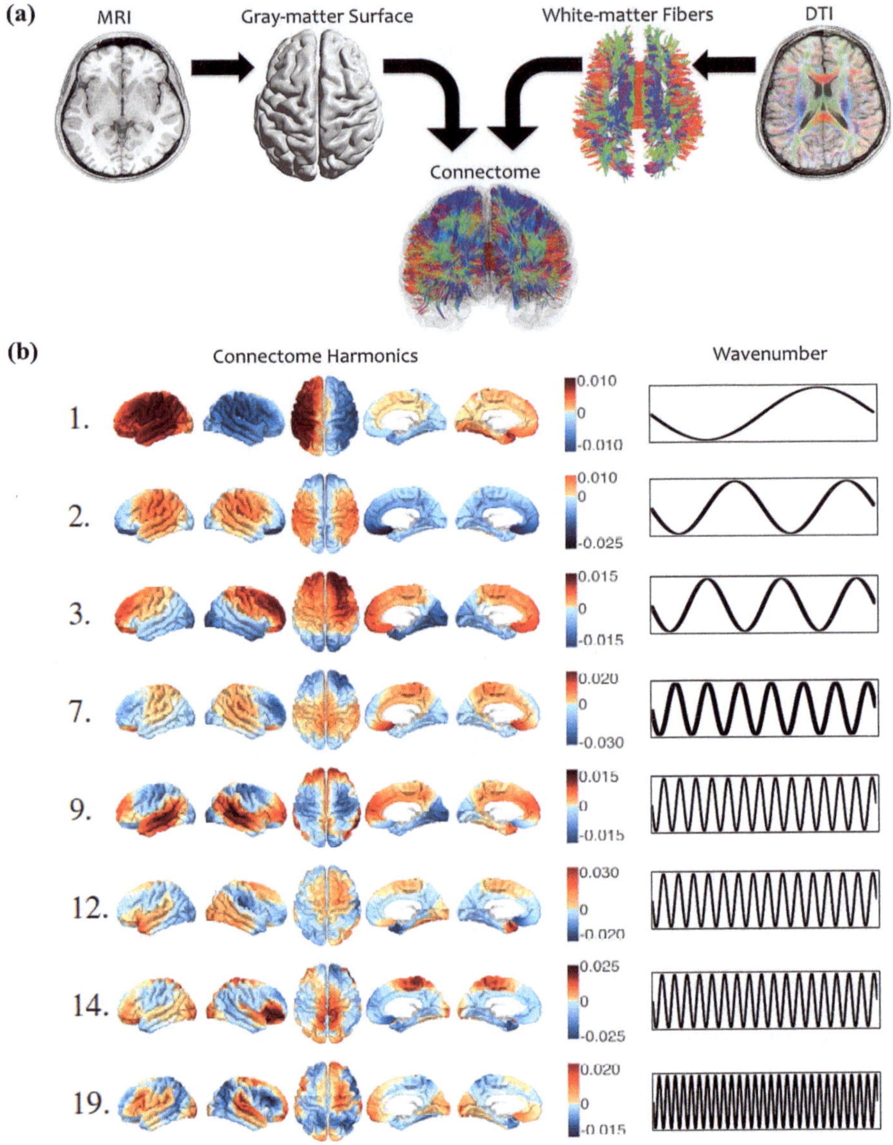

Fig. 2 Computation of connectome harmonics. **a** The human connectome was formed by combining the local cortical (gray matter) connections, which encode the particular geometry of the human cortex, with the long-distance white matter connections. **b** Connectome harmonics computed as the eigenvectors of the graph Laplacian—analogous of the Laplace operator in graph theory [33]—applied to the human connectome. The figure has been adapted from [18]

A unique feature of connectome harmonic representation is that it enables the investigation of the temporal evolution of different frequency brain states separately [19]. Intuitively, equating each brain state to a connectome harmonic with a unique temporal frequency and a spatial wavelength and expressing the brain dynamics as the temporally evolving co-activation of these harmonic brain states, is theoretically analogous to hearing a musical piece and decomposing it into its musical notes. As next, we first discuss how synchrony emerges in biological systems and biological oscillators and then turn our attention to how one can infer what type of dynamical regime the brain operates from the simple question of what kind of music is your brain playing.

3 Synchrony in Biological Oscillators

Synchrony does not only occurs in simple physical systems, such as vibrating metal plates or musical instruments. Remarkably, biological systems and biological oscillators also collectively synchronize their activity leading to the emergence of complex, global patterns. School of fish, or flock of birds synchronize their activity to effortlessly move as a group, thousands of fireflies orchestrate their flashing to flash on and off in synchrony and even thousands of neurons synchronize their firings to give rise to coherent oscillations throughout the cortex [34]. With its 100 billion neurons interconnected by trillions of synapses, studying the synchrony in the brain is clearly not as simple as studying the synchrony in vibrating metal plates. In particular, due to the cooperation and collaboration between the individual elements, the collective activity of biological systems is governed by non-linear equations; i.e. the dynamics of the whole system is different than the sum of its parts. However, here we discuss how simple mathematical tools describing harmonics in linear systems can be used to study the collective dynamics of macro-scale brain activity and how synchrony in populations of biological oscillators relates to the physical concepts of phase transitions and criticality.

Order emerges in the collective behaviour of large biological systems with no central control; i.e. no part directs the behaviour of any other part. For instance in school of fish, flock of birds, ant colonies or group of neurons, no individual directs the behaviour of another but the global pattern of collective activity emerges (self-organizes) from the interactions between the constituents of the group. How does such self-organization occur in large, collective biological systems, and in particular how do large groups of biological oscillators mutually synchronize their activity?

While studying the mutual synchrony in populations of biological systems as a model of coupled oscillators, Winfree and Kuramoto have separately discovered that there was a critical value of population diversity at which the amount of synchrony increased abruptly [34]. Remarkably, in physical systems such abrupt changes in organization occur at a phase transition; imagine cooling water. Until the temperature reaches 0°, there is no change in the structure of water, the water is liquid and the water molecules move rather freely. However, at 0°, at the freezing point, the phase transition occurs from liquid to solid and a new form of order is born in the form of ice. What Winfree discovered in his initial model of coupled oscillators, which then formed the base of the well-known Kuramoto model, was that for wide range of values for the diversity of oscillator frequencies, no synchrony occurred until a critical threshold was reached. Just like during a phase transition, at a critical threshold for diversity; i.e. sufficient homogeneity in the nature of individual oscillators; Winfree found that some oscillators spontaneously locked their frequencies and the narrower the distribution of frequencies became, the more oscillators joined the synchronized pack [34]. In dynamical systems, when this transition (e.g. freezing water) between a relatively disordered (e.g. water) and a more ordered state (e.g. ice) occurs at the critical value of the driving parameter (temperature), the system is said to be at criticality. Hence, Winfree has discovered that mutual synchronization in coupled oscillators is analogous to a phase transition, like the freezing of water into ice [34].

It is worth noting that although the example of melting ice provides an intuitive understanding of the remarkably sharp transition between an ordered and less ordered state, which is also a key signature of critical phenomenon, this particular example belongs to the class of first order (discontinuous) phase transition. Generally, the critical behavior observed in dynamical systems such as coupled oscillators is characterized as second-order (continuous) phase transition, where the transition between ordered and less-ordered states occurs continuously yet with a sharp transition; i.e. with infinite slope at the critical point [35]. In such critical phenomena, the correlation length becomes much larger than the range of microscopic interactions, enabling communication across all scales and leading to the emergence of a collective behavior that is independent of the system's microscopic detail [35–37]. Furthermore, this diverging correlation length gives rise to striking similarities in the behaviour near critical point among systems that are otherwise quite different in nature [35, 37].

In fact, the brain is also hypothesized to work near a phase transition because criticality enhances information processing capabilities and offers certain functional advantages [3, 38, 39]. As next, we discuss how whole-brain criticality and the concept of phase transition relate to the repertoire of (harmonic) brain states and oscillatory brain dynamics.

4 Whole-Brain Criticality and the Repertoire of Connectome Harmonics

In Winfree and Kuramoto's models, when the dispersion of frequencies is sufficiently reduced, the system abruptly crosses a phase transition threshold and synchronization emerges [34]. We now extend the concept of frequency spectrum for synchronous states of brain activity. To this end, we define the brain states—the building blocks of brain activity—as the connectome harmonics [19] and study the brain dynamics as the temporal evolution of the repertoire of these harmonic brain states.

Dynamical system is referred to a system whose state evolves over state space continuously over time according to a fixed rule.
Steady state in a dynamical system means the system does not change over time or continuously changes back to the same state. For instance, imagine placing a pendulum on a table. After swinging back and forth for a period, the pendulum will come to stillness hanging towards the table. This is an steady state of the pendulum.
Stability in a dynamical system means perturbations, i.e. slight disturbances, die out.
In the pendulum example, the state, where the pendulum is hanging with all its weight at the lowest possible point is also a stable state, as the small disturbances will die out and the pendulum will return to the same stable state after the transient period. However, now imagine that the pendulum was perfectly balanced upwards, with its weight at the top of its circle. Although this position of the pendulum is a steady state, it is unstable. The pendulum may stay there if not pushed. However any small perturbation, i.e. a breath of wind, will push it out of this steady state, making this state unstable.
Attractor in a dynamical system is a state to which the system converges after initial transient dynamics, for a wide variety of starting conditions of the system. When the system is close enough to an attractor it will remain close even if slightly disturbed.
Multistability in a dynamical system refers to the co-existence of multiple attractors [40].
Metastability in a dynamical system also refers to the co-existence of multiple attractors [41]. However, unlike in multistability, there are no stable attractors yielding a form of winnerless dynamics [40].
Bifurcation is an abrupt qualitative change in the system's dynamics occuring when one or more parameter pass through critical values, i.e, a small change in one or more parameters cause a qualitative change in system dynamics [41].
Criticality is the critical point (i.e. critical value of an order parameter in a dynamical system), at the brink of a bifurcation, where a stable state is becoming unstable and the system displays certain characteristic dynamical features, of which most are related to enhanced fluctuations [41]. At criticality the systems attractor(s) is(are) only weakly stable [40].

Similar to the temporal activity of generic dynamical systems, the functional complexity of brain activity can also be studied in terms of its *spatial* frequency content: when the connectome-harmonic repertoire (the dispersion of frequencies of harmonic brain states) becomes narrower, the brain activity becomes more synchronised (like the emergence of a synchronized pack in Kuramoto and Winfree's models) and resides in a more ordered regime (like ice in relation to water). At a critical expansion of the connectome-harmonic repertoire, the brain dynamics can approach criticality displaying more complex patterns of activity (combination between a synchronised pack as well as individual asynchronous oscillators in Kuramoto and Winfree's models) similar to the dynamics observed at a phase transition (transition of ice into water).

In computational neuroscience literature, it has already been proposed that the optimal exploration of the dynamical repertoire of brain states is achieved by maximizing *metastability*; i.e. the variability of the state repertoire [41]. Using a Kuramoto (coupled phase oscillator) model for whole-brain dynamics, the brain states have been previously defined as the oscillator phases of a set of brain areas and the metastability has been measured as the standard deviation of the Kuramoto order parameter across time, yielding how the synchronisation between the different brain areas fluctuates across time [41–43]. Notably, the concept of metastability is closely related to *criticality*. Equating each of these brain states to an attractor, *metastability* occurs when the brain dynamics approach criticality, when multiple brain states (attractors) become unstable (or weakly stable) [41] and co-exist within winnerless dynamics [40]. As the maximal metastability allows for a full exploration of the rich functional repertoire of brain states, it has been hypothesized to underlie healthy brain dynamics [41].

Here, we explore the ideas of metastability and criticality in terms of the repertoire of ***harmonic*** **brain states**; i.e. connectome harmonics [18, 19] and review the experimental evidence from the application of this method to fMRI data showing that the brain dynamics reside at the edge of criticality. Unlike previous definition of brain states, such as those in Kuramoto model [41], each *harmonic brain state* is associated with a unique frequency of oscillation and hence the connectome-harmonic decomposition of brain activity not only reveals the repertoire of brain states but also the repertoire of (fully synchronised) oscillator frequencies. Furthermore, due to the implicit link between the Fourier basis and the connectome harmonics [19], as also discussed in Sect. 2, this decomposition can yield the amount of synchronization and complexity of brain activity in terms of its frequency (harmonic) content. The fully synchronous oscillations (such as in resonance phenomena) and no-synchrony (such as in noisy or chaotic dynamics) become the two extreme tendencies of highly strong versus no (weak) structural organization in the connectome-harmonic language (see also the simple illustration of this relation in the temporal domain in Fig. 3).

Like in the temporal domain the Fourier decomposition of a pure oscillation corresponds to a narrow and sharp peak in the frequency spectrum, a strongly synchronous state of brain activity is associated with the activation of a narrow range of connectome harmonics. On the other extreme, the same way that the Fourier transform of a noisy or a chaotic signal yields a broad range of frequencies; a highly disorganized pattern of brain activity would require the activation of a broad range of connectome harmonics. Right at the edge of criticality; i.e. at the brink of a bifurcation, where the dynamics transition between the two extreme tendencies of order and disorder, the system reaches its optimum state repertoire, where structural organization and flexibility of dynamics are balanced. On the other hand, by definition, each connectome harmonic provides a brain state that exhibits long-range correlation, and provides an attractor for brain dynamics. A broad repertoire of connectome harmonics, yields a

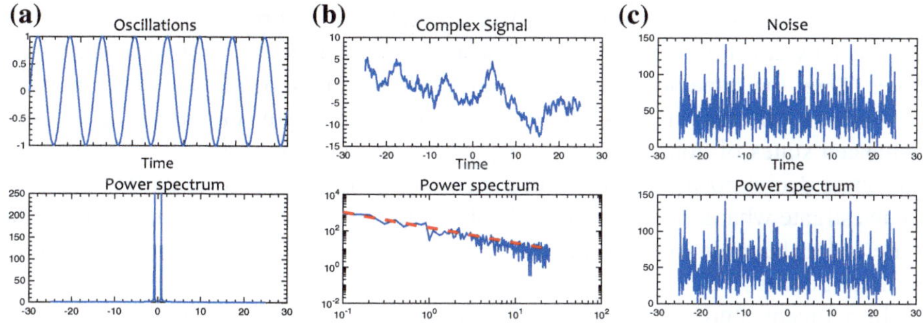

Fig. 3 Oscillations and complexity. **a** Simple harmonic oscillations are composed of a narrow peak in the frequency domain. **b** Frequency spectrum following power-laws yields a more complex signal in the temporal domain. **c** A more evenly or randomly distributed frequency content corresponds to simple noise. Hence, the power-law spectrum can be seen to provide the balance between structure and randomness, between oscillations and noise

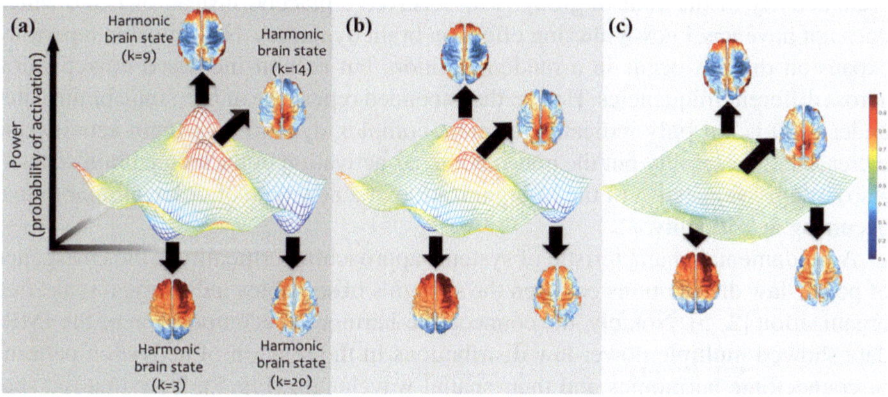

Fig. 4 Connectome-harmonic decomposition and criticality (metastability). Brain dynamics are proposed to evolve on a landscape of harmonic brain states—connectome harmonic brain states (or their combinations). The change from **a** to **c**; i.e. the flattening of the landscape of attractors of possible harmonic brain states—illustrates increased metastability occurring when brain dynamics are tuned further towards criticality. The more equally the probabilities of these harmonic brain states are distributed, the more likely it is that the brain activity resides at the metastable (winnerless) dynamics

landscape of multiple metastable attractors, which in turn leads to complex dynamics of brain activity. Hence, the metastability can be simply estimated as the richness of the connectome harmonic repertoire. Figure 4 illustrates the idea of metastable brain dynamics in terms of the attractor landscape of connectome harmonics.

In summary, unlike simple physical systems, such as vibrating metal plates, brain dynamics can be understood as constantly transitioning between the harmonic attractors over time instead of only oscillating with one of the natural frequencies. Using our musical analogy, the brain acts like an orchestra composing a complex musical piece by constantly activating and deactivating combinations of various musical notes. Thus, the same orchestra (the brain) could be said to be playing different types of music in different states such as sleep, conscious awake state and drug-induced enhanced states of consciousness, such as the psychedelic state. The connectome-harmonic decomposition of functional neuroimaging data, such as fMRI, allows us to investigate what kind of music the healthy brain is playing in the conscious awake state and if the brain dynamics exhibit signatures of a phase transition, i.e. of whole brain criticality.

In a recent study, we explored the dynamical repertoire of connectome harmonics in two different states of consciousness; conscious wakefulness and LSD-induced psychedelic state. It has been previously hypothesized that LSD, a powerful psychedelic compound, tunes the brain dynamics towards criticality [44]. Remarkably, we first found that LSD indeed induced an expansion of the connectome harmonic repertoire, when compared to the brain activity in the resting state; i.e. to the brain activity of the control group, who received a placebo instead of LSD, which does not have any known altering effect on brain dynamics. Notably, this repertoire expansion did not occur in a random fashion, but exhibit increased co-activation across different frequencies. Hence, the expended repertoire of harmonic brain states under LSD is not only indicative of more complex dynamics of brain activity due increased metastability but the non-random co-activation of harmonic brain states is also strongly suggestive of the re-organization of a new type of order, similar to that occurring at criticality.

A fundamental characteristic of systems approaching criticality is the emergence of power-law distributions between the system's observables indicating a scale-free organization [2, 5]. Notably, the connectome-harmonic decomposition of the fMRI data showed multiple power-law distributions in the relation of activation patterns of connectome harmonics and their spatial wavelength (Fig. 5). Note that the spatial wavelength of connectome harmonics also relate to the temporal frequency of oscillations as in the example of vibrating metal plates. The power-law distributions showed a drop-off at the tail, which indicates a deviation of the distribution from power-laws and suggests that the brain dynamics reside close to criticality with slight deviations towards the subcritical regime [12]. This finding is highly consistent with previous studies that explore criticality in brain activity using neural avalanches [6–12]. Remarkably, under LSD, the power (amount of activation)-frequency (spatial wavenumber) distribution of connectome harmonics fit power-laws more closely suggesting that the brain dynamics tune further towards criticality under the effect of LSD [45]. Hence, the LSD-induced variation in neural activity lead to alteration of the power-law exponent (steepness of the line) without destroying the power-law distributions. These findings not only provide supporting evidence for whole-brain criticality in the healthy resting brain, as also reported by previous studies [6–12], but also provide the first experimental evidence showing that a pharamcologically-induced repertoire expansion of harmonic brain states also leads to enhanced signatures of whole-brain criticality, such as power-law distributions emerging in the harmonic repertoire. Moreover, the definition of the brain states as connectome har-

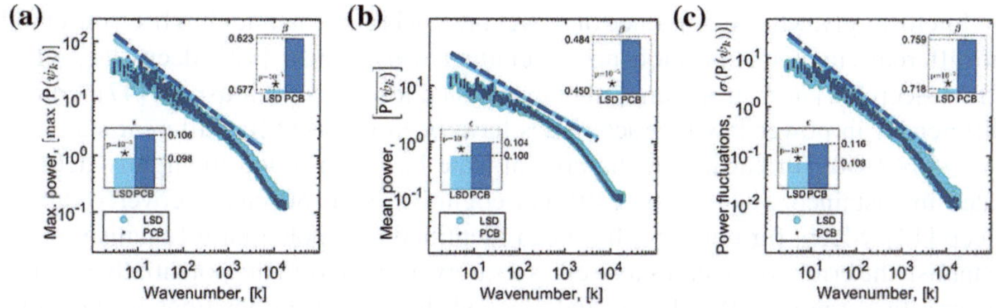

Fig. 5 Power-law distributions observed in the power-frequency relations of connectome harmonic decomposition of resting state functional magnetic resonance imaging data under LSD (cyan) and placebo (dark blue). Distribution of **a** maximum power, **b** mean power of connectome harmonics as well as **c** power fluctuations plotted in relation to the frequency (indicated by the spatial wavenumber of an harmonic pattern) follow power law distributions with a slight deterioration to the subcritical regime indicated by the cut-off observed in the tail of the distribution. The power-frequency relations in brain activity under LSD showed both a closer fit to power-laws (i.e. decreased error of fit (ϵ)) as well as a reduced slope—power-law exponent (i.e. decreased value of power-law exponent (β)). The figure has been adapted from [45]

monics and the exploration of critical dynamics in terms of the diversity of the harmonic repertoire, bridges the missing link between oscillatory brain dynamics and criticality.

5 Discussion

Criticality is poised right at the transition between order and disorder, and as an organizing principle could offer several functional advantages, which are necessary requirements for healthy brain dynamics. At criticality, a dynamical system is not yet as sensitive as in the chaotic regime, where even small disturbances get largely amplified and can lead to totally unpredictable behaviour of the system. It is also not as rigid as in the highly ordered state that the system is incapable of change. Hence, criticality provides a balancing point between these two opposing forces, between stability and flexibility.

It has been proposed that the critical regime, where two opposing processes are balanced [1], enables the essential dualism necessary for healthy brain dynamics; i.e. certain amount of stability (order) for coherent functioning and certain degree of disorder to enable flexibility [2], where the brain also may obtain the optimal balance between integration and segregation [46]. Notably, supporting this hypothesis, experimental studies demonstrated one of the key characteristic of critical dynamics—the power-law distributions—in large scale brain networks in fMRI [6, 7], EEG [8–10], MEG [6, 8, 10, 11] and intracranial depth recordings in humans [12] as well as in numerical simulations of computational models of brain dynamics [13, 14].

Interestingly, while emergence of power laws in brain dynamics has been observed in different states of consciousness including wakefulness [12], deep-sleep [12], REM sleep [12] and during anaesthetics induced loss of consciousness [47], crucial differences in power-law characteristics have been reported for different states [5, 12, 45, 48–50]. In particular, the critical state has been found to slightly deteriorated in sustained wakefulness [49] and cognitive load [50] and recovered during sleep [12]. Sleep, furthermore, has been identified as a potential mechanism to re-establish the critical dynamics as well as a safety margin from the potentially harmful supercritical regime [48]. These findings and theoretical propositions suggest that even though power-laws could be a feature of neural dynamics, which transcends levels of consciousness, differences in power-law distributions and the proximity of these states to critical dynamics are characteristic of different states. Furthermore, such deviations and subsequent re-emergence of power-laws with changing states of consciousness and cognitive-load strongly indicate that they originate from critical network dynamics, ruling out alternative explanations such as filtering or noise [5].

On the other hand, deviations of from the critical point have been also proposed to be symptomatic or causative for certain pathologies [5]. In particular, brain dynamics in depression [44, 48], addiction [44] and obsessive compulsive disorder (OCD) [44] have been associated with the subcritical regime [44, 48], whereas supercritical regime has been found to govern brain dynamics during epileptic seizures [4, 5, 48, 51] and in conditions such as autism [3]. In contrast to epileptic seizures, pharmacological alterations do not necessarily lead to the diminishment of power-laws. In particular, alterations in brain dynamics induced by dopamine agonist [52], dopamine antagonist [53], acetylcholine [54] and LSD [45] have been found to preserve the power-laws but change its critical exponents (steepness of power-laws). Take together, these findings strongly support that healthy brain dynamics reside sufficiently close to criticality and the power-law characteristics may be closely linked to the characteristics of brain dynamics.

In fact, these theoretical propositions and experimental findings become quite meaningful when considered in terms of the connectome harmonic repertoire attained at criticality. As the diversity of the repertoire of brain states is optimized at the edge of criticality [3], critical dynamics enable the emergence of maximally novel brain states. On the other hand, each harmonic brain state is by definition a fully synchronous state of oscillation and hence exhibits a high degree of organization (order). By expanding the repertoire of connectome harmonics, brain dynamics attained at criticality optimize the trade-off between stability and flexibility.

Interestingly, the balance between the complementary dynamics governing stability (ordered regime) and flexibility (chaotic regime), attained at criticality, has also been proposed to underlie the spatiotemporal brain activity in creative thinking [55]. As the diversity of the repertoire of brain states is maximized at the edge of criticality [3], maximum number of brain states becomes available for exploration, which in turn leads to maximum probability for novelty and creativity. In other words, critical dynamics enable the emergence of maximally novel brain states within an ordered (predictable) regime of brain dynamics. Studies revealing the network correlates of creativity provide further supporting evidence for this hypothesis [56]. The cooperation of two different types of brain networks; those linked to top-down control of attention and cognition; salience network (SAL), and the executive control network (ECN) associated with cognitive control together with the default mode network (DMN) associated with spontaneous thought [56] have been found to underlie divergent thinking and creative idea production. Hence, criticality, where the balance between stability (cognitive control) and flexibility (spontaneous thought) is attained, provides both of the necessary conditions required for creativity.

Surprising parallels emerge between creativity, critical dynamics and the connectome harmonic repertoire in light of the musical analogy once again. To answer 'what kind of music your brain is playing, when its dynamics approach criticality?', let us compare the LSD-induced psychedelic state, which shows enhanced signatures of criticality [45] and the wakeful resting state. As discussed above, LSD induces an expansion of the repertoire of connectome harmonics, which in the musical analogy corresponds to the use of an increased number of musical notes. Surprisingly, studies exploring neural basis of jazz improvisation [57] also report that the number of musical notes played during improvisation is significantly higher compared to memorized play of the same piece [57]. Hence, the critical tuning induced by LSD closely resembles jazz improvisation; just like improvising jazz musicians use many more musical notes in a spontaneous and non-random fashion, the brain combines many more of the harmonic brain states (connectome harmonics) spontaneously yet in a structured way, when approaching criticality. Beyond providing an analogy for intuitive understanding of dynamical changes in the brain, creativity and its underlying brain activity, such as observed during musical improvisation, remain yet to be explored in terms of the changes in the repertoire of connectome harmonics.

References

1. Chialvo, D.R.: Critical brain networks. Phys. A Stat. Mech. Appl. **340**, 756–765 (2004)
2. Plenz, D.: Viewpoint: the critical brain. Physics **6**, 47 (2013)
3. Shew, W.L., Plenz, D.: The functional benefits of criticality in the cortex. Neuroscientist **19**, 88–100 (2013)
4. Priesemann, V., et al.: Spike avalanches in vivo suggest a driven, slightly subcritical brain state. Front. Syst. Neurosci. **8**, 108 (2014)
5. Hesse, J., Gross, T.: Self-organized criticality as a fundamental property of neural systems. Front. Syst. Neurosci. **8**, 166 (2015)
6. Kitzbichler, M.G., Smith, M.L., Christensen, S.R., Bullmore, E.: Broadband criticality of human brain network synchronization. PLoS Comput. Biol. **5**, e1000314 (2009)
7. Tagliazucchi, E., et al.: Criticality in large-scale brain fMRI dynamics unveiled by a novel point process analysis. Front. Physiol. **3**, 15 (2012)
8. Linkenkaer-Hansen, K., Nikouline, V.V., Palva, J.M., Ilmoniemi, R.J.: Long-range temporal correlations and scaling behavior in human brain oscillations. J. Neurosci. **21**, 1370–1377 (2001)
9. Allegrini, P., Paradisi, P., Menicucci, D., Gemignani, A.: Fractal complexity in spontaneous EEG metastable-state transitions: new vistas on integrated neural dynamics. Front. Physiol. **1**, 128 (2010)
10. Palva, J.M., et al.: Neuronal long-range temporal correlations and avalanche dynamics are correlated with behavioral scaling laws. Proc. Natl. Acad. Sci. **110**, 3585–3590 (2013)

11. Shriki, O., et al.: Neuronal avalanches in the resting MEG of the human brain. J. Neurosci. **33**, 7079–7090 (2013)
12. Priesemann, V., Valderrama, M., Wibral, M., Le Van Quyen, M.: Neuronal avalanches differ from wakefulness to deep sleep-evidence from intracranial depth recordings in humans. PLoS Comput. Biol. **9**, e1002985 (2013)
13. Deco, G., Jirsa, V.K.: Ongoing cortical activity at rest: criticality, multistability, and ghost attractors. J. Neurosci. **32**, 3366–3375 (2012)
14. Haimovici, A., Tagliazucchi, E., Balenzuela, P., Chialvo, D.R.: Brain organization into resting state networks emerges at criticality on a model of the human connectome. Phys. Rev. Lett. **110**, 178101 (2013)
15. Stumpf, M.P., Porter, M.A.: Critical truths about power laws. Science **335**, 665–666 (2012)
16. Beggs, J.M., Timme, N.: Being critical of criticality in the brain. Front. Physiol. **3**, 163 (2012)
17. He, B.J.: Scale-free brain activity: past, present, and future. Trends Cogn. Sci. **18**, 480–487 (2014)
18. Atasoy, S., Donnelly, I., Pearson, J.: Human brain networks function in connectome-specific harmonic waves. Nat. Commun. **7**, 10340 (2016)
19. Atasoy, S., Deco, G., Kringelbach, M.L., Pearson, J.: Harmonic brain modes: a unifying framework for linking space and time in brain dynamics. Neuroscientist **24**, 277–293 (2017). https://doi.org/10.1177/1073858417728032
20. Buzsáki, G., Draguhn, A.: Neuronal oscillations in cortical networks. Science **304**, 1926–1929 (2004)
21. Fox, M.D., et al.: The human brain is intrinsically organized into dynamic, anticorrelated functional networks. Proc. Natl. Acad. Sci. USA **102**, 9673–9678 (2005)
22. Fox, M.D., Raichle, M.E.: Spontaneous fluctuations in brain activity observed with functional magnetic resonance imaging. Nat. Rev. Neurosci. **8**, 700–711 (2007)
23. Deco, G., Jirsa, V.K., McIntosh, A.R.: Emerging concepts for the dynamical organization of resting-state activity in the brain. Nat. Rev. Neurosci. **12**, 43 (2011)
24. Chladni, E.F.F.: Die akustik. Breitkopf & Härtel (1830)
25. Roos, C.: Quantum physics: simulating magnetism. Nature **484**, 461–462 (2012)
26. Britton, J.W., et al.: Engineered two-dimensional Ising interactions in a trapped-ion quantum simulator with hundreds of spins. Nature **484**, 489–492 (2012)
27. Schrödinger, E.: An undulatory theory of the mechanics of atoms and molecules. Phys. Rev. **28**, 1049 (1926)
28. Moon, C.R., et al.: Quantum phase extraction in isospectral electronic nanostructures. Science **319**, 782–787 (2008)
29. Turing, A.M.: The chemical basis of morphogenesis. Philos. Trans. R. Soc. Lond. B Biol. Sci. **237**, 37–72 (1952)
30. Kondo, S., Miura, T.: Reaction-diffusion model as a framework for understanding biological pattern formation. Science **329**, 1616–1620 (2010)
31. Robinson, P.A., et al.: Eigenmodes of brain activity: neural field theory predictions and comparison with experiment. NeuroImage **142**, 79–98 (2016)
32. Stewart, I.: Mathematics: holes and hot spots. Nature **401**, 863–865 (1999)
33. Chung, F.R.: Spectral Graph Theory, vol. 92. American Mathematical Soc., Providence (1997)
34. Strogatz, S.H.: Sync: How Order Emerges from Chaos in the Universe, Nature, and Daily Life. Hachette, UK (2012)
35. Stanley, H.E.: Scaling, universality, and renormalization: three pillars of modern critical phenomena. Rev. Mod. Phys. **71**, S358 (1999)
36. Kadanoff, L.P.: From Order to Chaos II: Essays: Critical, Chaotic and Otherwise. World Scientific (1999)
37. Marković, D., Gros, C.: Power laws and self-organized criticality in theory and nature. Phys. Rep. **536**, 41–74 (2014)
38. Beggs, J.M.: The criticality hypothesis: how local cortical networks might optimize information processing. Philos. Trans. R. Soc. Lond. A Math. Phys. Eng. Sci. **366**, 329–343 (2008)

39. Brochini, L., et al.: Phase transitions and self-organized criticality in networks of stochastic spiking neurons. Sci. Rep. **6**, 35831 (2016)
40. Breakspear, M.: Dynamic models of large-scale brain activity. Nat. Neurosci. **20**, 340 (2017)
41. Deco, G., Kringelbach, M.L.: Metastability and coherence: extending the communication through coherence hypothesis using a whole-brain computational perspective. Trends Neurosci **39**, 125–135 (2016)
42. Cabral, J., et al.: Exploring mechanisms of spontaneous functional connectivity in meg: how delayed network interactions lead to structured amplitude envelopes of band-pass filtered oscillations. Neuroimage **90**, 423–435 (2014)
43. Cabral, J., Kringelbach, M.L., Deco, G.: Functional connectivity dynamically evolves on multiple time-scales over a static structural connectome: models and mechanisms. NeuroImage **160**, 84–96 (2017)
44. Carhart-Harris, R.L., et al.: The entropic brain: a theory of conscious states informed by neuroimaging research with psychedelic drugs. Front. Hum. Neurosci. **8** (2014)
45. Atasoy, S., et al.: Connectome-harmonic decomposition of human brain activity reveals dynamical repertoire re-organization under LSD. Sci. Rep. **7**, 17661 (2017)
46. Deco, G., Tononi, G., Boly, M., Kringelbach, M.L.: Rethinking segregation and integration: contributions of whole-brain modelling. Nat. Rev. Neurosci. **16**, 430–439 (2015)
47. Lee, U., et al.: Brain networks maintain a scale-free organization across consciousness, anesthesia, and recoveryevidence for adaptive reconfiguration. J. Am. Soc. Anesthesiol. **113**, 1081–1091 (2010)
48. Pearlmutter, B.A., Houghton, C.J.: A new hypothesis for sleep: tuning for criticality. Neural Comput. **21**, 1622–1641 (2009)
49. Meisel, C., Olbrich, E., Shriki, O., Achermann, P.: Fading signatures of critical brain dynamics during sustained wakefulness in humans. J. Neurosci. **33**, 17363–17372 (2013)
50. Tinker, J., Perez Velazquez, J.: Power law scaling in synchronization of brain signals depends on cognitive load. Front. Syst. Neurosci. **8** (2015)
51. Meisel, C., Storch, A., Hallmeyer-Elgner, S., Bullmore, E., Gross, T.: Failure of adaptive self-organized criticality during epileptic seizure attacks. PLoS Comput. Biol. **8**, e1002312 (2012)
52. Stewart, C.V., Plenz, D.: Inverted-u profile of dopamine-nmda-mediated spontaneous avalanche recurrence in superficial layers of rat prefrontal cortex. J. Neurosci. **26**, 8148–8159 (2006)
53. Gireesh, E.D., Plenz, D.: Neuronal avalanches organize as nested theta-and beta/gamma-oscillations during development of cortical layer 2/3. Proc. Natl. Acad. Sci. **105**, 7576–7581 (2008)
54. Pasquale, V., Massobrio, P., Bologna, L., Chiappalone, M., Martinoia, S.: Self-organization and neuronal avalanches in networks of dissociated cortical neurons. Neuroscience **153**, 1354–1369 (2008)
55. Bilder, R.M., Knudsen, K.S.: Creative cognition and systems biology on the edge of chaos. Front. Psychol. **5** (2014)
56. Beaty, R.E., Benedek, M., Kaufman, S.B., Silvia, P.J.: Default and executive network coupling supports creative idea production. Sci. Rep. **5** (2015)
57. Limb, C.J., Braun, A.R.: Neural substrates of spontaneous musical performance: an fMRI study of jazz improvisation. PLoS One **3**, e1679 (2008)

Selen Atasoy obtained her Ph.D. in medical imaging jointly at the Technical Universtiy of Munich and Imperial College London in 2012. She then held a postdoctoral research position at the School of Psychology, University of New South Wales (UNSW), Sydney, where she started her work in experimental and computational neuroscience exploring neural correlates of consciousness. In 2015, she moved as postdoctoral research fellow to the Center for Brain and Cognition, Universitat Pompeu Fabra (UPF), Barcelona and started her work on the neural correlates of different states of consciousness. Currently, she is working as postdoctoral researcher at Hedonia Transnational Research Group at the Department of Psychiatry, University of Oxford. She has been awarded TUM Graduate School of Information Science in Health Ph.D. Fellowship, BMEP Biomedical Sciences Exchange Scholarship and Marie-Currie Postdoctoral Fellowship. Her current research focuses on exploring brain dynamics in consciousness, sleep, meditation, psychedelic states as well as in psychiatric disorders by analysing fMRI and MEG data within the mathematical framework of harmonic waves- a phenomenon ubiquitous in nature.

Gustavo Deco received his first Ph.D. in Physics in 1987 for his thesis on Relativistic Atomic Collisions. In 1987, he was a postdoc at the University of Bordeaux in France. From 1988 to 1990, he obtained a postdoc of the Alexander von Humboldt Foundation at the University of Giessen in Germany. From 1990 to 2003, he led the Computational Neuroscience Group at Siemens Corporate Research Center in Munich, Germany. He obtained his Habilitation (maximal academical degree in Germany) in Computer Science (Dr. rer. nat. habil.) in 1997 from the Technical University of Munich for his thesis on Neural Learning. In 2001, he received his Ph.D. in Psychology at the Ludwig-Maximilians-University of Munich. He is currently a Research Professor at the Institució Catalana de Recerca i Estudis Avançats (ICREA) and Professor (Catedrático) at the Pompeu Fabra University (UPF) where he leads the Computational Neuroscience group. He is also Director of the Center of Brain and Cognition (UPF).

His research interests include: Perceptions, memories, emotions, and everything that makes us human, demand the flexible integration of information represented and computed in a distributed manner. Normal brain functions require the integration of functionally specialized but widely distributed brain areas. The main aim of his research is to elucidate precisely the computational principles underlying higher brain functions and their breakdown in brain diseases. His research allows us to comprehend the mechanisms underlying brain functions bycomplementing structural and activation based analyses with dynamics. He integrates different levels of experimental investigation in cognitive neuroscience (from the operation of single neurons and neuroanatomy, neurophysiology, neuroimaging and neuropsychology to behaviour) via a unifying theoretical framework that captures the neural dynamics inherent in the computation of cognitive processes.

Professor Morten L. Kringelbach is based at the University of Oxford, UK, and University of Aarhus, Denmark. His prize-winning research focuses on reverse-engineering the human brain to elucidate the heuristics that allow us to survive and flourish. This research has grown organically over time, bringing together expertise in computer science, mathematics, complex systems, psychology, and neuroimaging to build precise whole-brain models of brain activity allowing for causal inferences to be made regarding the necessary and sufficient brain networks ensuring short- and long-term survival. He has published fourteen books, and over 300 scientific papers, chapters and other articles. He is a fellow of The Queen's College, Oxford, of the Association for Psychological Science, on the advisory board of Scientific American and a board member of the world's first Empathy Museum.

Complexity of Network Connectivity Promotes Self-organized Criticality in Cortical Ensembles

Paolo Massobrio and Valentina Pasquale

Abstract Large-scale dissociated in vitro cortical networks spontaneously exhibit recurrent events of propagating spiking and bursting activity, usually termed as *neuronal avalanches*, since their size and lifetime distributions can be described by a power law, as in critical sand pile models. Indeed, this spontaneous activity is originated by the synaptic interactions among neurons which are able to freely re-create networks exhibiting complex topological structures. However, experimental in vitro findings show that mature cortical assemblies not necessarily display a critical dynamics, but can follow two other different dynamic states, namely sub-critical and super-critical. Well-known factors that drive the network to these different states are the developmental stage and the excitation/inhibition balance. In this chapter, we will investigate the interplay between self-organized critical state and topological features of the underling cortical network. To investigate the role of connectivity in driving spontaneous activity towards critical, sub-critical or super-critical regimes, results achieved by combining both experimental and computational investigations will be presented and discussed.

1 Introduction

The spontaneous activity originated by the interactions of neuronal assemblies is a peculiar feature of the vertebrate nervous system [1]. In the cortex, it is characterized by oscillatory patterns which span different frequencies or rhythms [2], while in reduced neuronal systems, it is mainly characterized by a mixture of spikes and bursts lasting from tenths to hundreds of milliseconds [3]. Its analysis has revealed that

P. Massobrio (✉)
Department of Informatics, Bioengineering, Robotics, System Engineering (DIBRIS), University of Genova, Genova, Italy
e-mail: paolo.massobrio@unige.it

V. Pasquale
Optical Approaches to Brain Function Laboratory, Istituto Italiano di Tecnologia (IIT), Genova, Italy

N. Tomen et al. (eds.), *The Functional Role of Critical Dynamics in Neural Systems*, Springer Series on Bio- and Neurosystems 11,
https://doi.org/10.1007/978-3-030-20965-0_3

cortical networks generate scale-free activation patterns called *neuronal avalanches*, supporting the evidence of criticality in the brain. Such experimental findings are robust and can be found at different level of investigations: neuronal avalanches have been detected and identified for the first time in acute cortical slices [4] (Fig. 1a), and later also in organotypic slices [5]. The robustness of such dynamics was confirmed some years later when the same behavior was found in dissociated cortical cultures [6] (Fig. 1b). Finally, thanks to the advancements of neurotechnologies that allowed multi-site in vivo recordings, such a power-law dynamics was also detected in in vivo experimental models like awake monkeys [7] (Fig. 1c), anesthetized rats [8] and cats [9] (Fig. 1d), up to the human brain [10].

The panels of Fig. 1 display the avalanches size distributions in different in vitro and in vivo experimental models. The different colors individuate the avalanches size distributions as a function of the temporal bin use to detect avalanches. Indeed, it is relevant for correctly detecting neuronal avalanches, to reduce the time scale of observation, binning the data according to an optimal time bin derived from the inter-

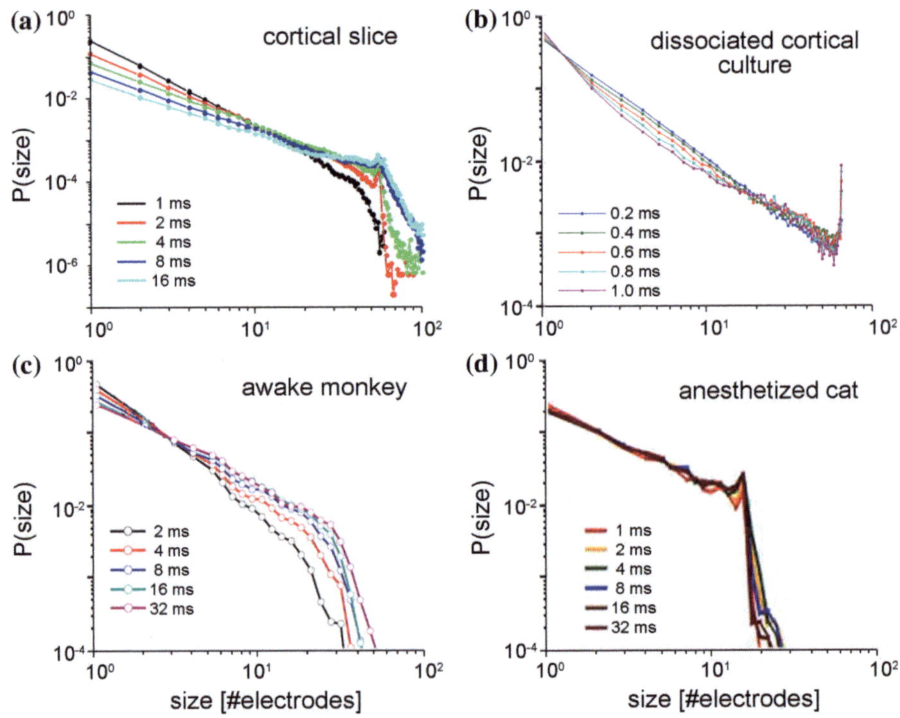

Fig. 1 Neuronal avalanches are a robust feature of cortical activity. Avalanche size distribution relative to: **a** cortical slice (adapted from [4]), **b** mature dissociated cortical culture (adapted from [6]), **c** awake monkey (adapted from [7]), **d** anesthetized cat (adapted from [9])

event interval (IEI) distribution.[1] It is worth to notice that the different experimental models show different dynamics in terms of temporal scale of observations: if for dissociated cultures and acute slices 0.2–0.3 and 4 ms are the optimal bin, respectively, such a values increase for in vivo preparations [7, 9]. In this chapter, we will apply the self-organize critical (SOC) approach to dissociated cultures. These preparations are relatively free of predefined constraints and allow neurons to self-organize during development, creating a network that exhibits complex spatio-temporal patterns of activity [11, 12]. Using this experimental framework, it is possible to study how the spontaneous electrophysiological activity of the network changes and matures during development [13, 14]. A marked sensitiveness of the spatio-temporal firing patterns to structural changes in the network during the in vitro maturation has been extensively demonstrated, showing variations in the burst patterns and also in the cross-correlation among all pairs of electrodes. Mature cultures (between 21 and 35 days in vitro (DIV)) exhibit a synchronized and distributed bursting activity, mixed to a highly variable spiking activity (Fig. 4). In 2008, we found that dissociated cortical cultures, although they are a simplified and reduced experimental model, are able to present a SOC behavior, typical of the complex systems. Among the possible factors that could support critical dynamics, the interplay between functional critical states and topological features of cortical networks remains poorly understood. Experimental evidences in vitro show that mature cortical assemblies coupled to Micro-Electrode Arrays (MEAs) not necessarily fall into a critical regime, but can also show sub-critical or super-critical states [6, 15].

2 Micro-electrode Array (MEA) Technology for Recording Electrophysiological Activity from Large-Scale Neuronal Ensembles

Microtransducers based on Micro-Electrodes Arrays (MEAs) have been demonstrated as powerful tools for recording the electrical activity of networks of neurons cultured in vitro [16]. Under this experimental condition, neurons are directly coupled to the microtransducer by a neuro-electronic junction, and the neuronal electrical activity is then extracellularly recorded [17].

The history of the microtransducer arrays as extracellular recording devices begins at the end of the 60s, when the first metal microelectrodes were adopted [18]. In 1980s, Gross et al. [19] and Pine [20] designed arrays made up of 32 electrodes able to record the electrophysiological activity of excitable cells, and validated this approach on

[1] The IEI is the probability density of time intervals between successive spikes of all the neurons. The average value of the IEI distribution gives an indication of the average time between two successive activations of any pair of neurons. Generally, the average IEI is calculated by averaging the values of the IEI distribution over a predefined time interval whose maximum value is determined as the average time interval corresponding to more the 99% of the area of the mean cross-correlogram (averaging cross-correlograms between all possible pairs of neurons).

neuronal networks. MEAs allow long-term neuron signal recording thanks to their non-invasive properties and, at the same time, allow applying external stimuli using the same electrodes. In 1992, Kawana et al. [21], began studying the plasticity in spontaneous and evoked activity patterns in dissociated rat cortical cultures [21–23]. They induced potentiation or depression in the activity by delivering a fast sequence of strong stimuli (tetanic stimulation). They demonstrated that it is possible to induce long-term potentiation (LTP) or depression (LTD) in a pathway-specific manner [22]. During the years, MEAs have been used for characterizing the response to electrical stimulation [24, 25], for generating peculiar patterns of activity [12], to study possible neurotoxic effects [26] up to closed-loop experiments [27, 28].

From a technological point of view, MEAs (Fig. 2) are made of cell-sized electrodes (10–100 μm diameter) placed onto a glass substrate (Fig. 2a). The electrodes, typically made of Au, Indium-Tin Oxide (ITO), Titanium Nitride (TiN), or black platinum, are biocompatible, long-term lasting, and have low impedance (less than 500 KΩ at 1 kHz) for low thermal noise. The MEA surface and electrode leads are coated with biocompatible insulators (e.g., polyamide or silicon nitride/oxide) which prevent short circuits with the electrolyte bath. These insulators, again coated with adhesion molecules such as poly-lysine and laminin, allow and help neuron coupling to the device surface. The low impedance of the electrodes, and the choice of a correct voltage range to avoid the generation of neurotoxic redox complexes, enable using them to deliver external stimuli. MEAs are equipped with an internal reference electrode used to minimize the possibility of pollution that should be caused by the introduction of an external reference electrode.

The results described in this chapter come from recordings with MEAs made up of 60 flat round electrodes made of TiN. Tracks and contact pads are made of Titanium and the insulation material is made of silicon nitride (Si_3N_4). The electrodes are positioned in an 8 × 8 layout grid (Fig. 2b) with four electrodes at the inactive corner. Electrode diameters are 30 μm with an inter-electrode distance of 200 μm.

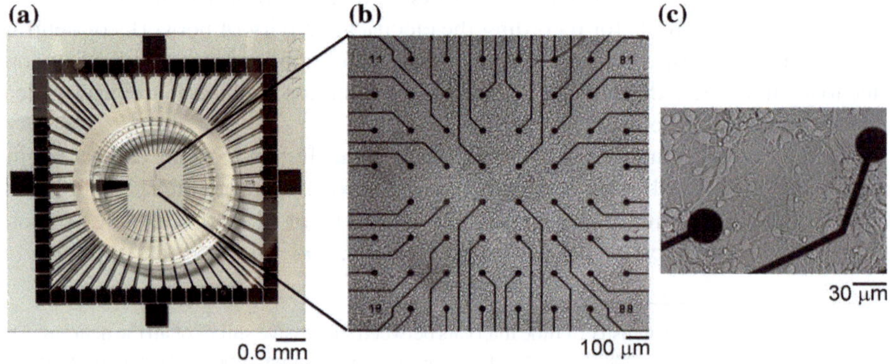

Fig. 2 In vitro experimental set-up based on Micro-Electrode Arrays (MEAs). **a** MEA device; **b** dissociated cortical culture coupled to a MEA. **c** Detail of two microelectrodes

Dissociated neuronal cultures were obtained from cerebral cortices of embryonic rats, at gestational day 18. Cells were then plated on MEA surface (Fig. 2b, c), pre-coated with adhesion promoting molecules (poly-lysine and laminin), at the final density of 1200–2000 cells/mm^2. They were maintained in culture dishes, each containing 1 ml of nutrient medium (i.e. serumfree Neurobasal medium supplemented with B27 and Glutamax-I) and placed in a humidified incubator having an atmosphere of 5% CO_2 and 95% O_2 at 37 °C. Half of the medium was changed weekly. Under these environmental conditions, cortical neurons showed excellent growth and robust synaptic connections that allowed to record spontaneous electrical activity from 7 DIV up to 5–6 weeks in vitro.

The extracellularly recorded signals are typically embedded in biological and thermal noise. Thus, the first pre-processing phase consists in identifying the "real" peaks corresponding to action potentials. Two main routs can be followed: *spike sorting* [29] or *spike detection* [30]. In this chapter, the presented analyses come from spike trains originated by applying to the raw data ad hoc algorithms of spike detection. In particular, the analyses presented in Sect. 3 come from spike trains detected by applying a simple hard threshold voltage (typically set as n-times the standard deviation of the noise) to the raw data. The results of Sect. 5 have been achieved by applying a more rigorous algorithm of spike detection based on a differential threshold, the peak lifetime period of the events, and the refractory period between two consecutive events [31].

3 In Vitro Cortical Cultures Show Different Dynamical States During Development

As revealed in the previous section, the use of in vitro dissociated cortical cultures coupled to MEA allows to monitor the natural development of an assembly. Generally, it is possible to record the first electrophysiological signals after 7 DIVs. At this stage, the global dynamics is essentially characterized by a random spiking activity and only a few short bursts can sporadically appear. Then, the establishment of strong synaptic connections as well as the maturation of the glutamatergic and GABAergic system allow to shape the emergent dynamics by increasing the bursting activity and the level of synchronization of the network [13, 14].

In 2010, Tetzlaff and co-workers characterized the development of in vitro cortical ensembles in terms of SOC [15]. Figure 3a shows the time course of the avalanches' size distribution relative to a dataset of 20 cultures. The plotted variable represent the normalized goodness-of-fit (GoF_{norm}) of the avalanches distributions; the ideal fit (corresponding to a power-law with an exponent $\alpha = -1.5$) is represented by the black dashed lines. Distributions that match or that are close to such a reference curve are "critical". Otherwise, if the curve displays a different trend, the exhibited dynamics can be *super-critical* when there is an excess of avalanches involving the whole network with respect to the expected number according to a power low distribution or

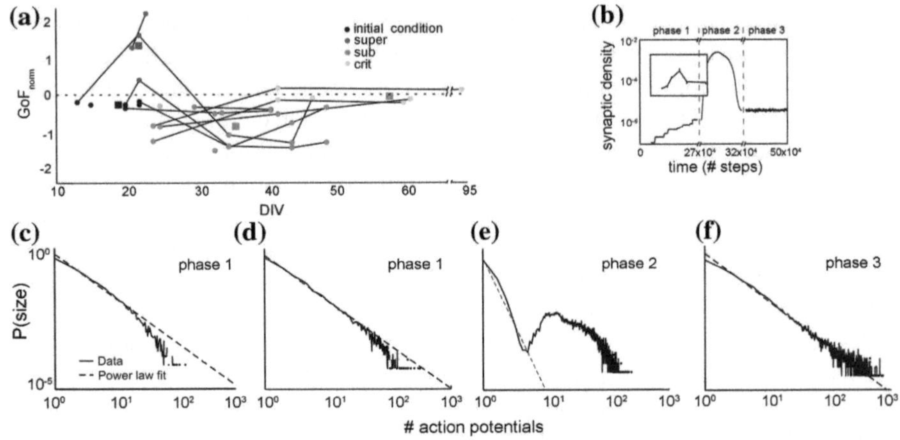

Fig. 3 During development cortical assemblies display different dynamical states. **a** Time course of the normalized goodness-of-fit (GoF_{norm}) for evaluating whether the avalanche distribution follows a power law with −1.5 exponent (dotted line). Red, green, and cyan points indicate super-critical, sub-critical, and critical dynamics. The initial conditions are marked in black. Squares indicate the mean values. Data coming from the same preparation are connected by means of a straight black line. **b** Morphological characteristics of the model as a function of the development split in three phases. Trend of the synaptic density (the inset show experimental results) as a function of the time course. **c–d** At early stage of development, the model foresees a slight sub-critical and critical regime. **e** In phase 2, when we observe a strong increase of the synaptic density, the network display a super-critical dynamics. **f** Once the network reaches a firing rate homeostasis (phase 3), a critical and stable dynamical state appears. Adapted from [15]

sub-critical if distribution of the avalanches follows an exponential decay. Although the initial state (black dots) presents GoF_{norm} in the range of [−0.19, −0.38] that remains up to 19 DIV, suggesting that cortical assemblies develop towards criticality, we also observed that this behavior is very unstable. After a couple of days, the avalanches distributions change dramatically towards a super-critical regime (about 22 DIV on average). After 36 DIV (on average) network activity decreases and a sub-critical regime appears. Finally, in the mature stage of development (at about 58 DIV, on average) the system becomes stable and ruled out by a power-law distribution, indication of a critical regime, since the deviation from the ideal −1.5 power law is close to zero ($GoF_{norm} \sim -0.06 \pm 0.17$). Such a trend is shared by most of the cultures of the dataset, although little deviations can be found: from initial condition (black) to super-critical (red) to sub-critical (green) and finally to a critical state (blue).

To support such an experimental findings, a computational model has been developed to verify which morphological variables (e.g., neuron morphology, synaptic density, connectivity degree, etc.) support this behavior. The model is based on two opposing mechanisms of axonal and dendritic growth which are regulated in order to reach the homeostasis of the network firing rate. The first mechanism regulates dendritic growth probabilities inversely to neuronal activity (firing rate), while the second is the axonal outgrowth promoted by the activity itself [32].

Simulating this model, it is feasible to reproduce the initial phase (phase 1) characterized by neuritic growth, followed by first a structural overshoot and pruning (phase 2), and then by a maturation phase (phase 3) which drives the network to a stable level of connectivity degree (Fig. 3b). Slowly growing connectivity in phase 1 leads over to the fast building of many synapses and a strong increase in activity in phase 2 (super-critical regime), while pruning leads to phase 3 with reduced number of synapses by lowering the activity (critical regime).

The panels c–f of Fig. 3 show four different snapshots of the avalanches distribution of the simulated network during development. Figure 3c, d are relative to the first phase of development (phase 1) when the synaptic density is relative low: the network dynamics evolve towards a non-stable critical regime that is destroyed (phase 2, Fig. 3e) during the maturation stage of the inhibitory system. The network drifts towards a super-critical state (about 22 DIV). Only in the mature stage (phase 3, Fig. 3f), when the homeostasis of the network firing rate has been reached thanks to the correct setting-up of the excitatory and inhibitory synaptic connections, the network becomes critical. The model foresees that developing inhibitory connections is an important factor to reach criticality in developing neuronal networks. Only if inhibition in the model is lowered (phase 3), the network displays critical dynamics.

4 Not All Mature Cortical Cultures Display Self-organized Criticality

In large-scale networks developing ex vivo and chronically coupled to MEAs neurons can freely form synaptic connections and, besides the fact that they grow on a rigid substrate, they are not constrained by any additional external cues.

As reported in the Introduction, these networks spontaneously exhibit complex spatio-temporal patterns of activity, characterized by synchronized and distributed bursting activity mixed with highly variable spiking activity [13]. The raster plots of Fig. 4, show 60 s of spontaneous activity of three representative cortical cultures during their mature stage of development.

The electrophysiological activity was recorded without any chemical or electrical stimulation (i.e., spontaneous activity). The recorded signals ranged from random spike activity to more complicated and synchronized burst signals. However, it is qualitatively evident from a visual inspection of the raster plots of Fig. 4 how the generated patterns of electrophysiological activity present different features in terms of burst duration, inter-burst interval distribution, level of synchronization, percentage of random spiking. This means that in vitro dissociated cortical cultures can display a rich repertoire of motifs [13, 33].

It is worth to notice that such different electrophysiological behaviors are independent from the cell preparation: the three snapshots of Fig. 4 are relative to three mature cortical cultures coming from the same preparation, plated to MEAs at the same final concentration (i.e., 1500 cells/mm^2), and recorded at the same degree of

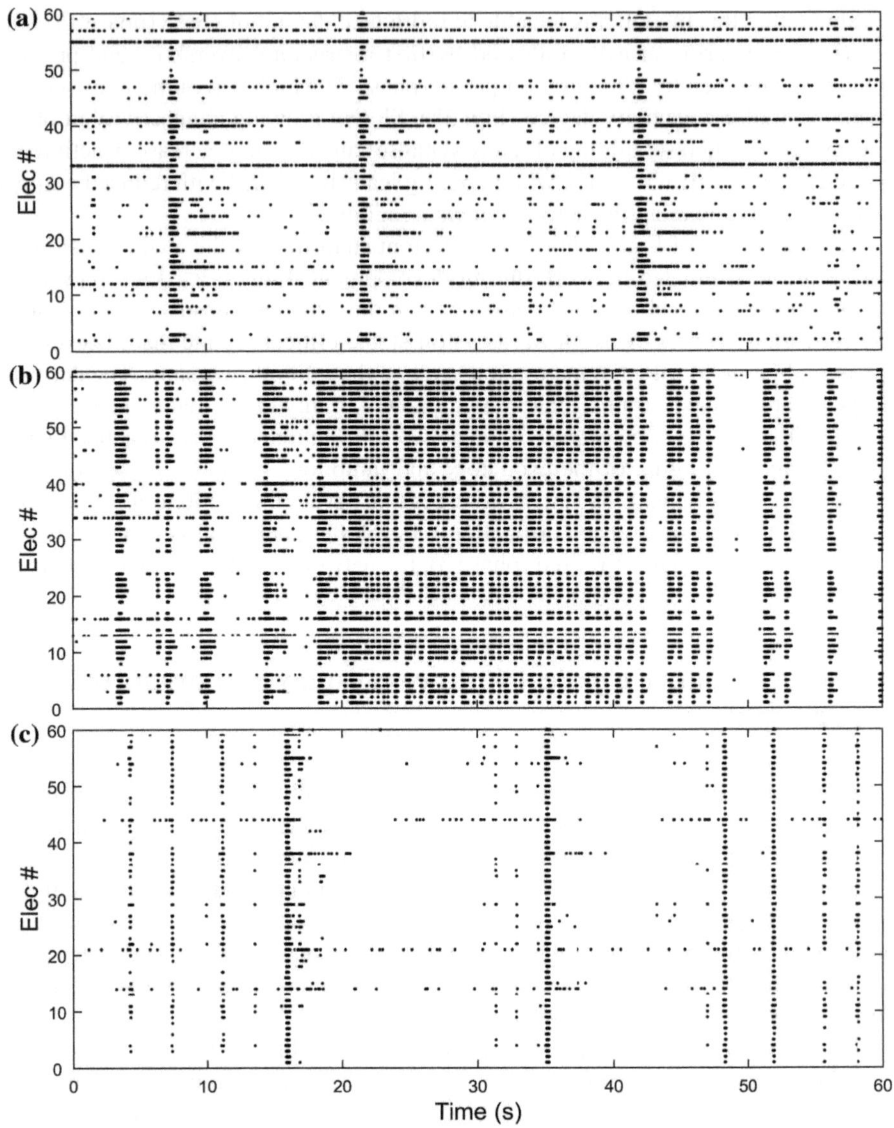

Fig. 4 Raster plots showing 1 min of spontaneous activity of three cortical cultures at DIV 27 recorded by a 60-microelectrodes MEA. Each dot represents a detected spike

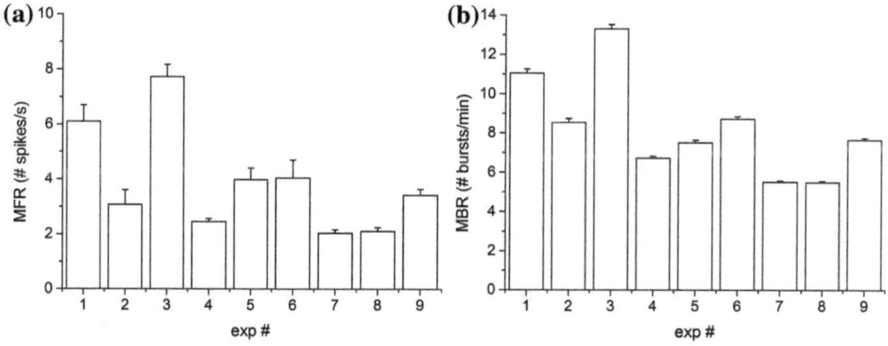

Fig. 5 Firing and bursting statistics relative to a dataset of n = 9 mature cortical cultures coupled to MEAs. **a** Mean firing rate, i.e., the firing rate relative to each electrode averaged over the all active electrodes of a MEA; **b** mean bursting rate, i.e., the bursting rate activity relative to each electrode averaged over the all active electrodes of a MEA. Data are expressed as mean ± standard error of the mean

development (i.e., 27 DIV). Figure 5 quantifies the macroscopic dynamics of these cultures (n = 9) by means of the evaluation of the mean firing rate (MFR) and the mean bursting rate (MBR) which quantify the frequency of spiking and bursting activity. Indeed, this wide variety of patterns of activity influence the dynamical states in which a cortical network can lie.

Once a culture reached the mature stage of development (i.e., fourth/fifth week in vitro), it shows a preferred dynamical state (critical or quasi-critical, sub-critical or super-critical). This means that not all cortical cultures evolve towards criticality during development [6].

In Fig. 6, we reported three examples of mature cortical cultures (4th and 5th week in vitro) which show a sub-critical behavior (blue line), critical (red line) and a strong super-critical (green line) behavior. Although the observed cultures did not share a common pathway of development, they demonstrate that networks of dissociated neurons can approach a critical state in the mature phase, giving rise to events of propagation of activity that corresponds to the description of neuronal avalanches.

Averaging the results of n = 4 different cultures, which displayed a critical behavior in the mature stage, we found an average slope of -1.60 ± 0.09 (mean ± std, RMSE < 10^{-3}) for the avalanche size and -1.86 ± 0.13 (RMSE < 10^{-2}) for the avalanche lifetime. Similarly, averaging the results of two sub-critical cultures, a slope of -2.03 ± 0.12 (RMSE < 10^{-3}) for the size and -2.45 ± 0.20 for the lifetime (RMSE < 10^{-3}) was achieved. Lastly, averaging the results of three super-critical cultures, we obtained -1.88 ± 0.17 (RMSE < 10^{-2}) and -2.19 ± 0.16 (RMSE < 10^{-2}), respectively for the size and the lifetime. Cultures have been clustered as critical, sub-critical, and super-critical, by the inspection of their avalanches' distributions and by the GoF values.

The question that arises after observing these different dynamical states that mature cortical assemblies can display is whether such different behaviors corre-

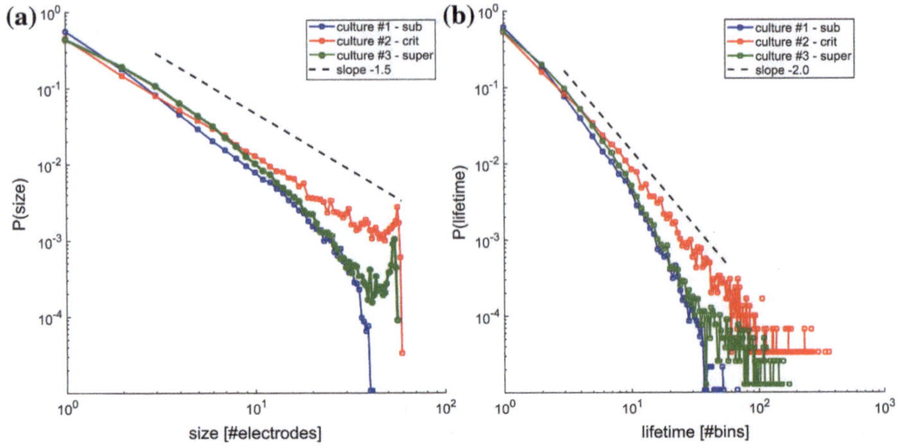

Fig. 6 Dissociated cortical cultures may display different dynamical states, namely sub-critical (blue), critical (red), and super-critical (green). **a** Avalanche size distributions. **b** Avalanche lifetime distributions. In black dashed, the reference power laws with exponents −1.5 (size), and −2.0 (lifetime) are reported

late with other parameters describing the electrophysiological patterns of activity. It was found that neuronal avalanches in cultures of neurons are associated with other parameters describing spiking and bursting dynamics, mainly the degree of synchronization of bursts among different channels and the proportion between spiking and bursting activity. Super-critical behavior is associated with a high degree of synchronization of bursts among all the electrodes, whereas sub-critical behavior is related to low synchronization and high percentage of non-clustered activity. Therefore, the critical state is achieved when spontaneous electrical activity is composed of both medium-synchronized network bursts [34] and a very small amount of random spikes. When the activity is highly synchronized, all neurons fire together and frequently originate avalanches involving the whole network: in this case, the distribution of avalanche sizes is bimodal, as in cortical slices after treatment with picrotoxin [4]. Conversely, when the electrical activity is poorly correlated, each electrode fires independently, and global avalanches occur with a lower probability. A medium-level synchronization usually corresponds to a nearly critical state, suggesting that criticality is strictly linked to the degree of connectivity, both in anatomical and functional terms [35, 36].

Figure 7 show the coincidence index[2] CI_0 between all pairs of electrodes and the percentage of random spikes (i.e. the fraction of spikes outside bursts). The former is a measure of the level of synchronization among all the electrodes, while the latter measures the proportion between spikes included within bursts and random spikes outside bursts (in other words, it is an indication of the level of burstiness of the net-

[2]The coincidence index is defined as the ratio of the integral of the cross-correlation function in a specified area (e.g., ±1 ms) around the zero bin to the integral of the total area [6].

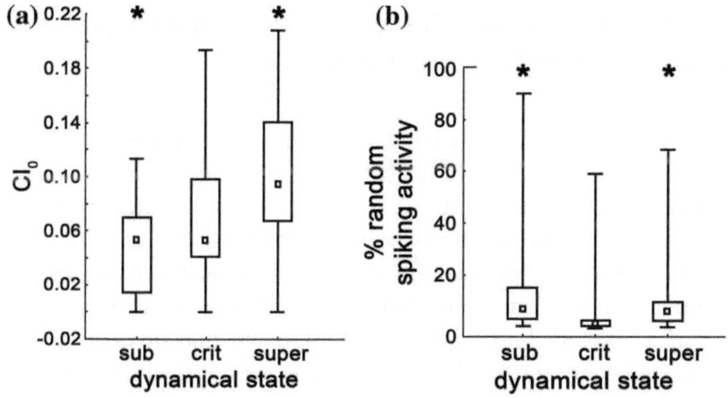

Fig. 7 Avalanches' distributions correlate with different macroscopic dynamics. **a** Box plots of the CI_0 (indication of the level of synchronization of the network) distributions for sub-critical, critical, and super-critical cultures; **b** box plots of the percentage of random spiking activity (i.e., spikes not belonging to bursts) for the same groups of assemblies. Stars indicate statistically significant differences ($p < 0.05$) with respect to the critical condition. Data are expressed as mean ± standard error of the mean. Adapted from [6]

work). These results correlate with the corresponding distributions of the percentage of random spiking activity and the coincidence index: in Fig. 7, we reported a box-plot representation obtained by considering all cultures (4 critical, 2 sub-critical and 3 super-critical) of the dataset. These statistical distributions are different as confirmed by statistical tests (ANOVA for ranks, Kruskal–Wallis test[3]: CI_0, H (N = 3356) = 1071.106, $p \leq 0.001$; percentage of random spiking activity, H (N = 3264) = 1112.986, $p \leq 0.001$).

Critical distributions of avalanche sizes and lifetimes correlate with average synchronization among electrodes, while sub-critical and super-critical distributions correspond to low-level synchronization and to high-level synchronization, respectively. In addition, observing Fig. 7b, we noticed that, in those cultures that tend to criticality, most spikes are concentrated within bursts (i.e. median and 25–75% percentile values of the proportion of random spikes are lower), whereas the other cultures have, on average, a less compact bursting activity, more evidently for sub-critical networks.

[3]Kruskal–Wallis non-parametric test was applied since the normality assumption was not verified by the considered dataset (Kolmogorov-Smirnov normality test).

5 Complex Network Topologies Promote Self-organized Criticality in Cortical Assemblies

The experimental findings described in the previous section show that in vitro mature cortical assemblies not necessarily fall into a critical regime, but can also show sub-critical or super-critical states (Fig. 6). Since dissociated cultures are free to grow without any physical or chemical constraint, their underlying structural connectivity is a relevant variable that can push the network towards different dynamical states. In 2006, Eytan and Marom found that the distribution of firing rates during the early phase of a network burst (i.e., a network event that involves all the neurons of a network) can be described by a power law, which is consistent with a scale-free topology of connectivity [37]. Indeed, this was the first observation relative to dissociated cortical cultures that postulated an interplay between the exhibited dynamics and the underling (functional) connectivity.

From that work, and by exploiting the tremendous advances in multichannel extra-cellular recording techniques that made possible the simultaneous recording of the electrophysiological activity of thousands of neurons [38–40], a more precise reconstruction of the "functional" network organization of neuronal assemblies has been reached. Nonetheless, also the development of high-computing methods for inferring the topological properties of neuronal assemblies [41, 42], as well as the support of computational models of large-scale neuronal assemblies allowed to find a reliable interplay between network dynamics and connectivity.

5.1 Scale-Free Networks with Small-Worldness Features Promote Self-organized Criticality

In this section, we present synthetic results achieved by developing a bio-inspired computational model reproducing the experimental electrophysiological patterns of activity of mature in vitro dissociated networks. In particular, we characterized the interplay between the different dynamic states (i.e. sub-critical, critical, super-critical) and the underlying network topology.

To achieve such an issue, the morphological network connectivity was modeled by means of the graph theory [43]. Graphs are made up of *nodes* which represent the neurons and *edges* which model the morphological connections among the neurons. The structure of the graph is described by the adjacency matrix, a square symmetric matrix of size equal to the number of nodes N with binary entries. If the element $a_{ij} = 1$, a connection between the node j to i is present, otherwise $a_{ij} = 0$ means no connection. All the auto-connections ($a_{ii} = 1$) are avoided. Then, the value 1 of the non-zero a_{ij} elements are substituted with numbers representing the different synaptic weights drawn from two normal distributions (one for excitatory and one for inhibitory weights). Each node (neuron) of the graph is "replaced" by a neuron model, whose dynamics is described by the Izhikevich equations [44]:

$$\frac{dv}{dt} = 0.04v^2 + 5v + 140 - u + I_{tot} \quad (1)$$

$$\frac{du}{dt} = a(bv - u) \quad (2)$$

with the after-spike resetting conditions:

$$if\ v \geq 40\,\text{mV} \rightarrow \begin{cases} v \leftarrow c \\ u \leftarrow u + d \end{cases} \quad (3)$$

In Eqs. (1–3), v is the membrane potential of the neuron, u is a membrane recovery variable which takes into account the activation of K^+ and inactivation of Na^+ channels. Equation (4) displays the used values for the four parameters.

$$a = \begin{bmatrix} 0.02 \\ 0.02 + 0.08r_i \end{bmatrix} b = \begin{bmatrix} 0.2 \\ 0.25 - 0.05r_i \end{bmatrix} c = \begin{bmatrix} -65 + 15r_i^2 \\ -65 \end{bmatrix} d = \begin{bmatrix} 8 - 6r_i^2 \\ 2 \end{bmatrix} \quad (4)$$

In Eq. (4), the first row are relative to the excitatory neurons (regular spiking, RS), while the second one to the inhibitory (fast spiking, FS) neurons. Although the network makes use of the class of RS and FS models, a random variable r_i (which spans from 0 to 1) was introduced for each neuron to introduce more variability in the neuron dynamics: a RS neuron is obtained if $r_i = 0$, whereas if $r_i = 1$, a bursting neuron is obtained. Such distribution is biased towards RS neuron. Figure 8a, b display the reproduced patterns of electrophysiological activity.

Figure 8c identifies the permitted connections among the neuronal populations: practically, excitatory neurons project to excitatory and inhibitory, while inhibitory neurons can establish synaptic connections only to excitatory neurons.

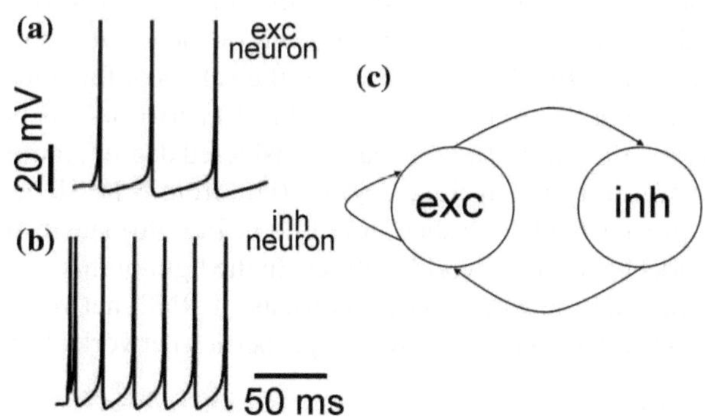

Fig. 8 Network model features. Electrophysiological patterns of **a** excitatory regular spiking and **b** inhibitory fast spiking neurons. **c** Sketch of the allowed connections among the excitatory and inhibitory populations

The "modes" of connectivity used in the simulations are random and scale-free. The model of random network (RND) follows the original derivation devised by Erdős and Rényi [45]. The fundamental assumption of random networks is that, despite the random placement of links, the correspondent graph is characterized by a uniform connection probability and a Poissonian/Gaussian degree distribution. The independent variables for building up a random graph are the number of nodes N and the total number of edges, with the condition that the minimum number of edges must be $N \cdot log\frac{N}{2}$. In scale-free networks (SF) [46] the degree distribution follows a power-law: thus, if m is the number of edges which incident to a node (i.e. the connectivity degree), the power-law distribution is given by [47]:

$$P(m) = m^{-\gamma} \tag{5}$$

where the characteristic exponent γ lies between 1.3 (slice recordings [7]) and 2 (fMRI recordings [5]).

For each RND and SF network topology, we generated 9 different configurations (labeled from N1 to N9), with an increasing average connectivity degree. Figure 9a shows the incoming degree (mean ± standard error) for SF (red) and RND (black) networks. On average, all networks have a comparable incoming degree (similar considerations can be also done for the outcoming degree); SF networks feature higher standard deviation values of the connectivity degree, given the presence of a small number of hub neurons. The inset of Fig. 9a quantifies the percentage of hub neurons in SF networks which spans from 1.8% in correspondence of the network N1 to 13.5% (N9). It is worth to notice that the composition of hub neurons maintains the same balance between excitation and inhibition of the entire network, i.e. 30% of inhibitory neurons.

Figure 9c, d show the degree distributions of SF and RND networks, respectively. For all SF networks, the degree distribution can be fitted by a power-law (Fig. 9c) and the corresponding exponent lies between −1.65 and −1.43, with no specific correlation to the average degree. The degree distributions of RND networks have been fitted by a Gaussian distribution (Fig. 9d), whose mean value corresponds to the network average degree. Finally, Fig. 9b shows the values of the Small-World Index (SWI)[4] for the different SF (red) and random (black) networks. As predicted by the theory, SF networks are always (i.e., for each considered degree) more clustered than RND ones. However, for high average degrees (i.e., from N4), SF networks present a value of SWI greater than 1 (dashed gray line in Fig. 9b) suggesting that such a pool of SF networks have small-world features. In the light of this topology analysis, the dataset of simulated network models consists of RND networks (N1–N9), SF networks (N1–N4) and SF with small-world properties networks (N5–N9).

[4]The Small-World Index (SWI) is defined as: $SWI = \frac{CC_{net}}{CC_{rnd}} / \frac{PL_{net}}{PL_{rnd}}$, where CC_{net} and PL_{net} are the cluster coefficient and the path length of the investigated network; while CC_{rnd} and PL_{rnd} correspond to the cluster coefficient and the path length of random networks equivalent to the original network (i.e., with the same number of nodes and links). A SWI higher than 1 suggests the emergence of a small-world topology.

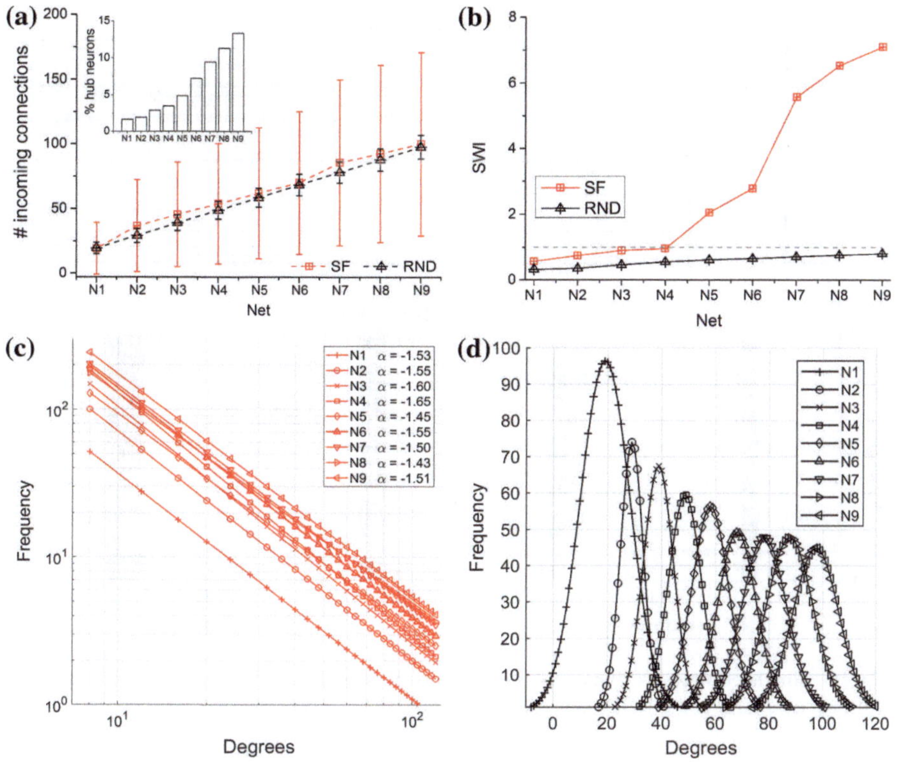

Fig. 9 Topologies features of the computational model mimicking dissociated neuronal network dynamics. **a** Incoming degree (mean ± standard error of the mean) for SF (red) and RND (black) networks. The inset shows the percentage of hub neurons in SF networks as a function of the degree (i.e., from N1 to N9). **b** Small-World Index (SWI) for SF (red) and RND (black) networks as a function of the degree. The dashed gray line (SWI = 1) indicate the threshold over that a network presents small-world features. Degree distribution of **c** SF and **d** RND networks for the different degree (i.e., from N1 to N9). Adapted from [36]

By simulating the aforementioned dataset, it was found that both RND and SF networks display a mixture of synchronous bursts and asynchronous spiking activity as in vitro biological preparations display (Fig. 4 and [34, 48]). Figure 10 displays 60 s of activity of two representative SF (Fig. 10a) and RND (Fig. 10b) simulations.

However only SF networks with SW features display either sub-critical (i.e. showing faster exponential decay) or critical (i.e. power-law) distributions of avalanche sizes depending on the value of the SWI (correlated with network degree). Intuitively, in SF networks with small-world features the coexistence of short- and long-range connections, promoting both segregation and integration of information through high clustering and short path length, should favor the generation of avalanches of all sizes, and thus a critical behavior [49]. On the other hand, in RND networks, where inte-

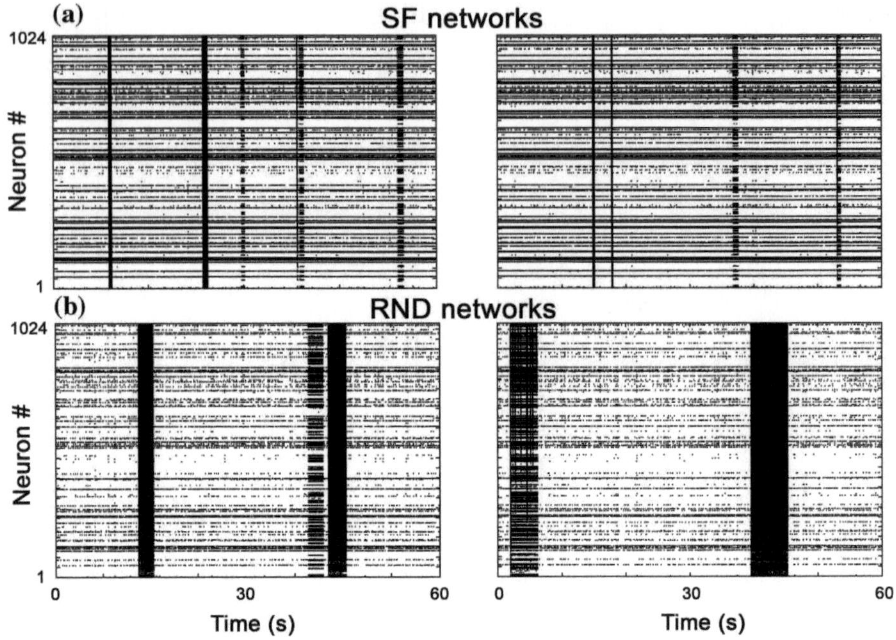

Fig. 10 Simulated network dynamics. Raster plots of 60 s of simulated activity relative to two representative **a** SF and **b** RND networks

gration of activity prevails and clusterization is low, only a super-critical behavior emerges at physiological mean firing rates.

Figure 11 displays four significant examples of avalanche size distributions of two SF (A) and two RND (B) networks, respectively. Figure 11 shows that by varying in a wide range the connectivity degree SF networks, the dynamic state varies as a function of the average degree. Figure 11a shows the avalanche sizes' distributions obtained from the simulation of a SF network with incoming average degree of 33.7 $\pm$ 19.0 (N2, blue line). The power-law behavior is not identifiable and an exponential drop emerges after about 1 decade. By increasing the average degree up to 93.6 $\pm$ 43.7 (N7, red line), the network shifts to a critical regime characterized by a power-law relationship ($\alpha = -1.61$). On the other way round, in RND networks the avalanche distribution always shows a peak in correspondence to the biggest avalanches after an exponential decay (Fig. 11b, green lines), indicative of a super-critical state.

To quantify whether SF and RND networks display power-law distributed avalanches size, we estimated the goodness-of-fit ($GOF_{\text{power law}}$) by means of the Kolmogorov-Smirnov (KS) distance.[5] For the considered simulations, a significance threshold of 0.1 was set [50]. *P*-values are higher than 0.1 only for SF networks with

[5]It is used to measure the distance between the empirical distribution and the fitted model. When the *p*-value is close to 1, the data set is considered to be drawn from the fitted distribution, otherwise it should be rejected.

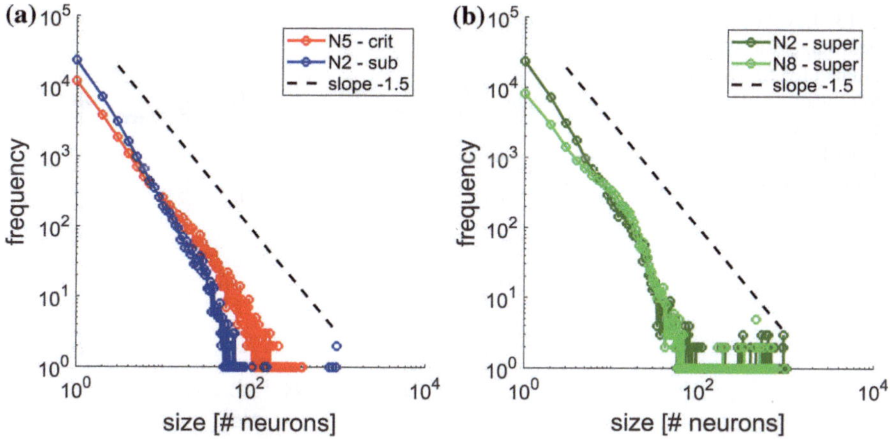

Fig. 11 Avalanche size distributions of SF and RND networks. **a** SF networks can display both sub-critical (blue line) and critical (red line) states depending on the average degree. **b** RND networks display always a super-critical dynamics (green lines) independently of the average network degree. The dashed black line identifies the reference slope of −1.5. Adapted from [36]

Fig. 12 GoF evaluated for SF (red) and RND (black) networks. The gray line indicates the p-level value used to assess power-law fitting. Data are expressed as mean ± standard error of the mean. Adapted from [36]

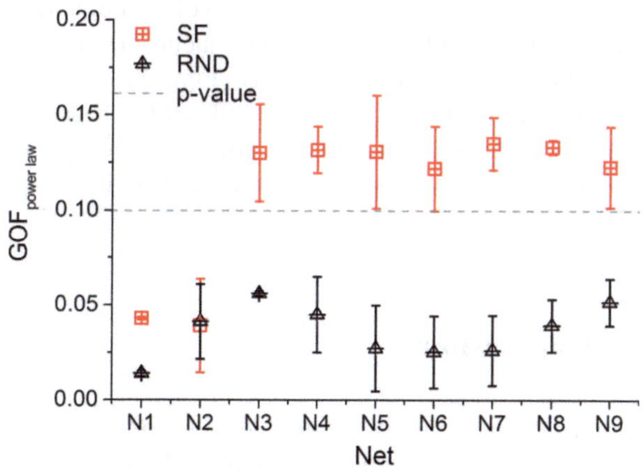

a medium/high degree (i.e., from N3 to N9, Fig. 12, red squares). On the contrary, the avalanche distributions of other SF and of all RND networks present a *p*-value below 0.1 and thus cannot be considered as ruled by a power-law relationship.

The results of Fig. 12 suggest two main observations: first, RND topologies, in the simulated physiological regimes, do not promote power-law avalanche size distribution, even by spanning the incoming average degree in a wide range (from 19.5 ± 4.4 to 97.7 ± 9.1). The second result is that SF networks display a dual behavior: for low connectivity degrees (N1 and N2), the fitting methods do not support a power-law model for the avalanche distribution (GoF < 0.1), while for higher connectivity, the avalanche size distribution approaches a power-law regime (GoF > 0.1). It is worth to notice that the SF network configurations supporting the

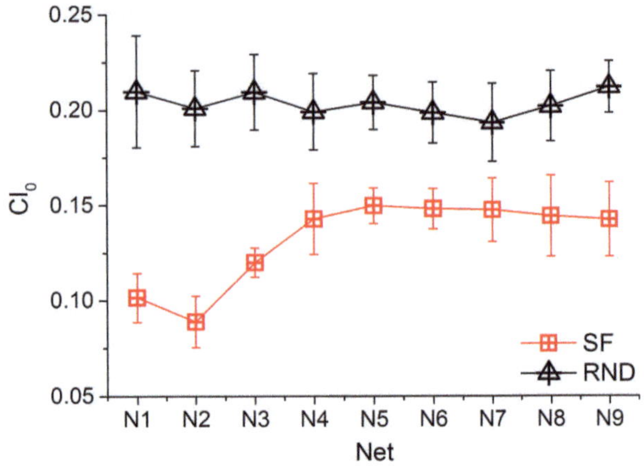

Fig. 13 Level of synchronization (CI_0) evaluated for SF (red) and RND (black) networks. Data are expressed as mean ± standard error of the mean. Adapted from [36]

avalanches' power law distributions are the ones that exhibit small-world features since SWI > 1 (Fig. 9b).

The presented simulated results correlate with the experimental findings of Fig. 7a regarding the level of synchronization of the network. Highly connected SF networks with small-world features are less synchronized than the corresponding RND ones, thus allowing to achieve criticality (Fig. 13).

The combination of SF and small-world properties is the necessary substrate for sustaining the network in an "optimal" regime of synchrony and thus allowing a smooth transition between dynamic states (from sub-criticality to criticality) [51, 52].

6 Conclusions

In this chapter, we presented results coming from the analysis of the spontaneous dynamics generated by large-scale dissociated neuronal assemblies coupled to Micro-Electrode Arrays (MEAs). As Sects. 3 and 4 show, neuronal avalanche size and lifetime distributions follow a power-law relationship in bi-logarithmic scale. This is an evidence of Self-Organized Criticality (SOC) that was found in some cultures at a mature stage of development supporting the idea that this behavior is a universal mechanism that is spontaneously implemented in many neuronal systems both at in vitro and in vivo level. Such an avalanche behavior is a general dynamics that applies at different time scales, reflects different hierarchical levels of organization. The results achieved considering dissociated cultures as biological substrate suggest that despite its simplicity, such an in vitro model is capable to reproduce network dynamic behaviors that resemble those found in other in vitro preparations in which the architecture is partly maintained (i.e. cultured and acute slices [4, 5]). However, not all the cultures necessarily reach a stable critical dynamics. As Fig. 6

shows, mature cortical ensembles can display two other possible avalanche distributions, namely sub-critical and super-critical. Since in this experimental model their development is free of predefined constraints, every network can be substantially different from the others, in terms of the underlying connectivity (not known a priori). Therefore, it is not surprising that not all cultures self-organize and reach a critical state.

The simulations of the model presented in Sect. 5.1 demonstrate that structural features like the presence of hubs, the physiological balance between excitation and inhibition, and the concurrent presence of scale-free and small-world features are necessary ingredients to induce SOC dynamics. The combination of these two topological properties is the necessary substrate for sustaining the network in an "optimal" regime of synchrony and permitting a smooth transition between dynamic states (from sub-criticality to criticality) [51, 52].

The use of high-density MEAs with a large (i.e., thousands) number of electrodes would certainly confirm the prediction of the model: the use of standard MEAs with tens of microelectrodes is not enough to correctly infer in a reliable way the topological properties of a functional network [53]. Future studies should use these new generation of MEAs to better reconstruct the network organization and study its role in shaping the emergent dynamics.

References

1. O'Donovan, M.J.: The origin of spontaneous activity in developing networks of the vertebrate nervous system. Curr. Opin. Neurobiol. **9**, 94–104 (1999)
2. Buzsaki, G., Draguhn, A.: Neuronal oscillations in cortical networks. Science **304**, 1926–1929 (2004)
3. Marom, S., Shahaf, G.: Development, learning and memory in large random networks of cortical neurons: lessons beyond anatomy. Q. Rev. Biophys. **35**, 63–87 (2002)
4. Beggs, J.M., Plenz, D.: Neuronal avalanches in neocortical circuits. J. Neurosci. **23**, 11167–11177 (2003)
5. Beggs, J.M., Plenz, D.: Neuronal avalanches are diverse and precise activity patterns that are stable for many hours in cortical slice cultures. J. Neurosci. **24**, 5216–5229 (2004)
6. Pasquale, V., Massobrio, P., Bologna, L.L., Chiappalone, M., Martinoia, S.: Self-organization and neuronal avalanches in networks of dissociated cortical neurons. Neuroscience **153**, 1354–1369 (2008)
7. Petermann, T., Thiagarajan, T.C., Lebedev, M.A., Nicolelis, M.A., Chialvo, D.R., Plenz, D.: Spontaneous cortical activity in awake monkeys composed of neuronal avalanches. Proc. Natl. Acad. Sci. U.S.A. **106**, 15921–15926 (2009)
8. Gireesh, E.D., Plenz, D.: Neuronal avalanches organize as nested theta- and beta/gamma oscillations during development of cortical layer 2/3. Proc. Natl. Acad. Sci. U.S.A. **105**, 7576–7581 (2008)
9. Hahn, G., Petermann, T., Havenith, M.N., Yu, S., Singer, W., Plenz, D., et al.: Neuronal avalanches in spontaneous activity in vivo. J. Neurophysiol. **104**, 3312–3322 (2010)
10. Poil, S.S., van Ooyen, A., Linkenkaer-Hansen, K.: Avalanche dynamics of human brain oscillations: relation to critical branching processes and temporal correlations. Hum. Brain Mapp. **29**, 770–777 (2008)

11. Van Pelt, J., Corner, M.A., Wolters, P.S., Rutten, W.L.C., Ramakers, G.J.A.: Long-term stability and developmental changes in spontaneous network burst firing patterns in dissociated rat cerebral cortex cell cultures on multi-electrode arrays. Neurosci. Lett. **361**, 86–89 (2004)
12. Rolston, J.D., Wagenaar, D.A., Potter, S.M.: Precisely timed spatiotemporal patterns of neural activity in dissociated cortical cultures. Neuroscience **148**, 294–303 (2007)
13. Wagenaar, D.A., Pine, J., Potter, S.M.: An extremely rich repertoire of bursting patterns during the development of cortical cultures. BMC Neurosci. **7** (2006)
14. Chiappalone, M., Bove, M., Vato, A., Tedesco, M., Martinoia, S.: Dissociated cortical networks show spontaneously correlated activity patterns during in vitro development. Brain Res. **1093**, 41–53 (2006)
15. Tetzlaff, C., Okujeni, S., Egert, U., Worgotter, F., Butz, M.: Self-organized criticality in developing neuronal networks. PLoS Comput. Biol. **6**, e1001013 (2010)
16. Taketani, M., Baudry, M. (eds.): Advances in Network Electrophysiology: Using Multi-Electrode Array. Springer, New York (2006)
17. Massobrio, P., Massobrio, G., Martinoia, S.: Interfacing cultured neurons to microtransducers arrays: a review of the neuro-electronic junction models. Front. Neurosci. **10** (2016)
18. Robinson, D.A.: The electrical properties of metal microelectrodes. Proc. IEEE **56**, 1065–1071 (1968)
19. Gross, G.W., Williams, A.N., Lucas, J.H.: Recording of spontaneous activity with photoetched microelectrode surfaces from mouse spinal neurons in culture. J. Neurosci. Methods **5**, 13–22 (1982)
20. Pine, J.: Recording action potentials from cultured neurons with extracellular microcircuit electrodes. J. Neurosci. Methods **2**, 19–31 (1980, February)
21. Canepari, M., Bove, M., Maeda, E., Cappello, M., Kawana, A.: Experimental analysis of neuronal dynamics in cultured cortical networks and transitions between different patterns of activity. Biol. Cybern. **77**, 153–162 (1997)
22. Jimbo, Y., Robinson, H.P.C., Kawana, A.: Strengthening of synchronized activity by tetanic stimulation in cortical cultures: application of planar electrode arrays. IEEE Trans. Biomed. Eng. **45**, 1297–1304 (1998)
23. Kamioka, H., Maeda, E., Jimbo, Y., Robinson, H.P.C., Kawana, A.: Spontaneous periodic synchronized bursting during formation of mature patterns of connections in cortical cultures. Neurosci. Lett. **206**, 109–112 (1996)
24. Le Feber, J., Stegenga, J., Rutten, W.L.C.: The effect of slow electrical stimuli to achieve learning in cultured networks of rat cortical neurons. PLoS ONE **5**, e8871 (2010)
25. Wagenaar, D.A., Pine, J., Potter, S.M.: Effective parameters for stimulation of dissociated cultures using multi-electrode arrays. J. Neurosci. Methods **138**, 27–37 (2004)
26. Defranchi, E., Novellino, A., Whelan, M., Vogel, S., Ramirez, T., van Ravenzwaay, B., et al.: Feasibility assessment of micro-electrode chip assay as a method of detecting neurotoxicity *in vitro*. Front. Neuroeng. **4**, 6 (2011)
27. Wagenaar, D.A., Madhavan, R., Pine, J., Potter, S.M.: Controlling bursting in cortical cultures with closed-loop multi-electrode stimulation. J. Neurosci. **25**, 680–688 (2005)
28. Martinoia, S., Sanguineti, V., Cozzi, L., Berdondini, L., Van Pelt, J., Tomas, J., et al.: Towards an embodied in-vitro electrophysiology: the NeuroBIT project. Neurocomputing **58–60**, 1065–1072 (2004)
29. Lewicki, M.S.: A review of methods for spike sorting: the detection and classification of neural action potentials. Netw. Comput. Neural Syst. **9**, R53–R78 (1998)
30. Wilson, S.B., Emerson, R.: Spike detection: a review and comparison of algorithms. Clin. Neurophysiol. **113**, 1873–1881 (2002)
31. Maccione, A., Gandolfo, M., Massobrio, P., Novellino, A., Martinoia, S., Chiappalone, M.: A novel algorithm for precise identification of spikes in extracellularly recorded neuronal signals. J. Neurosci. Methods **177**, 241–249 (2009)
32. Abbott, L.F., Rohrkemper, R.: A simple growth model constructs critical avalanche networks. Prog. Brain Res. **165**, 13–19 (2007)

33. Wagenaar, D.A., Nadasdy, Z., Potter, S.M.: Persistent dynamic attractors in activity patterns of cultured neuronal networks. Phys. Rev. E **73** (2006)
34. Van Pelt, J., Wolters, P.S., Corner, M.A., Rutten, W.L.C., Ramakers, G.J.A.: Long-term characterization of firing dynamics of spontaneous bursts in cultured neural networks. IEEE Trans. Biomed. Eng. **51**, 2051–2062 (2004)
35. Sporns, O., Chialvo, D.R., Kaiser, M., Hilgetag, C.C.: Organization, development and function of complex brain networks. Trends Cogn. Sci. **8**, 418–425 (2004)
36. Massobrio, P., Pasquale, V., Martinoia, S.: Self-organized criticality in cortical assemblies occurs in concurrent scale-free and small-world networks. Sci. Rep. **5** (2015)
37. Eytan, D., Marom, S.: Dynamics and effective topology underlying synchronization in networks of cortical neurons. J. Neurosci. **26**, 8465–8476 (2006)
38. Berdondini, L., Imfeld, K., Maccione, A., Tedesco, M., Neukom, S., Koudelka-Hep, M., et al.: Active pixel sensor array for high spatio-temporal resolution electrophysiological recordings from single cell to large scale neuronal networks. Lab Chip **9**, 2644–2651 (2009)
39. Eversmann, B., Jenker, M., Hofmann, F., Paulus, C., Brederlow, R., Holzapfl, B., et al.: A 128 x 128 CMOS biosensor array for extracellular recording of neural activity. IEEE J. Solid-State Circuits **38**, 2306–2317 (2003)
40. Frey, U., Egert, U., Heer, F., Hafizovic, S., Hierlemann, A.: Microelectronic system for high-resolution mapping of extracellular electric fields applied to brain slices. Biosens. Bioelectron. **24**, 2191–2198 (2009)
41. Pastore, V.P., Godjoski, A., Martinoia, S., Massobrio, P.: SpiCoDyn: a toolbox for the analysis of neuronal network dynamics and connectivity from multi-site spike signal recordings. Neuroinformatics **16**, 15–30 (2018)
42. Rubinov, M., Sporns, O.: Complex network measures of brain connectivity: uses and interpretations. NeuroImage **52**, 1059–1069 (2010)
43. Sporns, O.: Graph theory methods for the analysis of neural connectivity patterns. In: Kotter, R. (ed.) Neuroscience Databases. A Practical Guide. Klüwer (2002)
44. Izhikevich, E.M.: Simple model of spiking neurons. IEEE Trans. Neural Netw. **14**, 1569–1572 (2003)
45. Erdős, P., Rényi, A.: On random graphs I. Publ. Math. **6**, 290–297 (1959)
46. Barabasi, A.-L., Albert, R.: Emergence of scaling in random networks. Science **286**, 509–512 (1999)
47. Dorogovtsev, S., Mendes, J.: Evolution of networks. Adv. Phys. **51**, 1079–1187 (2002)
48. Timofeev, I., Grenier, F., Bazhenov, M., Sejnowski, T.J., Steriade, M.: Origin of slow cortical oscillations in deafferented cortical slabs. Cereb. Cortex **10**, 1185–1199 (2000)
49. Friedman, N., Ito, S., Brinkman, B.A., Shimono, M., DeVille, R.E., Dahmen, K.A., et al.: Universal critical dynamics in high resolution neuronal avalanche data. Phys. Rev. Lett. **108**, 208102 (2012, 18 May)
50. Clauset, A., Shalizi, C.R., Newman, M.E.J.: Power-law distributions in empirical data. SIAM Rev. **51**, 661–703 (2009)
51. Priesemann, V., Wibral, M., Valderrama, M., Pröpper, R., Le Van Quyen, M., Geisel, T., et al.: Spike avalanches in vivo suggest a driven, slightly subcritical brain state. Front. Syst. Neurosci. **8**, 2014-June-24 (2014)
52. Tomen, N., Rotermund, D., Ernst, U.: Marginally subcritical dynamics explain enhanced stimulus discriminability under attention. Front. Syst. Neurosci. **8**, 2014-August-25 (2014)
53. Pastore, V.P., Massobrio, P., Godjoski, A., Martinoia, S.: Identification of excitatory-inhibitory links and network topology in large scale neuronal assemblies from multi-electrode recordings. PLoS Comput. Biol. (2018)

Paolo Massobrio graduated in Biomedical Engineering at the University of Genova in 2004 and received his Ph.D. in Bioengineering from the same institution in 2008. Currently, he is associate professor of Bioengineering at the University of Genova where he teaches courses of bioelectronics and computational neuroscience. His research activities fit into the field of neuroengineering, by investigating neuronal network properties and neural interfaces. His current research interests include the study of the functional topological properties of large-scale cortical assemblies and the interplay between dynamics and connectivity.

Valentina Pasquale graduated in Bioengineering at the University of Genova in 2006 and received her Ph.D. in Humanoid Technologies from the same institution in 2010. She has worked on the development of innovative analysis methods for neuroscience data, ranging from electrophysiology to imaging. Her current research interests include thalamo-cortical circuits in slow-wave oscillations, and holographic optogenetic stimulation.

From Neurons to Networks: Critical Slowing Down Governs Information Processing Across Vigilance States

Christian Meisel

Abstract The general idea that computational capabilities are maximized at or nearby critical states related to phase transitions or bifurcations led to the hypothesis that neural systems in the brain operate at or close to a critical state. Near phase transitions, a system is expected to recover more slowly from small perturbations, a phenomenon called critical slowing down. In this chapter we will review and discuss recent studies that have identified critical slowing down as a pervasive feature in neural system functioning and information processing across different spatial scales from individual neurons to cortical networks. First, we will provide an easily accessible introduction into the theory of critical slowing down with an emphasis on its scaling laws. Second, we will review experimental work using the whole-cell patch clamp technique demonstrating how critical slowing down governs the onset of spiking in individual neurons. The associated scaling laws identify a saddle-node bifurcation underlying the transition to spiking in pyramidal neurons and fast-spiking interneurons. We will discuss implications for the integration of synaptic inputs and neuronal information processing in general. Third, we will review evidence for the existence of critical slowing down at the cortical network level. Recent studies in rodents and humans conclusively show that cortex is goverend by long dynamical timescales expected from critical slowing down that support temporal information integration but change as a function of vigilance state and time awake. The results provide novel mechanistic and functional links between behavioural manifestations of sleep, waking and sleep deprivation, and specific measurable changes in the network dynamics relevant for characterizing the brain's changing ability to integrate and process information over time and across vigilance states.

C. Meisel
University Clinic Carl Gustav Carus, Dresden, Germany
e-mail: christian@meisel.de

N. Tomen et al. (eds.), *The Functional Role of Critical Dynamics in Neural Systems*, Springer Series on Bio- and Neurosystems 11,
https://doi.org/10.1007/978-3-030-20965-0_4

1 Introduction

In recent years it has become apparent that the concept of phase transitions is not only applicable to the systems classically considered in physics. It applies to a much wider class of complex systems exhibiting phases, characterized by qualitatively different types of long-term behavior. The pertinence of such a framework for brain functioning goes back to the 1950s when Alan Turing suggested it as a way the nervous system is afforded the speed and flexibility required for instantaneous reaction to novelty [44]. Since then, the concept of phase transitions has been shown to accurately describe the behavior of neural systems along different scales, including individual neurons [15, 37], brain network function [2, 8], the transition between mutually exclusive motor programs [40] or the on- or offset of disease states such as epileptic seizures [20, 31].

In the terminology of physics, a system is said to exhibit a phase transition if it passes a threshold where the emergent macroscopic behavior changes qualitatively. To characterize this transition, one usually defines an 'order parameter' to distinguish different macroscopic behaviours of the system. Upon variation of an ambient property, called the 'control parameter', one then investigates how this order parameter changes. Generally, there is a smooth change in the order parameter as the control parameter is changed. At certain points, however, the values of the order parameters undergo a rapid transition, exhibiting a 'jump' or sharp turn. At these critical points, the phase transition occurs, which marks the boundary between different phases. If the system resides at or near this critical point, it is said to be poised at criticality. The mathematical analog of phase transitions are bifurcations. Bifurcation theory and physics have been shown to provide the appropriate tools to quantitatively describe the rapid transitions observed in many neural systems.

The relevance of phase transitions and bifurcations for neural systems is further supported by their advantageous functional roles. Being at or near a critical state is attractive because criticality has been argued to provide computational advantages in some instances [24]. In the critical states, small changes can have a large effect on the system. Therefore, a large dynamical repertoire can potentially be explored at minimal energy expense [12, 19, 41]. Furthermore, at criticality systems exhibit long-range spatial and temporal correlations which can subserve memory of events long in the past or across cortical areas far apart. Such properties of critical states may prove to be advantageous for computation and memory in the brain.

The systematic effects in dynamics that systems exhibit as they are brought closer to the critical point are usually referred to as *critical slowing down.* As will be reviewed in more detail below Sect. 2, critical slowing down in its broadest sense refers to the tendency of systems to recover more and more slowly from small perturbation the closer they get to the critical point [47]. Recent research has provided important new insights into critical slowing down in theory and experiment. At the theoretical level, the scaling laws underlying critical slowing down, i.e., how specifically certain system properties change as a function to the distance from the critical point, could be identified [22, 31]. At the experimental level, critical slowing down was observed to occur in different neural systems and across spatial scales. Importantly, these experimental observations closely matched the theoretical predictions and, collectively, have identified critical slowing down as a pervasive feature of neural systems ranging from individual neurons to large-scale cortical networks. As such, these results provide important insight into neural system functioning and how it subserves information processing in the brain.

The current chapter aims to provide an accessible overview over these new insights. The underlying motivation is that a thorough understanding of critical transitions and critical slowing down will be essential to grasp the principles of information processing in neural systems. We will start by providing an introduction into the theory of critical slowing down with a particular emphasis on its newly identified scaling laws. Next, we will summarize experimental evidence of critical slowing down in biological neurons and its implications for information processing and functioning. Finally, we will review experimental evidence for critical slowing down at the cortical network level. We will discuss how critical slowing down provides mechanistic and functional links between behavioural manifestations of sleep, waking and sleep deprivation, and specific measurable changes in the network dynamics relevant for characterizing the brain's changing information processing capabilities.

2 The Theory of Critical Slowing Down

The following section is intended to provide an intuitive and easily accessible introduction to critical slowing down and its scaling laws. Critical slowing down results from the ever slower recovery from small perturbations when a bifurcation or phase transition is approached [47]. For demonstration purposes, let us consider a ball in a two-well potential (Fig. 1, left). Initially the ball sits in one of the two wells. The bifurcation in this case corresponds to the transition of the ball to the other well, which is facilitated by slowly changing the potential landscape. Now consider a small perturbation, for example by kicking the ball a bit, upon which the ball will roll up and down in the left well, but will eventually return to the bottom of the first well exponentially fast (Fig. 1, right). As the bifurcation or phase transition, at which the ball driven by noise fluctuations transitions to the second well, approaches, this recovery from perturbations becomes progressively slower. This slowing of dynamics upon approaching the bifurcation or phase transition point is the essence of critical slowing down. Consequently, critical slowing down can be monitored by measuring the recovery rate of system variables after small perturbations but it also manifests itself by an increase in its fluctuations, i.e., variance, due to the longer relaxation times, as well as higher autocorrelation values near the bifurcation [6, 14].

The occurrence of critical slowing down is a natural consequence of a system at or near a critical phase transition. Over the last years, critical slowing down has attracted considerable attention in a wide range of systems outside of neuroscience where it was predominantly studied as a potential warning signal of an impending bifurcation or tipping point [38]. In many real-world settings, warning signals of impending

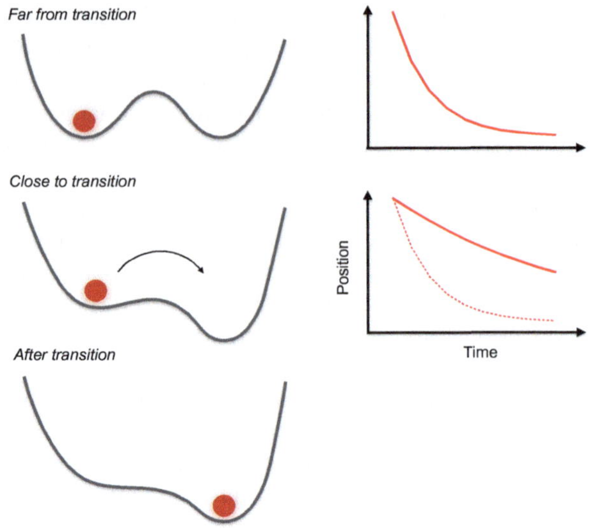

Fig. 1 Schematic demonstration of critical slowing down. The ball in the first well needs longer and longer to recover from small perturbations upon approaching the bifurcation point, at which the ball transitions to the second well. The solid and dashed lines on the right indicate the recovery upon perturbations

critical transitions are highly desirable because it is often difficult to revert a system to the previous state once a critical transition has occurred [23, 39].

Research over the last years has identified the exact scaling laws of critical slowing near a bifurcation by studying bifurcation normal forms. Recall that phase transitions in physics equate to bifurcations in mathematics. Normal forms are model systems associated with a bifurcation exemplifying the bifurcation type. These normal forms can be used to quantify how exactly quantities like recovery rates from small perturbations change as the bifurcation point is approached. Critical slowing down and its scaling can be derived for different bifurcations which correspond to different types of transitions. For demonstration purposes how these scaling laws can be derived, consider a saddle-node bifurcation which is a common type of bifurcation where a stable fixed point loses stability to an unstable fixed point. A normal form of a saddle-node bifurcation is given by

$$\frac{\mathrm{d}V}{\mathrm{d}t} = y + V^2. \tag{1}$$

Here, y is the parameter which controls the distance to the bifurcation and which is approached from the negative side. There are two equilibria $V = \pm\sqrt{-y}$ for $y < 0$ and the saddle-node bifurcation occurs for $y_c = 0$. The equilibrium $V^- := -\sqrt{-y}$ is stable because the linearized system around V^- is

$$\frac{\mathrm{d}X}{\mathrm{d}t} = (\mathrm{D}_V f)(V^-)X = -2V^-X = -2\sqrt{-y}X, \tag{2}$$

where D is the differential. If one assumes y to be quasi-stationary, one may solve (2) and obtain

$$X(t) = X(0)e^{-2\sqrt{-y}t}. \tag{3}$$

From Eq. (3) one can see that when dynamics is perturbed slightly away from the stable equilibrium V^- for $y < 0$ it will return to V^- exponentially fast. More specifically, the exponent by which it returns scales like $\mathcal{O}(\sqrt{-y})$ in terms of the y-variable as $y \nearrow 0$. Consequently, the closer one gets to the bifurcation point $y_c = 0$, the longer it takes to recover from a perturbation which is the essence of critical slowing down [11, 38, 47]. By studying these normal forms, one can thus derive the exact mathematical scaling laws of how quantities like recovery rate change as a function of distance to the bifurcation point and for different types of bifurcations [22, 31].

Conclusively, theory posits critical slowing down to govern the dynamics near transitions in neural systems and, thereby, to subserve an essential role in information processing. Rigorous experimental evidence of critical slowing down, however, has been sparse. In the following, we will discuss recent work demonstrating critical slowing down in neural systems and how it supports information processing.

3 Critical Slowing Down in Individual Neurons

Functionally, quiescence and spiking can be regarded as two different states or phases a neuron can be in. The transition between these two phases, i.e., the onset of spiking in individual neurons, has long been studied with the mathematical tools of bifurcation theory [15]. It is because of this proximity to a bifurcation that neurons are excitable and have the ability to exhibit a qualitative change in their dynamics. By injecting a current, a quiescent neuron can be gradually driven towards the spiking threshold (Fig. 2a, b). At the threshold, its qualitative behavior changes rapidly—it starts to fire action potentials. In the terminology of phase transitions, the injected current corresponds to the control parameter. A suitable order parameter can be defined by the neuron's firing frequency, for example, which becomes non-zero at the spiking onset. As is well-known, the onset of spiking is caused by the complex interplay between membrane ion channels, i.e., the interaction of many constituents. Seminal studies have provided evidence that this complex biological system can be accurately described as a bifurcation or phase transition [13, 15, 36, 37].

Although theory provides quantitative predictions on critical slowing down [42], direct experimental evidence in neurons approaching their spiking threshold has long been lacking. Specifically, bifurcation theory posits that variance and autocorrelation in subthreshold membrane potential activity should increase as a neuron comes closer to spiking, while the recovery from small perturbations should take increasingly longer [6, 14]. Importantly, the scaling laws by which these measures change is predicted by the bifurcation type [22, 31]. The experimental confirmation of critical slowing down has therefore long represented a missing link to theory.

Using whole-cell patch-clamp recordings, we developed a stimulation protocol to monitor markers of critical slowing down [14, 38, 39] while systematically driving

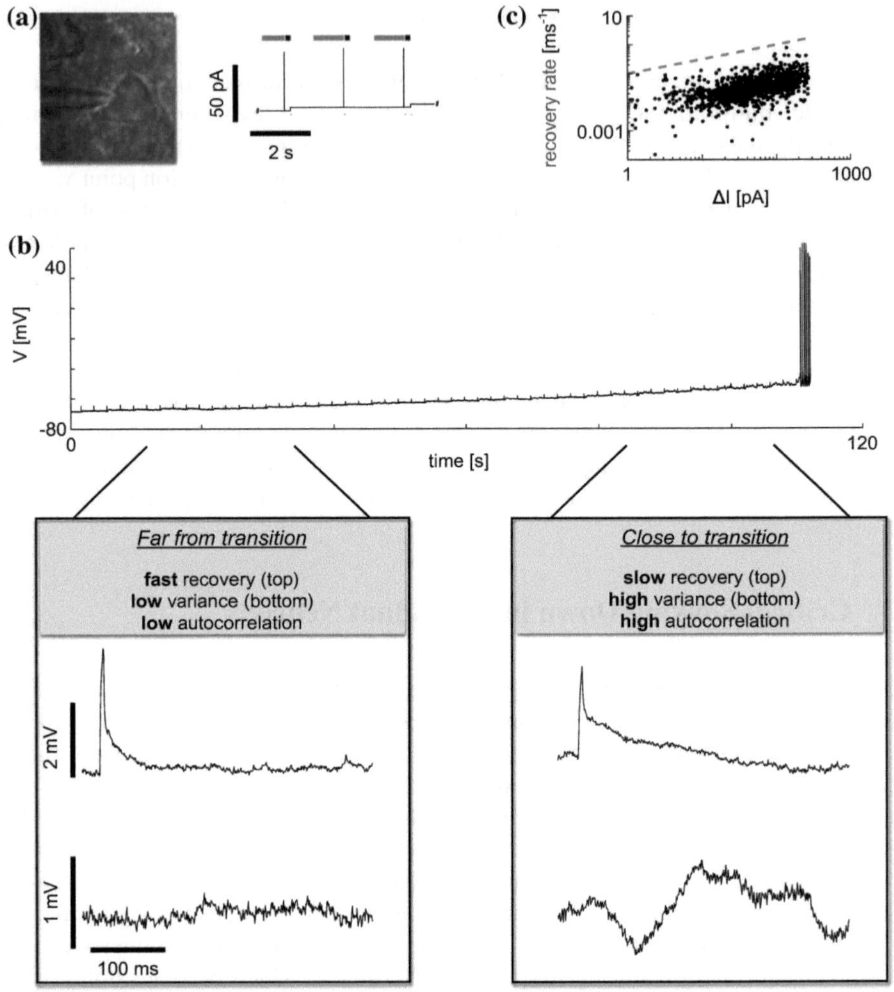

Fig. 2 Critical slowing down governs the transition to neuron spiking. **a**, Using the whole-cell patch-clamp technique, a pyramidal neuron is driven to its spiking threshold by injecting a step-wise increasing current with small perturbations on top. **b**, Subthreshold voltage exhibits clear signatures of critical slowing down upon reaching spiking onset: a progressively slower recovery from perturbations along with increasing variance and autocorrelation. These changes become apparent when comparing subthreshold dynamics far from (left) and close to the spiking onset (right). **c**, The experimental measures (black dots) follow predictions for a saddle-node bifurcation (grey dashed line). Specifically, in log-log coordinates, the dashed line has a slope of 0.5, which is the scaling exponent expected for a saddle-node bifurcation (see Eq. (3)). Measures of the recovery rate from small perturbation follow this scaling relationship predicted by theory [29]

the neuron towards its spiking threshold by increasing the injected current (Fig. 2a, b). Characteristics of critical slowing down, i.e., a slower recovery from perturbations along with increasing variance and autocorrelation, became evident as the neuron approached the spiking threshold [29]. The onset of spiking thus corresponds to a critical transition from a dynamical perspective. Furthermore, the scaling laws suggest a saddle-node bifurcation governing slowing down and the onset of spiking (Fig. 2c). It is interesting to note that, by measuring the scaling laws of slowing down, the type of transition (or, more specifically, bifurcation) can thus be determined purely from the subthreshold activity without observing the transition itself.

Collectively, these results further demonstrate the deep insights that the concept of critical transitions already provides at the individual neuron level. For example, it explains why subthreshold voltage variance [16] or shape of postsynaptic potential [9] are affected by a neuron's holding potential. The occurrence of critical slowing down near the spiking threshold has considerable implications on the integration of inputs and, more generally, information processing in neurons. Interestingly, to our knowledge, this particular aspect of neuron functioning has not been investigated in much detail yet. In fact, many neuron models, such as most leaky integrate-and-fire neuron models, do not take the effects of critical slowing down into account, by omitting the dynamical modeling of action potential generation [5].

The systematic decrease in recovery rate resulting from critical slowing down implies that inputs to a neuron, for example in the form of excitatory postsynaptic potentials (EPSPs), will become wider the closer the neuron is to the spiking threshold. For demonstration purposes, consider the two responses to the same input depicted in Fig. 2b, one far and one close to the onset of spiking. It is clear that the more wide or more narrow shape of the EPSP from other neurons' inputs will affect how subsequent inputs are integrated and, by consequence, affect the generation and timing of action potentials. Specifically, the widening of EPSPs due to critical slowing down should progressively facilitate the temporal integration of small inputs to a neuron the closer it gets to the spiking threshold since a longer memory of past inputs is maintained.

4 Critical Slowing Down in Cortical Networks Is Maintained by Sleep

A large body of research has identified bifurcations and, more recently, also critical slowing down as essential features of individual neuron functioning with implications to their ability to integrate and process information. At the neural network level, the maintenance and integration of information over extended periods of time is similarly thought to be important for information processing. For example, in decision-making and working memory tasks [7, 10, 17, 21], the ability to integrate information over time may increase the signal-to-noise ratio and help to maintain some memory of the past. Persistent network activity characterized by slowly decaying perturbation responses and autocorrelation functions, which are generic features of critical slowing down, as discussed above, may provide the basis for this integration in cortex. Consequently, neural network models, when poised at criticality, exhibit the slowest decaying autocorrelation functions [27, 30]. The likelihood of critical slowing down to play a role in cortical networks dynamics is further supported by the observation of other neuronal activity patterns consistently following power-law distributions, a hallmark of systems at a continuous phase transition [2, 4, 8, 48]. Slowly decaying autocorrelation functions have been reported in awake cortex [25, 34] and to disappear under anesthesia [3]. Critical slowing down and the long temporal memory effects it brings about is therefore an attractive and likely feature not only for individual neurons, but also for cortical network functioning. Monitoring critical slowing down might consequently provide insights into the brains changing ability to integrate and process information over time and across vigilance states.

Using a multi-modal experimental approach spanning work on humans and rodents, we recently asked whether there is indication of critical slowing down in cortical network activity and whether it changes across time and across different vigilance states [28, 30, 32]. Different vigilance states are known to impact cognitive capabilities and, more generally, cortical information processing. Specifically, it has long been a common observation that the brain's ability to process information declines during extended waking [1, 18, 33, 45]. Cortical information processing is also affected by sleep. It has been suggested that during non-rapid eye movement sleep (NREM sleep, a form of deep sleep) the brain loses its ability to effectively integrate information [26]. This drastic decline of information integration capabilities during NREM sleep has also been proposed to underlie the concomitant loss of consciousness [43]. The neural correlates underlying the decline of information processing capabilities during extended wake and sleep, however, have been difficult to identify.

We focused on the decay of autocorrelation functions as a likely prerequisite for cortical network information integration with direct relationship to critical slowing down. The analysis of data from electroencephalogram in humans [28, 32] as well as neuron and local field potential activity in rodents [30] demonstrated that long-range temporal correlations related to slowly decaying autocorrelation functions in cortical network activity are dependent on vigilance states. While wake and REM sleep exhibit long timescales suitable for information integration, these long-range temporal correlations break down during NREM sleep. Furthermore, extended wake leads to a progressively faster decay of autocorrelation functions suggestive of a progressive decline of information processing capabilities in cortical networks (Fig. 3).

The slow autocorrelation function decays observed during normal wake and also during REM sleep are in line with the idea that cortex dynamics is poised in the vicinity of a phase transition where critical slowing down governs activity. Critical slowing down occurs upon approaching a critical transition; it can therefore be expected to govern dynamics even when networks are not exactly at the critical point but, for example, in a slightly subcritical regime [35]. At or near criticality, cortex dynamics can thus benefit from long temporal information integration capabilities

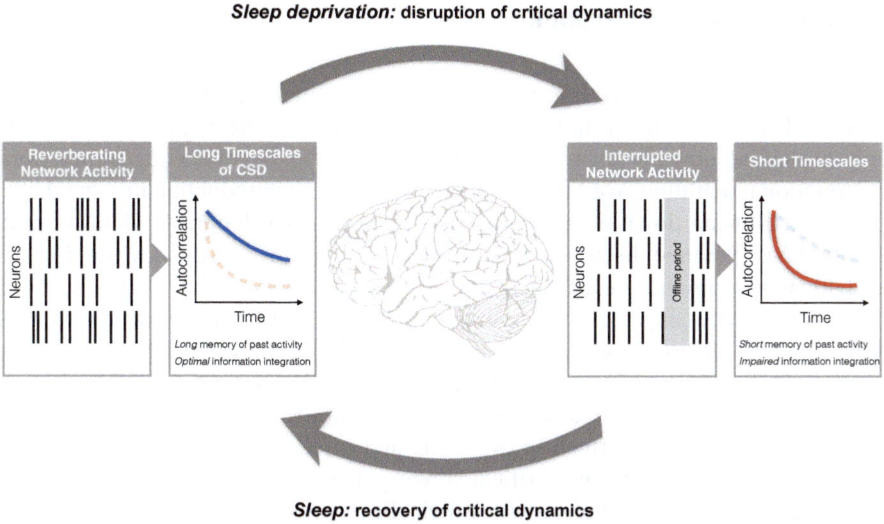

Fig. 3 Sleep recovers critical network dynamics for optimal information processing. Normal cortex activity is characterized by reverberating, critical dynamics (left). The long timescales associated with critical slowing down (CSD) afford the memory maintenance of past activity and thereby contribute to optimal information integration. During extended wake, these long timescales are increasingly disrupted by neuronal offline periods, which may lead to a decline of optimal information integration and cognition (right). Sleep, conversely, reestablishes critical dynamics and its associated long timescales for optimal function during wake [28, 30, 32]

which may help to, for example, increase the signal-to-noise ratio in decision-making and working memory tasks. Conversely, during NREM sleep and extended wake the long-range temporal correlations were found to be reduced suggesting an impaired ability to integrate information over time. But what causes this disruption of critical dynamics during NREM sleep and also during extended wake? In detailed analyses we found that the tendency of neurons to go into a down state leads to this disruption. These periods where neurons do not participate in ongoing network dynamics anymore, and so to say go offline for a brief amount of time, are well known to occur in NREM sleep, but have also been observed as local phenomena during extended wake [46]. The occurrence of such offline periods disrupts the normally occurring reverberating cortical network activity and shuts down any memory it might have of the past (Fig. 3, [30]). The increased disruption of critical slowing down and its associated long timescales during extended wake complements other findings that have also reported fading signatures of critical dynamics during sleep deprivation [32]. Collectively, these observations thus support an intriguing hypothesis for a function of sleep, to re-organize cortical networks towards critical dynamics for optimal information processing during wake.

5 Summary and Outlook

The existence of a critical transition and, by consequence, critical slowing down in individual neuron function is now well established. At the neural network level, critical slowing down is attractive as it may provide the basis for effective temporal information integration. We here provided an accessible introduction into the theory of critical slowing down with an emphasis on recent developments identifying its scaling laws. We then discussed experimental evidence of critical slowing down in individual neurons and at the cortical network level and how it relates to and supports information processing at each spatial scale. It became apparent how critical slowing down can provide novel mechanistic and functional links between behavioural manifestations of sleep, waking and sleep deprivation, and specific measurable changes in the network dynamics relevant for characterising the brain's changing ability to integrate and process information over time and across vigilance states. As shown for the individual neuron case, critical slowing down is best monitored by slightly changing a suitable control parameter which allows the rigorous identification of the transition mechanism from subthreshold dynamics. Novel genetic and optogenetic tools might afford a similar experiment at the cortical network level and, for example by careful titration of the excitation/inhibition axis, to characterize critical slowing down with similar rigor to identify and map out the critical transition governing cortex dynamics.

References

1. Banks, S., Dinges, D.F.: Behavioral and physiological consequences of sleep restriction. J. Clin. Sleep Med. **3**, 519–528 (2007)
2. Beggs, J.M., Plenz, D.: Neuronal avalanches in neocortical circuits. J. Neurosci. **23**, 11167–11177 (2003)
3. Bellay, T., Klaus, A., Seshadri, S., Plenz, D.: Irregular spiking of pyramidal neurons organizes as scale-invariant neuronal avalanches in the awake state. Elife **4**, e07,224 (2015)
4. Benayoun, M., Kohrman, M., Cowan, J., van Drongelen, W.: EEG, temporal correlations, and avalanches. J. Clin. Neurophysiol. **27**, 458–464 (2010)
5. Burkitt, A.: A review of the integrate-and-fire neuron model: I. homogeneous synaptic input. Biol. Cybern. **95**, 1–19 (2006)
6. Carpenter, S.R., Brock, W.A.: Rising variance: a leading indicator of ecological transition. Ecol. Lett. **9**, 311318 (2006)
7. Chaudhuri, R., Knoblauch, K., Gariel, M.A., Kennedy, H., Wang, X.J.: A large-scale circuit mechanism for hierarchical dynamical processing in the primate cortex. Neuron **88**(2), 419–431 (2015)
8. Chialvo, D.R.: Emergent complex neural dynamics. Nat. Phys. **6**, 744–750 (2010)
9. Fricker, D., Miles, R.: EPSP amplification and the precision of spike timing in hippocampal neurons. Neuron **28**(2), 559–569 (2000)
10. Friston, K., Breakspear, M., Deco, G.: Perception and self-organized instability. Front. Comput. Neurosci. **6**, 44 (2012)
11. Guckenheimer, J., Holmes, P.: Nonlinear Oscillations, Dynamical Systems, and Bifurcations of Vector Fields. Springer, New York, NY (1983)

12. Haldeman, C., Beggs, J.: Critical branching captures activity in living neural networks and maximizes the number of metastable states. Phys. Rev. Lett. **94**, 058, 101 (2005)
13. Hodgkin, A.L., Huxley, A.F.: A quantitative description of membrane current and its application to conduction and excitation in nerve. J. Physiol. (Lond.) **10**, 500 (1952)
14. Ives, A.R.: Measuring resilience in stochastic systems. Ecol. Monogr. **65**, 217233 (1995)
15. Izhikevich, E.: Neural excitability, spiking, and bursting. Int. J. Bif. Chaos **10**, 1171–1266 (2000)
16. Jacobson, G.A., Diba, K., Yaron-Jakoubovitch, A., Oz, Y., Koch, C., Segev, I., Yarom, Y.: Subthreshold voltage noise of rat neocortical pyramidal neurones. J. Physiol. **564**, 145160 (2005)
17. Kiebel, S.J., Daunizeau, J., Friston, K.J.: A hierarchy of time-scales and the brain. PLoS Comput. Biol. **4**(11), e1000,209 (2008)
18. Killgore, W.D.: Effects of sleep deprivation on cognition. Prog. Brain Res. **185**, 105–129 (2010)
19. Kinouchi, O., Copelli, M.: Optimal dynamical range of excitable networks at criticality. Nat. Phys. **2**, 348–351 (2006)
20. Kramer, M.A., Truccolo, W., Eden, U.T., Lepage, K.Q., Hochberg, L.R., Eskandar, E.N., Madsen, J.R., Lee, J.W., Maheshwari, A., Halgren, E., Chu, C.J., Cash, S.S.: Human seizures self-terminate across spatial scales via a critical transition. Proc. Natl. Acad. Sci. U.S.A. **109**(51), 21116–21121 (2012)
21. Kringelbach, M.L., McIntosh, A.R., Ritter, P., Jirsa, V.K., Deco, G.: The rediscovery of slowness: exploring the timing of cognition. Trends Cogn. Sci. (Regul. Ed.) **19**(10), 616–628 (2015)
22. Kuehn, C.: A mathematical framework for critical transitions: bifurcations, fast-slow systems and stochastic dynamics. Physica D **240**(12), 1020–1035 (2011)
23. Lade, S.J., Gross, T.: Early warning signals for critical transitions: a generalized modeling approach. PLoS Comput. Biol. **8**(2), e1002,360 (2012)
24. Langton, C.G.: Computation at the edge of chaos: phase transitions and emergent computation. Physica D **42**, 12–37 (1990)
25. Linkenkaer-Hansen, K., Nikouline, V.V., Palva, J.M., Ilmoniemi, R.J.: Long-range temporal correlations and scaling behavior in human brain oscillations. J. Neurosci. **21**, 13701377 (2001)
26. Massimini, M., Ferrarelli, F., Huber, R., Esser, S.K., Singh, H., Tononi, G.: Breakdown of cortical effective connectivity during sleep. Science **309**(5744), 2228–2232 (2005)
27. Meisel, C.: Linking cortical network synchrony and excitability. Commun. Integr. Biol. **9**(1), e1128,598 (2016)
28. Meisel, C., Bailey, K., Achermann, P., Plenz, D.: Decline of long-range temporal correlations in the human brain during sustained wakefulness. Sci. Rep. **7**(1), 11, 825 (2017)
29. Meisel, C., Klaus, A., Kuehn, C., Plenz, D.: Critical slowing down governs the transition to neuron spiking. PLoS Comput. Biol. **11**(2), e1004,097 (2015)
30. Meisel, C., Klaus, A., Vyazovskiy, V.V., Plenz, D.: The interplay between long- and short-range temporal correlations shapes cortex dynamics across vigilance states. J. Neurosci. **37**(42), 10114–10124 (2017)
31. Meisel, C., Kuehn, C.: Scaling effects and spatio-temporal multilevel dynamics in epileptic seizures. PLoS One **7**(2), e30,371 (2012)
32. Meisel, C., Olbrich, E., Shriki, O., Achermann, P.: Fading signatures of critical brain dynamics during sustained wakefulness in humans. J. Neurosci. **33**(44), 17363–17372 (2013)
33. Mignot, E.: Why we sleep: the temporal organization of recovery. PLoS Biol. **6**(4), e106 (2008)
34. Petermann, T., Thiagarajan, T.C., Lebedev, M.A., Nicolelis, M.A.L., Chialvo, D.R., Plenz, D.: Spontaneous cortical activity in awake monkeys composed of neuronal avalanches. Proc. Natl. Acad. Sci. U.S.A. **106**, 15921–15926 (2009)
35. Priesemann, V., Wibral, M., Valderrama, M., Propper, R., Le Van Quyen, M., Geisel, T., Triesch, J., Nikoli?, D., Munk, M.H.: Spike avalanches in vivo suggest a driven, slightly subcritical brain state. Front. Syst. Neurosci. **8**, 108 (2014)
36. Rinzel, J.: A formal classification of bursting mechanisms in excitable systems. In: Proceedings of International Congress Math, Berkeley (1986)

37. Rinzel, J., Ermentrout, B.: Analysis of neural excitability and oscillations. In: Koch, C., Segev, I. (eds.) Methods in Neuronal Modeling: From Ions to Networks, p. 251. MIT Press, Cambridge, MA (1989)
38. Scheffer, M., Bascompte, J., Brock, W.A., Brovkin, V., Carpenter, S.R., Dakos, V., Held, H., van Nes, E.H., Rietkerk, M., Sugihara, G.: Early-warning signals for critical transitions. Nature **461**, 5359 (2009)
39. Scheffer, M., Carpenter, S.R., Lenton, T.M., Bascompte, J., Brock, W., Dakos, V., van de Koppel, J., van de Leemput, I.A., Levin, S.A., van Nes, E.H., Pascual, M., Vandermeer, J.: Anticipating critical transitions. Science **338**, 344–348 (2012)
40. Schöner, G., Kelso, J.A.S.: Dynamic pattern generation in behavioral and neural systems. Science **239**, 1513–1520 (1988)
41. Shew, W.L., Yang, H., Petermann, T., Roy, R., Plenz, D.: Neuronal avalanches imply maximum dynamic range in cortical networks at criticality. J. Neurosci. **9**, 15595–15600 (2009)
42. Steyn-Ross, D.A., Steyn-Ross, M.L., Wilson, M.T., Sleigh, J.W.: White-noise susceptibility and critical slowing in neurons near spiking threshold. Phys. Rev. E **74**, 051, 920 (2006)
43. Tononi, G.: Consciousness as integrated information: a provisional manifesto. Biol. Bull. **215**(3), 216–242 (2008)
44. Turing, A.: Computing machinery and intelligence. Mind **59**, 433–460 (1950)
45. Van Dongen, H.P., Maislin, G., Mullington, J.M., Dinges, D.F., Van Dongen, H.P., Mullington, J.M., Dinges, D.F.: The cumulative cost of additional wakefulness: dose-response effects on neurobehavioral functions and sleep physiology from chronic sleep restriction and total sleep deprivation. Sleep **26**(2), 117–126 (2003)
46. Vyazovskiy, V.V., Olcese, U., Hanlon, E.C., Nir, Y., Cirelli, C., Tononi, G.: Local sleep in awake rats. Nature **472**(7344), 443–447 (2011)
47. Wissel, C.: A universal law of the characteristic return time near thresholds. Oecologia **65**, 101–107 (1984)
48. Worrell, G.A., Cranstoun, S.D., Echauz, J., Litt, B.: Evidence for self-organized criticality in human epileptic hippocampus. Neuroreport **13**, 2017–2021 (2002)

Christian Meisel studied physics and medicine at the University of Freiburg, Germany. He then went on to complete his residency training in neurology at the University Clinic in Dresden. Research fellowships led him to the Max Planck Institute for the Physics of Complex Systems and the National Institutes of Health, USA. He currently holds a dual faculty appointment at Harvard Medical School and the University of Dresden. His research interests relate to physics approaches towards understanding cortical network function and information processing to facilitate better diagnostics and treatments in epilepsy, sleep medicine and beyond.

The Challenge of Taming a Latching Network Near Criticality

Chol Jun Kang and Alessandro Treves

Abstract The Potts associative memory can be regarded as a model of extended cortical networks characterized, near a critical line, by spontaneous latching dynamics, i.e. the unguided hopping from one attractor to the next. Can Potts dynamics also be guided, and follow specific instructed transitions between attractors? In this paper, we study to what extent instructions, given via an additional *hetero*-associative learning rule, determine latching sequences in an adaptive Potts neural network. Each global activity pattern is both stored as an attractor and associated with a certain strength λ to D randomly generated and a priori selected other patterns. Increasing either the strength λ of *hetero*-couplings or D leads to longer latching sequences, but also to lower retrieval quality. Further, while the fraction of transitions that follow the instructions initially increases with λ, beyond a certain value it drops, the more rapidly the larger D, as spontaneous dynamics ride on top of instructed transitions, taking them off course. This is shown to be due to the (random) instructions not reflecting the structure of correlations among memories, which drives spontaneous dynamics. In fact, when the number of instructed options D is large, the network appears to choose, among them, those with the same correlations as the spontaneous transitions.

Keywords Neural network · Potts model · Latching · Hetero-associative learning rule

---

C. J. Kang · A. Treves (✉)
SISSA - International School for Advanced Studies, Via Bonomea 265,
34136 Trieste, Italy
e-mail: ale@sissa.it

C. J. Kang
The Abdus Salam International Centre for Theoretical Physics, Strada Costiera 11,
34151 Trieste, Italy

Department of Physics, Kim Il Sung University, RyongNam Dong,
Taesong District, Pyongyang, Democratic People's Republic of Korea

A. Treves
Kavli Institute for Systems Neuroscience/Centre for Neural Computation,
Norwegian University of Science and Technology, Trondheim, Norway

N. Tomen et al. (eds.), *The Functional Role of Critical Dynamics in Neural Systems*, Springer Series on Bio- and Neurosystems 11,
https://doi.org/10.1007/978-3-030-20965-0_5

1 Introduction

The Potts associative memory network, originally studied merely as a variant of mathematical or potentially applied interest [1–5], can be interpreted as a reduced model of cortical dynamics, in which local patches of cortex are modelled as Potts units [6]. Thus, it offers one approach to model spontaneous dynamics in extended cortical systems, in particular if simple mechanisms of temporal adaptation are taken into account [7]. It can be subject to rigorous analyses of e.g. their storage capacity [8], or of the mechanics of saltatory transitions between states [9] and is amenable to a description in terms of distinct 'thermodynamic' phases [10, 11]. The dynamic modification of thresholds (temporal adaptation) with timescales separate from that of retrieval, together with the correlation between cortical states, are key features characterizing cortical operations, and Potts network models may contribute to elucidate their roles. Adaptation and its role in semantic priming [12] have been linked to the instability manifested in schizophrenia [13]. This model of spontaneous behaviour may also be relevant to elucidate the neural dynamics underlying confabulation [14] and perhaps free association and creativity [10, 15]. As opposed to mental processes that clearly require executive control [16], these are situations where the dynamics can be thought of as largely self-organizing.

Saltatory or *latching* dynamics can be regarded as a critical phenomenon in that a suitably defined measure of its quality is observed to diverge close to a critical line. As discussed in previous studies [17], below the line latching dynamics is short lasting, while above the line it does not distinguish well the currently active state from the rest. Latching quality, defined below as the product of duration and discriminability, is shown in [17] and in Fig. 2 to take high values only in a narrow band in phase space.

In previous papers, such spontaneous latching dynamics has been extensively studied [17, 18] in two distinct limits: the slowly adapting regime, where inhibition is taken to act very slowly, and the fast adapting regime, where its time scale is faster than those of excitation and of rate adaptation. In reality GABA_A and GABA_B mechanisms provide cortical circuits with inhibition near both limits [19], but the analysis is clearer if they are considered separately. Such analysis has been largely limited to the case of randomly correlated memory patterns, and we have seen (see e.g. Fig. 15 in [17]) that, particularly in a narrow band along the critical line where infinite latching arises and is of good quality, in the slowly adapting regime, it is the (random) positive fluctuations in the degree of correlation between memory patterns that determine which transitions occur. In the fast adapting regime transitions appear to be more random, and in fact to avoid pairs of correlated patterns, so we focus on the slowly adapting limit as a more interesting case of self-organizing dynamics. Note that this is self-organizing dynamics near criticality, not self-organized criticality. To explore the latter, in this context, one could study a model of how cortical states come to be stored in memory, possibly through a learning process that itself involves spontaneous trajectories. Here, we limit ourselves to a scenario where learning is taken to have occurred already, to have established specific cortical states as memory attractors, and we study the endogenous dynamics among such attractors, after external inputs have subsided.

One can ask, within this scenario, to what extent one may superimpose on spontaneous dynamics explicit *instructions*, that is, a list of transitions that the network is instructed or encouraged to go through, by encoding them in the connection weights. The answer is not straightforward, because the instructions may interfere with a collective behaviour, latching, which is well expressed only in a narrow region near criticality. Following [20], who proposed that transitions may be obtained in an associative memory by adding specific couplings between each activity configuration and its intended successor, here we introduce therefore a *hetero*-associative additional component to the previously purely *auto*-associative Hebbian weights. We focus on how effectively are these instructed transitions, encoded in the learning rule, followed during latching, and on how much these instructions alter the spontaneous behaviour of the network, i.e., the one determined by the correlational structure of its memories.

2 The Model

Consider an attractor neural network model comprised of Potts units. The rationale for the model is that each unit represents a local network of many neurons, with its own attractor dynamics [21, 22], but in a simplified/integrated manner, regardless of detailed local dynamics. Local attractor states are represented by $S + 1$ Potts states: S active ones and one quiescent state (intended to describe a situation of no retrieval in the local network), σ_i^k, $k = 0, 1, \ldots, S$, with the constraint that $\sum_{k=0}^{S} \sigma_i^k \equiv 1$.

2.1 *Interactions*

The 'synaptic' connection between two Potts units is a tensor, summarizing the effect of very many actual connections between neurons in the two local networks, but still, following the Hebbian learning rule [23], the connection weight between unit i in state k and unit j in state l is written as [8]

$$J_{ij}^{kl} = \frac{c_{ij}}{Ca(1 - a/S)} \sum_{\mu=1}^{p} \left(\delta_{\xi_i^\mu,k} - \frac{a}{S}\right) \left(\delta_{\xi_j^\mu,l} - \frac{a}{S}\right) (1 - \delta_{k0})(1 - \delta_{l0}), \quad (1)$$

where c_{ij} is 1 if two units i and j have a connection and 0 otherwise, C is the average number of connections per unit, a is the sparsity parameter, i.e. the fraction of active units in every stored global activity pattern ($\{\xi_i^\mu\}$, $\mu = 1, 2, \ldots, p$) and p is the number of stored patterns. The last two delta functions imply that the learned connection matrix does not affect the quiescent states. We will use the indices i, j

for units, k, l for (local) states and μ, ν for patterns (which are a subset of all global activity states).

Units are dynamically set in their activity states in the following way:

$$\sigma_i^k(t) = \frac{\exp(\beta r_i^k(t))}{\sum_{l=1}^{S} \exp(\beta r_i^l(t)) + \exp[\beta(\theta_i^0(t) + U)]} \tag{2}$$

and

$$\sigma_i^0(t) = \frac{\exp[\beta(\theta_i^0(t) + U)]}{\sum_{l=1}^{S} \exp(\beta r_i^l(t)) + \exp[\beta(\theta_i^0(t) + U)]}, \tag{3}$$

where $r_i^k(t)$ is the input to (active) state k of unit i integrated over a time scale τ_1, while U and $\theta_i^0(t)$ are, respectively, the constant and time-varying component of the effective overall threshold for unit i, which in practice act as inverse thresholds on its quiescent state. $\theta_i^0(t)$ varies with time constant τ_3, to describe local network adaptation produced by inhibition. The stiffness of the local dynamics is parametrized by the inverse "temperature" β (or T^{-1}), which is then distinct from the standard notion of thermodynamic noise. The input-output relations (2) and (3) ensure that $\sum_k \sigma_i^k(t) \equiv 1$.

In addition to the overall threshold, $\theta_i^k(t)$ is the threshold for unit i specific to state k, and it varies with time constant τ_2, representing adaptation of the individual neurons active in that state, i.e. their neural or even synaptic fatigue.

2.2 *Dynamics*

The time evolution of the network is governed by the following equations of motion

$$\tau_1 \frac{dr_i^k(t)}{dt} = h_i^k(t) - \theta_i^k(t) - r_i^k(t) \tag{4}$$

$$\tau_2 \frac{d\theta_i^k(t)}{dt} = \sigma_i^k(t) - \theta_i^k(t) \tag{5}$$

$$\tau_3 \frac{d\theta_i^0(t)}{dt} = \sum_{k=1}^{S} \sigma_i^k(t) - \theta_i^0(t). \tag{6}$$

Global dynamics crucially depend on the relative magnitude of the time constants τ_1, τ_2 and τ_3 for collective neuronal activity and for the thresholds for active and inactive states. In this paper, we stay in what we call the slowly adapting regime, characterized by $\tau_1 \ll \tau_2 < \tau_3$, where activity changes rapidly, compared to the adaptation of the active states (τ_2) and to the extremely slow inhibition effects (τ_3). The latter therefore is taken to model only GABA_B inhibition in the cortex [19]; to include also GABA_A action one would have to consider an additional time scale $\tau_3' \ll \tau_1 < \tau_2 \ll \tau_3$, which would make the analysis cumbersome. Simulations in a combined fast and slowly adapting regime are left for future work.

The field experienced by the unit i in state k reads

$$h_i^k = \sum_{j \neq i}^{N} \sum_{l=1}^{S} J_{ij}^{kl} \sigma_j^l + w \left(\sigma_i^k - \frac{1}{S} \sum_{l=1}^{S} \sigma_i^l \right). \tag{7}$$

The *local feedback term* w is an important parameter that modulates the stability of the local attractors and leads the network to speedily reach one of the attractors. For further details and a discussion of these modelling choices, we refer to [8–10, 17, 18, 24, 25].

2.3 Memories

Randomly correlated patterns are generated by following the probability distribution

$$\begin{aligned} P(\xi_i^\mu = k) &= a/S \\ P(\xi_i^\mu = 0) &= 1 - a \end{aligned} \tag{8}$$

as in the simplest of the paradigms discussed in [7]. Other paradigms include algorithms for generating correlated patterns, which are not considered in this paper.

The overlap or correlation of the current state of the network with the stored patterns $\left\{ \xi_j^\mu |_{\mu=1,\dots,p} \right\}$ can be measured as

$$m_\mu(t) = \frac{1}{Na(1 - a/S)} \sum_{j \neq i}^{N} \sum_{l \neq 0}^{S} \left(\delta_{\xi_j^\mu l} - \frac{a}{S} \right) \sigma_j^l(t). \tag{9}$$

to describe the activity of the network in memory spacc. In practice, however, we are not interested in all such overlap values, but only in those, if any, that stand out before the rest. As a simple assessment of this *discriminability* of retrieval, i.e., whether the network is prevailingly correlated with a single stored pattern, we use d_{12}, defined in terms of two highest overlap values at each time t, $m_1(t)$ and $m_2(t)$, as

$$d_{12} = \langle \int dt \, (m_1(t) - m_2(t)) \rangle_{\text{cue}}, \tag{10}$$

where angle brackets mean the (quenched) average over different initial cues, after the temporal average obtained by integration.

Finally, combining such average discriminability with the length l over which latching extends, we introduce the latching quality Q, defined as

$$Q = d_{12} \cdot l \cdot \eta, \tag{11}$$

where η is 0 when the network becomes trapped in the initially cued pattern during simulation time, and 1 otherwise.

3 Imparting Latching Instructions with an Associative Learning Rule

How can the suggestion by Kanter and Sompolinsky [20] be implemented in the Potts network? The tensor connection between unit i in state k and unit j in state l is generalized by adding a *hetero*-associative component of strength λ, to

$$J_{ij}^{kl} = \frac{c_{ij}}{Ca(1-a/S)} \sum_{\mu=1}^{p} \left(\delta_{\xi_i^\mu,k} - \frac{a}{S}\right) \left\{ \left(\delta_{\xi_j^\mu,l} - \frac{a}{S}\right) + \lambda \sum_{d=1}^{D} \left(\delta_{\xi_j^{\nu_d},l} - \frac{a}{S}\right) \right\} (1-\delta_{k0})(1-\delta_{l0}), \tag{12}$$

where the λ term effectively guides or instructs each pattern in the direction of D other patterns. At each stage in the latching sequence, the network may follow one of the D instructions, or proceed of its own to a different transition (or the sequence may stop).

We provide the network with a table in which each memory pattern ($\{\xi^\mu|_{\mu=1,2,\ldots,p}\}$) is associated with its own set of D instructed patterns ($\{\xi^\nu|_{\nu=1,2,\cdots,D}\}$) that are selected randomly among the p that are stored auto-associatively.

An example of latching partially governed by instructions is shown in Fig. 1. A latching sequence 1-5-3-7-$\cdots$-15-$\cdots$ is indicated by solid lines while the patterns *hetero*-associated to each pattern in the sequence are denoted by dashed lines. In the example, patterns (4,5,7) are associated to pattern 1 and latching proceeds towards pattern 5. For patterns 1, 5, 7, 15, dynamics flows along the associated patterns. Only for pattern 3 spontaneous latching occurs.

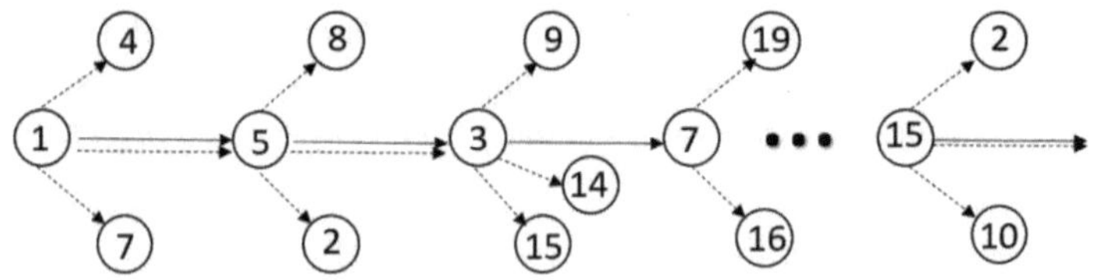

Fig. 1 An example of latching sequence (1-5-3-7-$\cdots$-15-$\cdots$) and the corresponding instructions ((4, 5, 7) to 1, (2, 3, 8) to 5, (9, 14, 15) to 3, ...). Instructed transitions are denoted by dashed lines, while solid lines denote those, instructed or not, occurring in the latching sequence

It should be noted that the unambiguous identification of which patterns occur at distinct stages in the sequence is only possible when latching is of sufficient quality. If not, it may for example happen that an instructed transition does occur, but is *masked* by a spontaneous transition to a different pattern, occurring simultaneously and with slightly larger overlap. It is therefore appropriate to first assess the quality of latching dynamics, in the presence of instructions.

4 The Effect of *Hetero*-Associative Instructions on Latching Dynamics

The character of latching dynamics in a Potts network may be quantified in terms of several different measures, including the latching length, l, the difference between the two highest overlaps, d_{12}, and the crossover in overlap between two successive patterns, m_{cross} [9, 10, 17, 18]. The quality of latching, Q, combines the first two of these measures to give a visual impression of where robust latching occurs in phase space.

We first address the question of what is the effect on latching behavior, in terms of the above quantities, when *hetero*-associations supplement the original *auto*-associative learning rule, with relative strength λ.

We keep the parameters $N = 600$, $C = 90$, $p = 200$, $S = 7, a = 0.25$, $U = 0.1$, $\beta = 12.5, w = 0.45, \tau_1 = 3.3, \tau_2 = 100.0$ and $\tau_3 = 10^6$, corresponding to the slowly adapting regime, throughout the chapter. Simulations are terminated after $6 \cdot 10^5$ updates and repeated over 1000 cued patterns. To see the influence of instructions on latching, we focus on the $D = 2$ case, where each pattern is *hetero*-associated with two other patterns at the learning stage.

We show examples of latching behavior with and without the λ term in Fig. 2a, b, c. Figure 2b shows that adding a small *hetero*-associative component to the connection

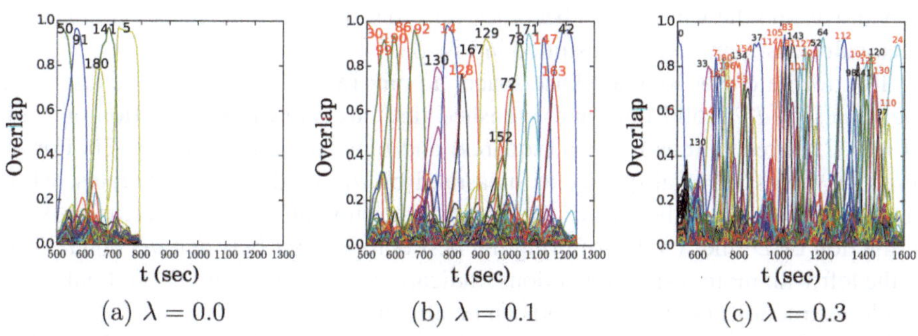

Fig. 2 Retrieval dynamics with $\lambda = 0$, 0.1 and 0.3 in **a**, **b** and **c**. Numbers indicate the patterns with the highest overlap that compose the retrieved sequence, and those in red (light black) denote instructed patterns. In these examples, $D = 2$

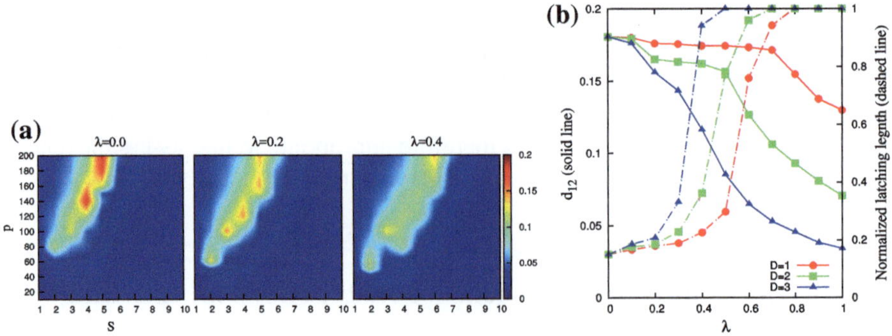

Fig. 3 **a** Phase space of $Q(S, p)$ with *hetero*-coupling strength $\lambda = 0.0, 0.2$ and 0.4. $D = 2$; **b** λ dependence of d_{12} (solid line) and l (dashed line), with $S = 7$ and $p = 200$. Red (circles), green (squares) and blue (triangles) stand for $D = 1, 2, 3$

weights has the main effect of lengthening the latching sequence, which however also becomes less distinct. In the next panel, with larger λ, retrieval quality has deteriorated much further, and one begins to notice collective instabilities, or bursts of nearly simultaneously retrieved patterns, that stand in marked contrast to the relatively clean sequence of the purely spontaneous latching in Fig. 2a.

Numbers on top of the largest overlaps in Fig. 2 comprise the sequence, and those in red indicate instructed transitions. Even before a quantitative analysis, the panel suggest that tripling the λ value does not succeed in eliminating spontaneous transitions. In fact, we show below that the opposite is the case.

The bright regions, or *bands*, where relatively high-Q latching occurs are shown, for different values of λ (0.0, 0.2 and 0.4) in Fig. 3a. The number of *hetero*-associative instructions at each stage is still $D = 2$. As already discussed in [17], the area right to the band, with relatively large S and small p, shows good *quality* retrieval, measured by relatively high d_{12} (and m_{cross}), but short latching length. Instead, in the area left to the band, with relatively small S and large p, latching extends indefinitely but is very noisy, and the value of d_{12} becomes very low. The band decreases gradually in peak values as λ grows, as illustrated in Fig. 3a.

To afford a closer look at phase space, a point ($S = 7$, $p = 200$) is chosen and the values of d_{12} (solid line) and l (dashed line) are shown as a function of λ for $D = 1, 2, 3$ (red (circles), green (squares) and blue (triangles)) in Fig. 3b. For all three D values retrieval quality as measured by d_{12} gradually deteriorates, while latching duration rapidly reaches the length of the simulations, with increasing λ. As a function of D, the λ value offering the best compromise between d_{12} and l shifts to the left with more instructed options, indicating that in the large D limit only very gentle instructions (small λ values) can be effective.

We can interpret these observations in the following way. For a given value of D, a strong *hetero*-associative coupling λ enhances the network tendency to latch, resulting in prolonged sequences, but it also disrupts *auto*-associative retrieval, making the process noisier, and d_{12} lower. These effects are amplified for larger D values, since it becomes easier to latch in one of many instructed directions, but noise is also larger and it becomes difficult to retrieve any clean pattern. As a result, λ and D produce similar effects, in the sense that they both degrade latching quality while increasing latching length.

5 Instructed Versus Spontaneous Latching Transitions

How often does the Potts network follow the instructions it is given? Can we obtain an obedient network, that only does what it is told to do, in terms of transitions?

To measure the fraction of transitions that comply with the instructions given at the learning stage, we introduce a *compliance* index f as

$$f = \frac{T_{instruct}}{T_{tot}}, \tag{13}$$

where $T_{instruct}$ is the number of transitions, i.e. pairs of successively retrieved patterns, with overlap above 0.5, that follow the instructions, and T_{tot} is the total number of pairs of successive patterns in the latching sequence. f is 1 if the network completely follows the instructions it is given, and 0 if it never does.

For convenience, we introduce some abbreviations; FP denotes a pair of patterns that follows the instructions, SP a spontaneous transition, LP a generic latching pair, spontaneous or instructed, and AP any possible pair, whether occurring in a latching sequence or not.

Simulations are run with the same parameters as in the previous section.

The fraction of FPs, f, is shown for $D = 1, 2, 3$ (red (circles), green (squares), blue (triangles)) in Fig. 4. From the figure we see that the network initially follows the instructions to an extent proportional to λ, but it quickly reaches a maximum degree of compliance, around $\lambda = 0.15$, at values $f > 0.5$ (which increase mildly with D). For larger values of λ, the compliance f drops, all the more rapidly the larger is D.

This may be because the network is effectively accompanied towards an instructed pattern only with a gentle fillip, i.e., at small enough λ, whereas larger values of λ push the network with a shove that perhaps drives it to the instructed pattern, but then often also past it and onward to an immediate further transition, that steps beyond the instructed path. For larger values of D, the concurrent shove in several directions accelerates the decrease in compliance f.

The correlations between latching pairs in the absence of the *hetero*-associative term are shown in Fig. 5. C_1 and C_2 are the fractions of units that are active in the same (different) states in the two patterns of a pair. As discussed in [10, 17], latching occurs mostly for positively correlated pattern pairs, i.e., when C_1 is larger than its mean value, while C_2 is smaller. Indeed, in the scatterplot LPs (red crosses) are printed on top of APs (blue circles) around the bottom right portion of the distribution of the latter in the (C_1, C_2) plane. The vertical and horizontal lines which cross the

Fig. 4 λ dependence of f for $D = 1, 2, 3$ (red (circles), green (squares), blue (triangles)

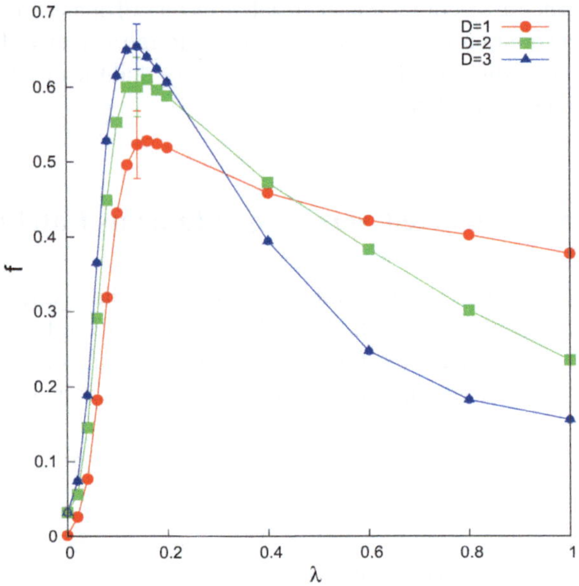

Fig. 5 C_1–C_2 correlation scatterplot for $D = 2$, $\lambda = 0$. Blue circles are for APs and red crosses for LPs. $S = 7$ and $p = 200$

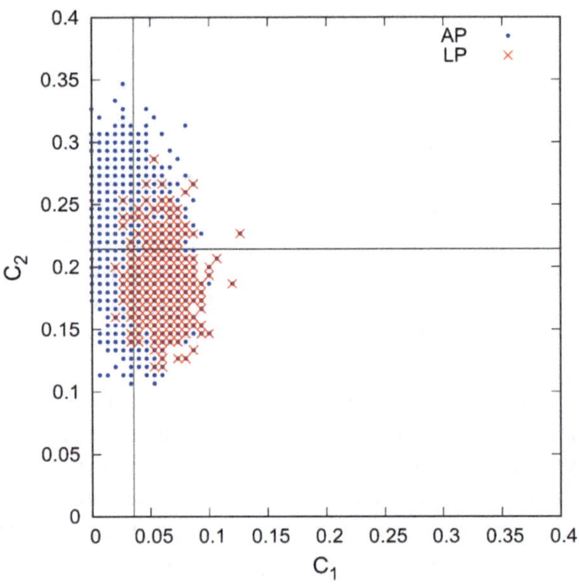

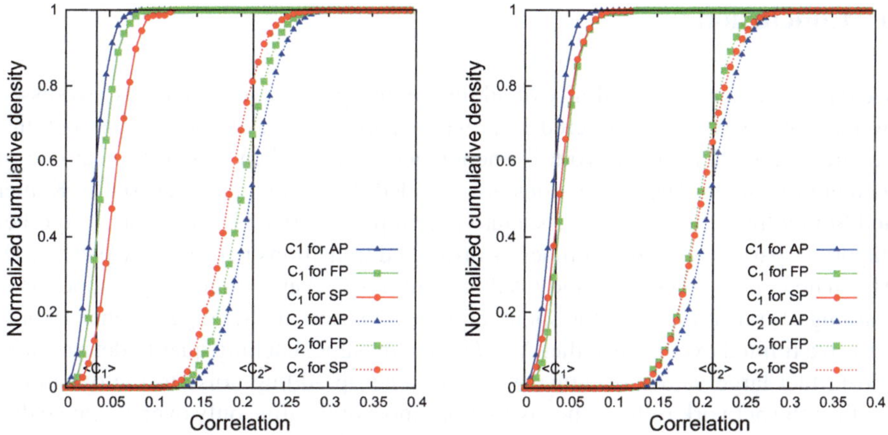

(a) $D = 2$ (b) $D = 10$

Fig. 6 Cumulative density of pattern pairs (AP in blue triangles, FP in green squares, SP in red circles) for increasing values of correlation, as measured by C_1 and C_2. Solid lines are for C_1 and dashed lines are for C_2. $\lambda = 0.1$, $S = 7$ and $p = 200$

figure represent $\langle C_1 \rangle \equiv a/S$ and $\langle C_2 \rangle \equiv a(S-1)/S$, the average values of C_1 and C_2.

As soon as we introduce the λ coupling term, the group of red crosses begins to diffuse towards the centre of the blue circle distribution, but in order to understand the change we need to separate FPs from SPs, among the full set of LPs. In Fig. 6, the cumulative density of APs, FPs and SPs (in blue triangles, green squares and red circles, respectively) are shown with respect to C_1 (solid lines) and C_2 (dashed lines). $\lambda = 0.1$, $S = 7$, $p = 200$ and both $D = 2$ and $D = 10$ cases are considered. The number of APs, FPs and SPs are of course normalized by their own total numbers in the latching sequence.

In Fig. 6a, for $D = 2$, there is a clear separation of FPs and SPs from APs along both C_1 and C_2 axes. Moreover, we see that SPs (red circles) are distributed towards larger C_1 and lower C_2 values, as already shown in Fig. 5. FPs (green squares) are closer to APs (blue (triangular) dots) in their scatter of correlation values, but still their cumulative density does not coincide with theirs. This is because at $D = 2$ there two options to follow at each stage and, even though the instructions are imparted at random, when the network does follow one of the two it tends to choose one with the correlations that more resemble those of SPs. The separation between SPs and APs is less marked when D is high ($D = 10$ in Fig. 6b), because at fixed λ increasing D increases the noise, and spontaneous transitions tend to occur more randomly; but even more imperceptible is the separation between SPs and FPs since, among the many options, the network apparently picks the instruction that moves it in a direction it would take anyway, spontaneously.

6 Conclusion

We have assessed the possibility of adding to the spontaneous dynamics, expressed by the Potts network considered in previous papers, a set of specific instructions, i.e., transitions that the network is encouraged to take when it is in or near one of its memory states. The instructions are encoded, following a suggestion by Kanter and Sompolinsky, in a *hetero*-associative term parametrized by a factor λ. Another important parameter is the number of instructed transitions per memory pattern, D. The main conclusion of the study is that combining what is effectively a supervised learning of transitions with the spontaneous expression of latching sequences works only to a limited extent. As either λ or D grow in value, latching quality deteriorates, and in fact large values of either parameter end up adding noise to the dynamics. Further, the network follows the instructions most of the time only over a λ range that narrows down around $\lambda = 0.15$ as D increases, while the transitions are increasingly indistinct. The ultimate reason for the difficulty of imparting instructions, in the model, is that these are arbitrary, while latching dynamics in the Potts network, especially in the critical band studied before, in the slowly adaptive regime, favors transitions between correlated pairs of patterns. If D is large, the network can choose among many options the ones that are more correlated to its current state, but even then the presence of all the other options, with a sufficient λ factor, generates noise.

Instead of self-organized criticality, therefore, we have an example where criticality *prevents* the subjugation of self-organized dynamics to an arbitrary command (in this case given with an additional term in the recurrent weights, but similar results are expected with an external drive). Away from criticality, the endogenous dynamics is much less constrained—even though degraded in quality or limited in duration—and easier to control.

In conclusion, it is difficult to harness the Potts network, an interesting simple model of complex spontaneous dynamical behaviour, to externally-determined goals via supervised learning. To explore the capability of the model to approach concrete problems where latching dynamics may be relevant, it is critical of course to include structure in the so far unstructured homogeneous network, and to allow the dynamics to harmoniously reflect such structure, whether explicit or implicit, without attempting to force it to follow a prescribed course.

Acknowledgements Extensive discussions with several colleagues and in particular with Vezha Boboeva and Michelangelo Naim are gratefully acknowledged, as well as support from Human Frontier Science Program grant RGP0057/2016.

References

1. Kanter, I.: Potts-glass models of neural networks. Phys. Rev. A **37**(7), 2739 (1988)
2. Bollé, D., Dupont, P., Mourik, J.V.: Stability properties of Potts neural networks with biased patterns and low loading. J. Phy.s A: Math. Gen. **24**(5), 1065 (1991)

3. Bollé, D., Dupont, P., Huyghebaert, J.: Thermodynamic properties of the q-state Potts-glass neural network. Phys. Rev. A **45**(6), 4194 (1992)
4. Bollé, D., Vinck, B., Zagrebnov, V.A.: On the parallel dynamics of the q-state Potts and q-Ising neural networks. J. Stat. Phys. **70**(5), 1099–1119 (1993)
5. Bollé, D., Cools, R., Dupont, P., Huyghebaert, J.: Mean-field theory for the q-state Potts-glass neural network with biased patterns. J Phys A: Math. Gen. **26**(3), 549 (1993)
6. Naim, M., Boboeva, V., Kang, C.J.: Treves A (2018) Reducing a cortical network to a Potts model yields storage capacity estimates. J. Stat. Mech.: Theory Exp. **4**, 043304 (2018)
7. Treves, A.: Frontal latching networks: a possible neural basis for infinite recursion. Cognit. Neuropsychol. **22**(3–4), 276–291 (2005)
8. Kropff, E.: Treves A (2005) the storage capacity of Potts models for semantic memory retrieval. J. Stat. Mech. Theory Exp. **08**, P08010 (2005)
9. Russo, E., Namboodiri, V.M.K., Treves, A., Kropff, E.: Free association transitions in models of cortical latching dynamics. New J. Phys. **10**(1), 015008 (2008)
10. Russo, E., Treves, A.: Cortical free-association dynamics: distinct phases of a latching network. Phys. Rev. E **85**(5), 051920 (2012)
11. Abdollah-nia, M.F., Saeedghalati, M.: Abbassian A (2012) Optimal region of latching activity in an adaptive Potts model for networks of neurons. J. Stat. Mech. Theory Exp. **02**, P02018 (2012)
12. Lerner, I., Bentin, S., Shriki, O.: Spreading activation in an attractor network with latching dynamics: automatic semantic priming revisited. Cognit. Sci. **36**(8), 1339–1382 (2012)
13. Lerner, I., Bentin, S., Shriki, O.: Excessive attractor instability accounts for semantic priming in schizophrenia. PLoS ONE **7**(7), e40663 (2012)
14. Burgess, P.W., Shallice, T.: Confabulation and the control of recollection. Memory **4**(4), 359–412 (1996)
15. Iyer, L.R., Doboli, S., Minai, A.A., Brown, V.R., Levine, D.S., Paulus, P.B.: Neural dynamics of idea generation and the effects of priming. Neural Netw. **22**(5), 674–686 (2009)
16. Shallice, T.: Specific impairments of planning. Philos. Trans. R Soc. Lond. B Biol. Sci. **298**(1089), 199–209 (1982)
17. Kang, C.J., Naim, M., Boboeva, V., Treves, A.: Life on the edge: latching dynamics in a Potts neural network. Entropy **19**(9) (2017)
18. Russo, E., Pirmoradian, S., Treves, A.: Associative latching dynamics vs. syntax. In: Advances in Cognitive Neurodynamics (II), pp. 111–115. Springer, Dordrecht (2011)
19. Connors, B.W., Malenka, R.C., Silva, L.R.: Two inhibitory postsynaptic potentials, and $GABA_A$ and $GABA_B$ receptor-mediated responses in neocortex of rat and cat. J. Physiol. **406**(1), 443–468 (1988)
20. Sompolinsky, H., Kanter, I.: Temporal association in asymmetric neural networks. Phys. Rev. Lett. **57**, 2861–2864 (1986)
21. Hopfield, J.: Neural networks and physical systems with emergent collective computational abilities. Proc. Natl. Acad. Sci. USA **79**(8), 2554–2558 (1982)
22. Amit, D.J.: Modeling Brain Function: The World of Attractor Neural Networks. Cambridge University Press, Cambridge (1992)
23. Hebb, D.: The Organization of Behavior: A Neuropsychological Theory. Wiley, New York (1949)
24. Kropff, E., Treves, A.: The complexity of latching transitions in large scale cortical networks. Nat. Comput. **6**(2), 169–185 (2007)
25. Kropff, E.: Full solution for the storage of correlated memories in an autoassociative memory. In: Heinke, D., Mavritsaki, E. (eds.) Computational Modelling in Behavioural Neuroscience: Closing the Gap Between Neurophysiology and Behaviour, 2:225. Psychology Press, USA and Canada (2009)

Chol Jun Kang has studied physics at Kim Il Sung University, Pyongyang, DPRK and obtained a researcher position in the same university, until in 2013 he came to Trieste, Italy, within a program supported by the International Center for Theoretical Physics and by SISSA. In December 2017 he was awarded a Ph.D. in Cognitive Neuroscience, with a thesis on latching dynamics in Potts neural networks, advised by Alessandro Treves. He has then moved back to his position at Kim Il Sung University, where he hopes to establish the first theoretical neuroscience research group in the DPRK.

Alessandro Treves has studied physics at La Sapienza University in Rome, Italy, and was awarded a Ph.D. from the Hebrew University of Jerusalem, Israel, with a thesis on the onset of order in associative networks, advised by Daniel Amit. After a postdoc in the neurophysiology lab of Edmund Rolls at the University of Oxford, he has been at SISSA—the International School for Advanced Studies since 1992, where he le leads LIMBO—Liminar Investigations in Memory and Brain Organization.

Fading Memory, Plasticity, and Criticality in Recurrent Networks

Bruno Del Papa, Viola Priesemann and Jochen Triesch

Abstract Criticality signatures, in the form of power-law distributed neuronal avalanches, have been widely measured in vitro and provide the foundation for the so-called critical brain hypothesis, which proposes that healthy neural circuits operate near a phase transition state with maximal information processing capabilities. Here, we revisit a recently published analysis on the occurrence of those signatures in the activity of a recurrent neural network model that self-organizes due to biologically inspired plasticity rules. Interestingly, the criticality signatures are input dependent: they transiently break down due to onset of random external input, but do not appear under repeating input sequences during learning tasks. Additionally, we show that an important information processing ability, the fading memory time scale, is improved when criticality signatures appear, potentially facilitating complex computations. Taken together, the results suggest that a combination of plasticity mechanisms that improves the network's spatio-temporal learning abilities and memory time scale also yields power-law distributed neuronal avalanches under particular input conditions, thus suggesting a link between such abilities and avalanche criticality.

B. Del Papa · J. Triesch (✉)
Frankfurt Institute for Advanced Studies, Ruth-Moufang-Str. 1, 60438 Frankfurt, Germany
e-mail: triesch@fias.uni-frankfurt.de

B. Del Papa
e-mail: delpapa@fias.uni-frankfurt.de

V. Priesemann
Department of Non-linear Dynamics, Max Planck Institute for Dynamics and Self-Organization, Am Fassberg 17, 37077 Göttingen, Germany
e-mail: viola@nld.ds.mpg.de

N. Tomen et al. (eds.), *The Functional Role of Critical Dynamics in Neural Systems*, Springer Series on Bio- and Neurosystems 11,
https://doi.org/10.1007/978-3-030-20965-0_6

1 Introduction

In recent years, the critical brain hypothesis has emerged as an important candidate to understand the functioning of biological neural networks, and has been the focus of a number of reviews [7, 12, 22, 36]. This hypothesis states that healthy neural circuits operate near a second order phase transition with maximal information processing capabilities. In the brain, supporting experimental evidence typically comes from measurements of indirect signatures of criticality in the form of scale-free activation patterns, called neuronal avalanches. One of the most prominent scale-free properties are approximate power-law distributions of duration and size of avalanches, i.e. cascades of neural activity estimated from neuronal spikes and local field potentials (LFP) [6]. These power-law distributions have been widely observed in vitro [6, 18, 33, 53, 57] and in coarse measures of neural activity in vivo (such as LFP, EEG, MEG and BOLD activity) [40, 43, 47, 49, 54, 56], but in vivo spiking activity resembles instead a driven subcritical regime [5, 45, 46, 67]. Such disparity suggests that healthy neural networks are able to self-organize towards different dynamical regimes with potentially different functions, emphasizing the importance of understanding their particular self-organization mechanisms. However, even though power-law distributions provide the bulk of evidence supporting critical dynamics in the brain, including its adaptation under external input [50], the mechanisms underlying the network's continuous adaptation resulting in these distributions, as well as their particular functions, are still not fully understood.

The continuous network adaptation towards criticality is commonly associated with the self-organization observed in self-organized critical (SOC) systems [4, 6], which evolve to eventually reach a critical point without tuning of control parameters [7, 46]. In neural network models, SOC has been proposed to arise from ongoing synaptic plasticity action. Power-law distributions of avalanche sizes have been observed in networks of different complexity and plasticity mechanisms, ranging from simple activity-dependent dynamic synapses [2, 31] to spike-timing-dependent plasticity [37, 60] and a combination of short- and long-term plasticity [55]. Such studies have suggested that SOC is indeed achievable via continuous adaptation under biologically inspired plasticity mechanisms. Interestingly, SOC and other states (for example, a non-equilibrium Widom line [64]) can be achieved by networks with different tuning conditions, such as excitatory and inhibitory balance [33, 53], but a unified theory linking self-organization mechanisms with their biologically relevant functions, such as learning and memory, is still to be developed [41].

From the brain's perspective, critical dynamics could be highly desirable due to its functions: computational studies have shown a number of functional benefits of criticality [51]. Specifically, networks tuned to criticality show maximal dynamical range [26], maximal information capacity [53] and number of metastable states [20], while computation at the edge-of-chaos (i.e., at a phase transition point separating non-chaotic from chaotic dynamics) is known to increase performance in classification tasks [30] and maximize information transfer and storage [11] in recurrent neural networks, allowing them to perform complex computational tasks due to an increased fading memory capacity [8]. Nonetheless, the definition of "criticality" used in different studies is not necessarily the same: edge-of-chaos criticality (measured via perturbation analysis and Lyapunov exponents) and avalanche criticality (usually measured via criticality signatures and power-law distributions of relevant events) are commonly assumed to coexist, but they might refer to different phenomena and do not necessarily co-occur in neural networks [14, 15, 24, 25]. As a general link between edge-of-chaos criticality and neuronal avalanches is still missing, a deeper understanding of the self-organization mechanisms that lead to a supposed SOC regime is necessary in order to describe how useful information processing capacities might also arise from them.

An essential information processing property of recurrent neural networks is their fading memory capacity (sometimes referred to as working memory): their ability to retain information about recent inputs in their activity [8, 35]. Such property is a determinant for a network's total memory capacity, allowing them to learn temporal sequences [27]. Interestingly, the fading memory time scale is known to benefit from a combination of Hebbian and homeostatic plasticity in random reservoirs [27] and scales approximately logarithmically with the network size for systems tuned to edge-of-chaos criticality (but slower for other systems) [8]. This result suggests another potential functional role for criticality in the brain: an improvement in its fading memory capacity, which in turn results in enhanced temporal learning abilities. Any such link, however, relies on the assumption that edge-of-chaos and avalanche criticality co-exist, to some extent, under particular input conditions and network states. Notably, no study reports the presence of avalanche criticality in the activity of those reservoirs, although it can be speculated that long distance correlations should underlie both criticality phenomena simultaneously. Thus, the co-occurrence of neuronal avalanches and logarithmic scaling of the fading memory capacity in the same system might shed light on the functions of the first and the underlying mechanisms connecting both phenomena in neural circuits.

In this chapter, we start by reviewing a previous study on self-organizing recurrent neural networks (SORNs) that show criticality signatures, in the form of neuronal avalanches with power-law distributed durations and sizes, in their spontaneous activity [16]. Such models are of particular interest because in contrast to the majority of critical neural networks and SOC models, they were not developed to display criticality, but instead to study spatio-temporal learning [28, 29]. Moreover, they reproduce experimentally measured statistics of synaptic strengths [69] and variability of neural responses [21]. Nonetheless, the models self-organize due to biologically inspired plasticity mechanisms and can naturally display signatures of criticality. We continue by discussing the link between neuronal avalanches, learning and fading memory, showing how these phenomena may arise from the same plasticity mechanisms. In particular, our results suggest that the occurrence of criticality signatures in the networks' spontaneous activity might be linked to an increase in their fading memory time scale, therefore being beneficial for their learning abilities. We finish by arguing that the implications of this novel link could help to clarify how self-organization towards criticality or other states occurs in the brain and propose that the existence of neuronal avalanches might be tied to fundamental brain functions such as learning and memory.

2 Self-Organizing Recurrent Networks

We use models from the Self-Organizing Recurrent Neural Network (SORN) family [28, 69]. SORNs are reservoirs of perceptron-like neurons with dynamic synaptic weights evolving according to biologically inspired plasticity mechanisms. Initially developed to study sequence learning tasks [28], the original SORN model has been able to reproduce a wide range of findings on spontaneous brain activity and the variability of neural responses [21]. When combined with a simplified form of structural plasticity, the SORN is also able to reproduce the distribution and fluctuation patterns of synaptic efficacies observed in the cortex and hippocampus [69], while spontaneously generating synfire chains [70]. Additionally, structural plasticity allows SORNs to learn complex tasks, such as artificial grammars [17].

In this section, we describe in detail all the SORN variants used in this study. The main python code for our simulations is available at https://github.com/delpapa/SORN.

2.1 Model Dynamics

The SORN models, as described in [16], consist of a reservoir of N^E excitatory and $N^I = 0.2 \times N^E$ inhibitory threshold neurons, whose state at each discrete time step t is described by the binary vectors $\mathbf{x}(t) \in \{0, 1\}^{N^E}$ and $\mathbf{y}(t) \in \{0, 1\}^{N^I}$, corresponding to the activity of excitatory and inhibitory neurons, respectively. Biologically, each discrete time step corresponds to the characteristic size of the spike-timing-dependent plasticity window (roughly 20 ms [10]). Both neuron types can be active ("1" state) or inactive ("0" state) at each time step depending on their input, membrane noise and firing threshold. Connections between neurons are represented by synaptic weights W_{ij} (from neuron j to i) and may exist between different excitatory neurons (W^{EE}), from excitatory to inhibitory neurons (W^{IE}) and from inhibitory to excitatory neurons (W^{EI}). Connections between inhibitory neurons and self-connections are not included in the models. At each time step, the network state is updated according to the input each neuron receives and its current threshold:

$$x_i(t+1) = \Theta\left[\sum_{j=1}^{N^E} W_{ij}^{EE}(t)x_j(t) - \sum_{k=1}^{N^I} W_{ik}^{EI} y_k(t) + u_i^{\text{Ext}}(t) + \xi_i^E(t) - T_i^E(t)\right] \tag{1}$$

$$y_k(t+1) = \Theta\left[\sum_{j=1}^{N^E} W_{kj}^{IE} x_j(t+1) + \xi_i^I(t) - T_k^I\right], \tag{2}$$

where Θ is the Heaviside step function, $u_i^{\mathrm{Ext}}(t)$ is the external input received by neuron i at time step t and T_i^E and T_k^I are the excitatory and inhibitory neuronal thresholds, respectively. $\xi^E(t)$ and $\xi^I(t)$ represent the neuronal membrane noise level, assumed to be a Gaussian random variable (with zero mean and variance $\sigma^2 = 0.05$, unless stated otherwise).

The models are initialized as random reservoirs where 10% (20%) of randomly chosen excitatory (inhibitory) synapses W^{EE} (W^{EI}) are present, while the remaining weights W^{IE} are dense matrices, in which all weights are present. All individual weights are drawn from a uniform distribution over the interval [0, 0.1] and normalized so that the incoming excitatory and inhibitory synapses separately sum up to 1 for all neurons. The excitatory and inhibitory thresholds are initially randomly drawn from the uniform distributions $[0, T_{\max}^E]$ and $[0, T_{\max}^I]$, respectively, and the initial network state $x(0)$ and $y(0)$ is randomly selected. Following a previous implementation [69], the initialization parameters are $T_{\max}^E = 1$ and $T_{\max}^I = 0.5$.

2.2 Plasticity Mechanisms

The synaptic weights W^{EE}, W^{IE} and W^{EI} and the excitatory thresholds T^E are subject to plasticity at each time step t, while the inhibitory thresholds remain fixed. There are five different types of plasticity mechanisms acting in SORN models:

Spike-timing-dependent plasticity (STDP)

As a biologically inspired form of Hebbian learning, a discrete model of STDP [10] acts on all active excitatory to excitatory connections, increasing the weight W_{ij}^{EE} (from neuron j to neuron i) by a fixed value η_{STDP} every time neuron i fires exactly one time step after j. Conversely, the weight is decreased by the same value if neuron j fires one time step before i. Very small weights ($<10^{-6}$) are pruned after each update. Formally, STDP can be written as:

$$\Delta W_{ij}^{EE}(t) = \eta_{\mathrm{STDP}}\left[x_i(t)x_j(t-1) - x_j(t)x_i(t-1)\right]. \tag{3}$$

Inhibitory spike-timing-dependent plasticity (iSTDP)

Similarly to STDP, iSTDP acts on the inhibitory to excitatory synaptic weights. When an inhibition is unsuccessful, i.e., an inhibitory neuron k firing does not prevent an excitatory neuron i firing at the next time step, the connection W_{ik}^{EI}, if present, is increased by $\eta_{\mathrm{inh}}/\mu_{\mathrm{IP}}$, in which $\mu_{\mathrm{IP}} < 1$ represents the mean target firing rate of the network. If successful, i.e., if i is silent one time step after k firing, W_{ik}^{EI} is reduced by a smaller value η_{inh}. In practice, this plasticity rule balances the increase of activity due to STDP, regulating the overall network activity, and can be simply written as:

$$\Delta W_{ik}^{EI}(t) = \eta_{\text{inh}} y_k(t-1)\left[1 - x_i(t)(1 + 1/\mu_{\text{IP}})\right]. \tag{4}$$

Structural plasticity (SP)

A structural plasticity rule adds new synapses between previously unconnected excitatory neurons, representing the creation of new synapses in the cortex. At each time step, a synapse is added between a previously unconnected neuron pair with a small probability $p_{\text{SP}} = 0.1$. The weights of new synapses are set to a small value $\eta_{\text{SP}} = 0.001$, and the majority of them quickly decay below the STDP pruning threshold ($<10^{-6}$) and are eliminated [69].

Synaptic normalization (SN)

At each time step, after the STDP, iSTDP and SP updates, SN normalizes the incoming connections of every excitatory neuron, thus regulating the total amount of input it receives while keeping the relative strengths of the connections. This rule is separately applied to W^{EE} and W^{EI}:

$$W_{ij}(t) \leftarrow \frac{W_{ij}(t)}{\sum_j W_{ij}(t)}. \tag{5}$$

Intrinsic plasticity (IP)

Last, as a form of homeostatic plasticity, IP regulates the excitatory neurons' firing thresholds T^E at each time step, driving them towards a fixed target firing rate $\mu_{\text{IP}i}$ by a small learning rate η_{IP}:

$$\Delta T_i^E = \eta_{\text{IP}}\left[x_i(t) - \mu_{\text{IP}i}\right]. \tag{6}$$

Unless stated otherwise, the plasticity parameters for our simulations are: $\eta_{\text{STDP}} = 0.001$, $\eta_{\text{IP}} = 0.001$, $\eta_{\text{inh}} = 0.001$ and $\mu_{\text{IP}} = \mu_{\text{IP}i} = 0.1$.

2.3 External Input

External input is included in the model via the parameter $u_i^{\text{Ext}}(t)$ (Eq. 1), and consists of a sequence of symbols of length L, randomly chosen from an alphabet of size A (typically in the [10, 200] interval) and kept fixed throughout network training, with one symbol presented per time step. By definition, the same input symbol can occur twice in a given sequence. Each symbol provides strong extra input to a fixed subset of $N^U = 0.02 \times N^E$ neurons randomly chosen from the excitatory unit pool, forcing them to spike, while the remaining neurons are unaffected.

2.4 Readout Layer and Performance Evaluation

During learning or memory tasks, an additional readout layer is trained in a supervised fashion (via logistic regression) in order to evaluate the performance of the SORN models, similarly to the readout training process commonly used for static reservoirs [34]. In practice, such readout layer acts as a classifier and has the same size as the input's alphabet size A. The readout is trained on the network excitatory activity vector $\mathbf{x}(t)$ for T_{train} time steps while all plasticity mechanisms are turned off. Subsequently, the performance of the model is evaluated for T_{test} time steps, in a task dependent manner, but again with plasticity mechanisms off. First, for learning tasks whose input consists of repeating sequences of symbols (i.e., without any other input in between), the performance is defined as the percent of correct predictions of the next input symbol. In order to capture the performance for every symbol, the condition $T_{\text{test}} \gg L$ must be satisfied. Second, for memory tasks whose input consists of non-repeating random sequences ($L \rightarrow \infty$), the performance is evaluated on past time steps, i.e., at each time step t, the performance is defined as the normalized number of correct classifications of input symbols received t_{p} time steps in the past.

2.5 Fading Memory Time Scale Estimation

In general terms, fading memory is the ability of a neural network to store information about past inputs in its current activity, i.e., the network state $x(t)$ can be mostly determined by a finite number of past inputs [8]. (For a more formal description of the fading memory time scale in recurrent neural networks, see [35]. This property is also referred to as short-term memory in echo state networks, in an independent work [23]). Here, we estimate such capacity via the model's performance on classifying past inputs based on the current activity (see previous subsection for details). We define the fading memory time scale M_τ as the average memory of recent symbols stored in the SORN's excitatory activity, i.e., the average number of past time steps in which past input symbols can be correctly classified with a performance of at least 90%.

2.6 SORN Variants

For clarity, we distinguish between the two main variants of the SORN model given the plasticity mechanisms they incorporate. The original SORN model [28], henceforward referred to as SORN_O, includes STDP, SN and IP, but excludes membrane noise ($\xi_i^E(t) = \xi_k^I(t) = 0$). The extended SORN model [69], henceforward referred to as SORN_E, includes all five plasticity mechanisms listed in this section (STDP, iSTDP, SN, SP and IP) and Gaussian membrane noise (mean zero, variance

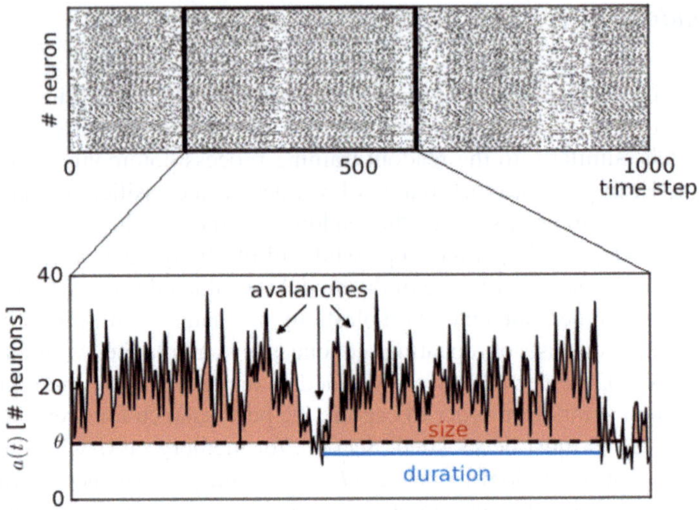

Fig. 1 Spontaneous activity and neuronal avalanches. Raster plot of the SORN$_E$ spontaneous activity after self-organization and neuronal avalanches definition via activity threshold θ. A single avalanche event starts when the activity goes above θ, and lasts as long as it stays about this threshold. Shaded red areas indicate the size of the avalanches and the blue line indicates the duration of a single avalanche event

$\sigma^2 = 0.05$) for both excitatory and inhibitory neurons. Any further modifications in these variants are detailed when necessary.

3 Spontaneous Activity: Self-Organization Towards Avalanche Criticality

In this section, we review one of the main findings from a previous study [16], by analyzing the occurrence of power-law distributed neuronal avalanches in the spontaneous activity and their stability (Fig. 1). Due to the combination of plasticity mechanisms, the SORN$_E$ displayed three distinct self-organization phases (see [16], Fig. 1a for an example of connectivity evolution). After random initialization, in the *decay phase*, the number of present W^{EE} synapses quickly decayed as a result of STDP pruning actions, reaching a minimum at around 10^5 time steps, when the *growth phase* starts. In this second phase, due to SP action, some new added connections remained part of the network, and the number of connections grew until around 2×10^6 time steps, when the fraction of connections stabilized (*stable phase*), but small deviations continued to be observed. During the *stable phase*, after the aforementioned transient period, bursts of asynchronous activity could be observed, suggesting potential avalanche events (Fig. 2).

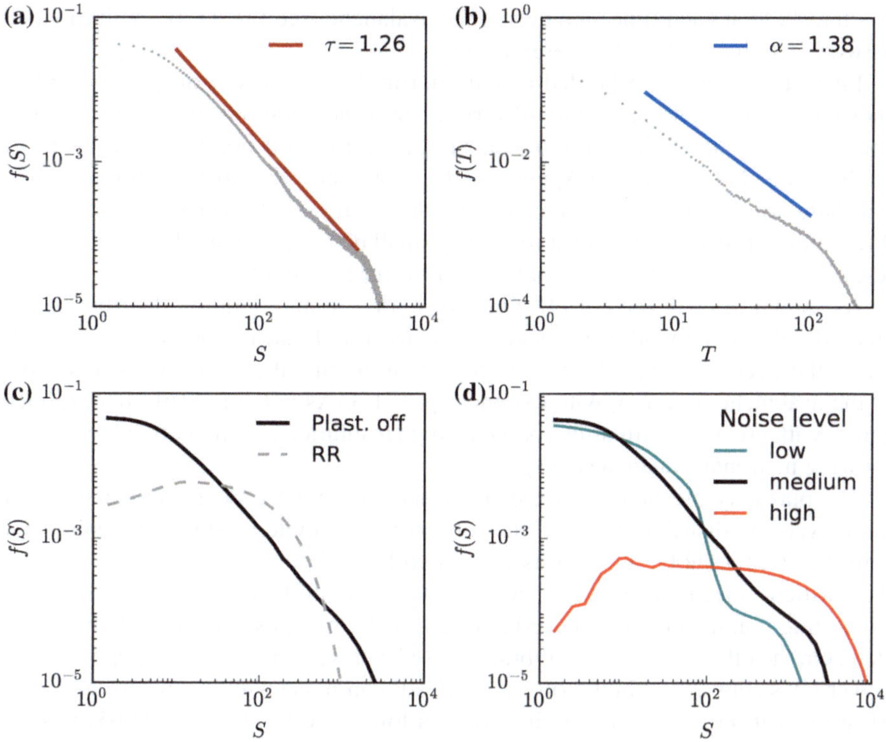

Fig. 2 Neuronal avalanches after self-organization, adapted from [16]. **a**, **b** Probability distributions of avalanches' sizes (S) and durations (T) and power-law fits with exponents τ and α, respectively. Gray points show the raw data of 50 independent simulations. **c** Distributions of avalanche sizes for the $SORN_E$ with all plasticity mechanisms turned off after self-organization (black) and for a random reservoir (RR, dashed gray). **d** For different levels of Gaussian neuronal membrane noise, different distributions of events appear (see text). Only a medium noise level ($\sigma^2 \approx 0.05$) results in distributions resembling power-laws. All curves are for avalanche events in the activity of networks of size $N^E = 200$

Although the $SORN_E$ self-organized towards a dynamical regime in which neuronal avalanches appeared, their definition was fundamentally different from classical SOC models such as the Bak–Tang–Wiesenfeld model [4] due to the lack of separation of time scales. In the $SORN_E$, as in neural activity in vivo, no complete pause exists between two consecutive avalanches, resulting in an overlap between one or more events. This entanglement could be partially solved via a process of activity thresholding [19, 42], which introduces a threshold θ above which avalanche events are measured. In general terms, an avalanche started at the time step when the network activity $a(t)$ went above the threshold and lasted while it remained above θ. This process allowed for an alternative definition of avalanche duration and size: the duration T of an avalanche was the number of time steps in which the activity remained above θ; the size S of an avalanche was equal to the sum of spikes exceeding this threshold at each time step during the avalanche (see Fig. 1). By definition, the network activity, θ, T and S were integers.

Interestingly, the thresholded activity during the *stable* self-organization phase showed bursts of spiking activity of various sizes and durations, whose distributions could be fit by power-laws up to a size dependent cut-off point (Fig. 2, for details on the power-law fitting, see Appendix). Here, we show distributions for an activity threshold equal to half of the mean network activity ($\theta = \langle a(t) \rangle /2$). However, the thresholding process was robust regarding small deviations from this value (see the results and supplementary material in [16] for further details).

As shown in [16], the cut-off point for both size and duration distributions increased with the network size, as also expected for classical SOC models. Furthermore, the average avalanche size followed an empirical power-law as a function of the avalanche duration, with exponent $\gamma \approx 1.3$. As noted in [16], this exponent agrees with previous estimations based on high resolution in-vitro cortical recordings showing neuronal avalanches [18].

The plasticity mechanisms were necessary for the network self-organization, as the power-law distributions were not observed in simple random reservoirs (RR) (Fig. 2c). Remarkably, as soon as the network reached the *stable phase*, none of the plasticity mechanisms were necessary to maintain the power-law distributions, suggesting that the network indeed reached a different, possibly critical, dynamical state. On the other hand, the membrane noise level was crucial for the appearance of power-laws (Fig. 2d): a high noise level (Gaussian noise with mean zero and $\sigma^2 \approx 5$) resulted in Poisson-like activity, while a low noise level ($\sigma^2 \approx 0.005$) resulted in distributions resembling a combination of exponentials. As similar conclusions could be drawn from different types of noise, the authors argued that the noise level regulated the dynamical regime after self-organization and potentially indicated that criticality signatures appeared only in a phase transition state.

In summary, the spontaneous activity of the SORN_E showed robust criticality signatures. The model self-organized towards a special "critical" dynamical state (following the avalanche criticality definition), which did not rely on the self-organization mechanisms to be maintained. The membrane noise level acted as a control parameter: neuronal avalanches only appeared at a transition point, poised between two distinct dynamical states. As the membrane noise could also represent input from other sources not included in the model (such as other brain areas), we continue the analysis by describing how the model behaves under external input.

4 External Input: Readaptation and Learning

Neural circuits in awake, behaving animals typically receive a large number of different external inputs (i.e., sensory inputs). For the critical brain hypothesis to hold true, adaptation mechanisms must tune neural networks towards criticality under various external input regimes. Experimental evidence has shown that the onset of external input breaks down the power-laws [50]. Specifically, that study showed that cortical ex-vivo activity measured in the turtle brain is not critical immediately after the onset of strong external input, but critical dynamics quickly reappears due to the system's readaptation. In [16], we showed similar results for the SORN_E using random external input: the criticality signatures transiently broke down after the onset of external input, but were rapidly brought back by action of plasticity. However, using structured input of simple learning tasks such readaptation did not occur and power-laws were not observed. We briefly contrast here both input conditions, emphasizing that the same model can achieve distinct dynamics under distinct external drives.

In a first experiment, random external input (sequence of length $L \to \infty$ and alphabet size $A = 10$) was given to the network after it reached the stable phase, and distributions of avalanche size and duration were measured immediately after its onset (avalanche events starting in the first 10 time steps after stimulus) and after readaptation due to plasticity, with all plasticity mechanisms on (2×10^6 time steps). The external input consisted of a random sequence of symbols, where each symbol provided strong input to a subset of N^U excitatory neurons (see the *External Input* subsection for further details). In agreement with the ex-vivo recordings, power-laws did not appear in the transient activity, but were brought back by the network's readaptation mechanisms (Fig. 3a, duration distributions not shown), supporting the argument that plasticity tunes the dynamical regime towards avalanche criticality.

In contrast to random external input, structured input is used in spatio-temporal learning tasks [21, 28] and is commonly associated with more realistic sensory input in behaving neural networks. Surprisingly, during such tasks, the repeating structure of the input was enough to destroy the power-law distributions in the SORN_E activity (Fig. 3b). Yet, longer sequences resulted in smoother distributions, qualitatively closer to power-laws, as long repeating input sequences resembled random input, suggesting that additionally to plasticity, the structure of the external drive fundamentally controls the model's dynamical regime.

The action of the plasticity mechanisms in the SORN_E abolished the criticality signatures under structured input. The same plasticity mechanisms, however, improved performance on a sequence learning task in the SORN_O [28], as well as in the SORN_E, compared to a randomly initialized reservoir (RR, Fig. 3c). Interestingly, the performance in the SORN_E was better under low membrane noise than under medium noise, and decreased to chance level for high noise (not shown). This, on the one hand is to be expected, as noise masks the input sequences. On the other hand, such result may appear surprising at first sight, as the SORN_E showed signatures of criticality under medium, but not low or high noise levels. What does this mean for the relation between criticality and information processing? First, one may conclude that performance is maximized at a state that does not show power-laws. However, as none of the tested conditions showed power laws, it is still conceivable that there may exist a state with power-law scaling and even better performance. Second, the task is fairly simple, as it predicted the pattern at $t + 1$ from the activity at t. Maximal performance at this task may not require critical dynamics. Note that criticality *maximizes* certain properties, such as susceptibility, correlation length and time, and pattern diversity. Maximization of these properties fosters performance in tasks that rely on them, e.g. tasks that require maintaining information about past input in their

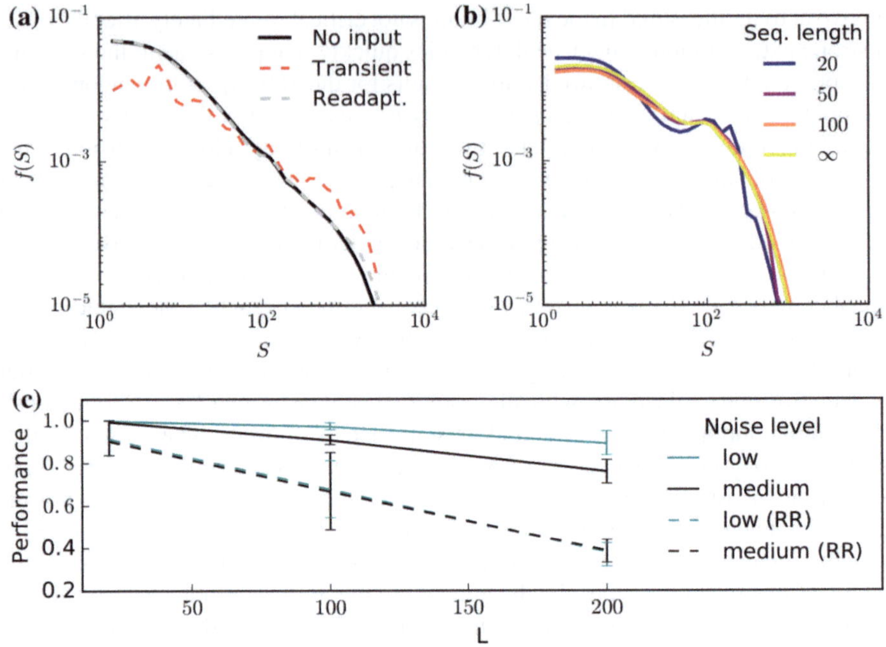

Fig. 3 **Externally driven neuronal avalanches, adapted from** [16]. **a** Probability distributions of avalanches' sizes (S) before the onset of random external input (black), immediately after (red) and after readaptation (gray), for a network of $N^E = 200$ neurons. Such distributions match experimental recordings from [50]. **b** Corresponding probability distributions after the onset of *structured* input of learning tasks (repeating sequences of different lengths L, with alphabet size $A = 10$). $L \to \infty$ indicates a random, non repeating sequence. **c** Performance of the SORN_E for the sequence learning task depends on the membrane noise level. This is not the case for random reservoirs (RR). Curves show averages of 50 simulations and error bars indicate the 25–75% percentiles interval

activity for long time (reservoir properties). However, for simple tasks as the one used here, fast forgetting might be of advantage [11]. Hence, the higher performance in the simple task under low membrane noise is expected.

To directly tackle the memory properties of the SORN_E, i.e. the ability of a network to store information about past inputs, we quantify in the next section how long information can be read out from network activity by assessing its fading memory time scale.

5 A Link Between Neuronal Avalanches and Fading Memory

Fading memory is the ability of a model to retain and combine information about previous inputs in its current activity, at any given time step. This property allows

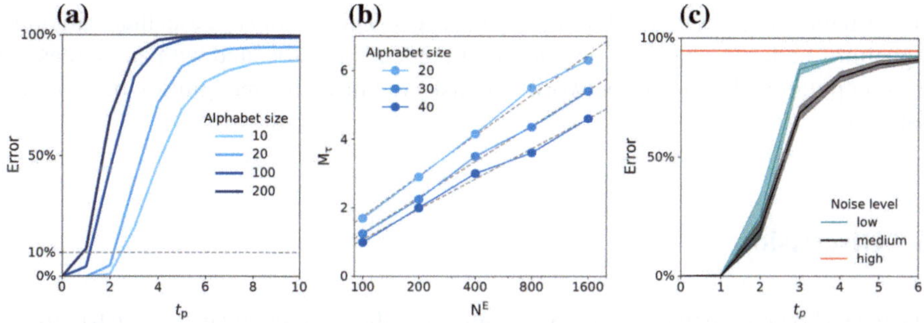

Fig. 4 Fading memory time scale in the SORN. a Error in input recall, for a $SORN_O$ of size $N^E = 200$ (no membrane noise). Curves show the average classification error for a symbol presented t_p time steps in the past. The dashed line shows the 90% performance (10% error) threshold used for the fading memory time scale estimation. **b** Fading memory time scale (M_τ) scaling with network size. Curves show the average M_τ for different alphabet sizes. Gray dashed lines show logarithmic functions, $a \cdot log(N^E) + b$ ($A = 20$: $(a, b) \approx (1.70, -6.09)$; $A = 30$: $(a, b) \approx (1.50, -5.64)$; $A = 40$: $(a, b) \approx (1.27, -4.78)$. **c** Error in input classification, for a $SORN_E$ of size $N^E = 200$ and $A = 20$. The membrane noise level greatly affects the model's performance, and the smallest error was observed for medium noise levels. All curves are averages of 20 SORN simulations

recurrent neural networks to perform complex temporal learning tasks [8] because it enables the networks to combine past and present input. In this section, we measure the fading memory time scale (M_τ) of the SORN models as the time for which readout of past inputs drops below 90%. We study M_τ under different input driven dynamics, and identify an increase in such capacity at the regime in which avalanche criticality occurs.

In order to estimate this fading memory time scale, we evaluated the $SORN_O$'s performance on random inputs received in previous time steps (see *Fading memory time scale estimation*). Each input symbol was independently chosen at random from an alphabet of size A, thus excluding any temporal correlation learning effect, as the input sequence was never repeated. The error when classifying past input symbols received t_p time steps before the current time step quickly increased with t_p, and the model's performance eventually reached chance level (Fig. 4a), for all tested alphabet sizes (qualitatively, bigger alphabets resulted in a "shorter" memory time capacity). Remarkably, the fading memory time scale increased approximately logarithmically with the network size (Fig. 4b), independent of the alphabet size, reaching $t_p \approx 6$ for $N^E = 1600$ and $A = 20$. Such scaling was comparable to the one observed in recurrent networks operating at the edge-of-chaos [8].

Last, we tested whether the power-law distributed neuronal avalanches present in the $SORN_E$'s spontaneous activity were associated with an improvement in its fading memory time scale by comparing different levels of membrane noise (Fig. 4c). As expected, high levels of membrane noise abolished the network's learning and memory capacity, where performance remained slightly above chance level. However, intermediate noise levels were not only associated with neuronal avalanches, but also resulted in an improved fading memory time scale, with a slightly higher performance when compared to low noise levels. This result suggests that criticality signatures in the spontaneous activity might indeed be associated with an increase in the computational power of networks for tasks requiring the recognition of temporal patterns.

6 Discussion

We have shown here that criticality signatures in the form of power-law distributions of duration and size of bursts of neural activity can be found in the spontaneous activity of a recurrent neural network subject to plasticity action. These signatures resulted from self-organization due to biologically inspired plasticity mechanisms and resembled neuronal avalanches observed in neural circuits [6, 18, 33, 40]. Notably, the power-law exponents (size: $\tau \approx 1.3$; duration: $\alpha \approx 1.4$) were smaller than those for a random-neighbor network (2 and 1.5, respectively), potentially reflecting the complex topology emerging after network self-organization. Our model also showed that such power-laws are input dependent and appeared under unstructured external input after a transient break down, a phenomenon that has also been observed in ex-vivo preparations [50]. Additionally, the power-laws also required a suitable intermediate level of membrane noise to occur in the model's spontaneous activity. This finding suggests that in cortex the strength of input received from other connected areas could act as a control parameter maintaining the local network circuit in a reverberating state, thus within the subcritical regime, as detailed by Zierenberg and colleagues [71]. Such result highlights the role of self-organization in driving a network towards a regime in which criticality signatures appear, but suggests that such regime is only achievable under particular input conditions.

Importantly, we refer here to self-organization *towards* criticality instead of typical self-organized criticality (SOC), as the SORN is not a pure SOC model such as the Bak–Tang–Wiesenfeld sandpile model [4]. In the SORN, as opposed to typical SOC models but in agreement with neural activity in-vivo, input and internal dynamics occur at the same time and separation of time scales is absent (i.e., avalanche events are not separated by pauses) [46]. The resulting overlap of avalanches was handled via activity thresholding [42], which excluded very small events and potentially induced left cut-offs in the power-law distributions (see Fig. 2). Although such cut-offs are not a property of SOC systems, they have been observed in cortical data [46]. Furthermore, power-law scaling alone is insufficient evidence for phase transitions, as non-critical stochastic systems have been shown to display such scaling [48, 59]. We argue, therefore, that even though SORN models might not operate at a critical point, their plasticity mechanisms drive them towards a regime in which experimentally measured criticality signatures appear, and suggest that future work should focus on understanding the link between criticality signatures (i.e., avalanche criticality) and phase transitions (i.e., edge-of-chaos criticality) in systems without separation of time scales. Furthermore, we note that in systems with non-zero external input, the structure of the input may generate apparent criticality in non-critical systems [44,

59]. Hence, whenever in a input-driven state, the input should be disentangled from the internal model dynamics when drawing conclusions about criticality.

The differences between the two types of phase transitions generically referred to as "critical points" have been highlighted before, and simulations showed that they are not equivalent in neural networks [24, 25]. In fact, our results suggest that the same occurs in SORN models: neuronal avalanches are present in their spontaneous activity, but perturbation analysis has previously shown that the SORN_O displays subcritical behavior, at least under specific input conditions [28]. Interestingly, even though criticality signatures are not associated with maximum performance in *simple* learning tasks [16], we note here that they coincide with improvement of the *time scale* of memory persistence in the model. Furthermore, such capacity scaled logarithmically with the network size, a property so far observed only in reservoirs tuned to edge-of-chaos criticality [8]. These findings imply that the presence of criticality signatures might not only be beneficial to a network's memory, but also linked to existence of edge-of-chaos critical dynamics, as both phenomena result from an increase in the long-range correlations across units and in time. Importantly, this implies that the dynamical state of the cortex or of a network could and should be adapted to task requirements. Such tuning is particularly sensitive if the network operates in a reverberating state in the vicinity of criticality [65, 66]. A formal description of the relation between these properties, however, is beyond of the scope of this manuscript and should be the focus of future research.

Whereas in models the link between processing capacity and criticality has been widely investigated, experimental evidence is scarce, potentially because it is harder to obtain. The two main challenges are first to tune the experimental system precisely form sub- to supercritical states, ideally in a manner that does not impede its natural processing capacities in a given state. Second, the processing capacities need to be quantified. A classical approach is to make use of pharmacological interventions, e.g. applying AP5 and DNQX to reduce excitation and thereby render a network subcritical, and applying TTX to reduce inhibition and render a system more supercritical [53]. Indeed, the dynamic range obtained from LFP recordings in-vitro under electrical stimulation is maximized in the unperturbed system and diminished when excitation or inhibition is reduced pharmacologically [52]. The same holds for the entropy of evoked patterns, and for the mutual information between stimulus strength and response pattern [51, 53]. Recently, Wibral and colleagues showed that the mutual information (MI) between the past activity of two neurons and their future spiking increases with maturation in-vitro [61]. Given that with maturation neural networks in-vitro approach a critical state [32], this clearly indicates that more information about the past can be read out in the future, as the network self-organizes towards a critical point. By characterizing contributions from the source neurons [9, 62, 63], the relative contribution of synergy to MI increased, whereas the unique contributions from each source decreased during the first four weeks. This indicates that the network develops information modification capabilities [61]. In the fifth week, however, the redundant or shared contribution dominated. Hence, information processing became highly similar across neurons, possibly due to a lack of external inputs [61]. Together, these studies in-vitro show that favorable information processing capabilities increase around criticality and thereby are very much in line with modeling results. Thus, the same relation is also expected to hold in-vivo.

Linking criticality or the deviation from it to performance in a particular task in experiments is challenging. Hence, theoretical work is crucial to drive a systematic understanding [11, 16, 58]. Experimentally, a large body of work focuses on testing whether recorded neural activity in-vivo or in-vitro complies with the criticality hypothesis in showing avalanche distributions without a direct link to function [3, 5, 6, 18, 39, 43, 45, 46, 49, 50, 54, 56, 68]. An elegant approach is to track the performance of a single task and monitor how the performance in this task relates to alterations in network state, as studied systematically by the group of Palva and Palva [38, 39]. The challenges when comparing between different tasks are twofold. First, tasks typically come with altered input for each condition. The input makes it very difficult to disentangle whether an observed difference is indeed caused by a deviation from criticality or any general state change, whether it is induced by transiently changing (non-stationary) input without underlying state change, or a combination of both. Developing approaches to disentangle the two scenarios is an important future challenge. The second challenge is that avalanche analysis requires tens of minutes, or even hours, of recordings to be able to detect differences. This is because an avalanche distribution that extends over the required two orders of magnitude comprises thousands of avalanches for sufficient statistics. Assuming a rate of one avalanche per second, the analysis requires at least 1000 s or 20 min of recording—per condition. A fine temporal resolution in state change is, therefore, difficult and data-costly. Novel approaches, such as the estimator derived recently by Wilting and Priesemann [66] and detailed in this book (for details, see the chapter by Priesemann, Levina and Wilting) now enables quantifying the distance to criticality in a much more data efficient manner, requiring at most a few minutes. Moreover, the estimator's invariance to subsampling is derived from first principles. With such an estimator, changes in dynamical state can be tracked with minute-resolution, or even down to seconds given a trial-based experimental design. Such novel approaches can pave the path to new experimental insights when linking dynamical state, criticality and task processing.

Taken separately, the results presented here seem to suggest an apparent contradiction: avalanche criticality is linked to an increase in the fading memory time scale of recurrent networks, but not to maximum performance in simple learning tasks with structured input. Although the understanding of the relation between both quantities is object of ongoing research, this discrepancy can be clarified by the nature of the learning task and the readout training procedure. Medium membrane noise levels resulted in an increased number of internal states, suggested by the increase in the fading memory time scale and power-law distributions, but such effect could not be exploited by a supervised readout layer as the noisy network activity slightly increased the error of the classifier.

Additionally, as the power-law distributions seem to require unstructured input in the SORN, their absence under repeating input sequences was not surprising. Despite the absence of the most common criticality signature, the network can achieve an internal structure that is beneficial for learning. Such results could be interpreted as a deviation from avalanche criticality due to structured input, while other network information processing abilities, including the fading memory time scale, remained unaffected. Therefore, criticality signatures might still indicate a favorable, although not unique, dynamical regime for recurrent networks.

The contrast between the presence of power-law distributed neuronal avalanches in the model's spontaneous activity and their absence under structured input also suggests an analogy between in-vivo and in-vitro activity in the brain. We have shown that the *same* plasticity mechanisms might result in the occurrence or absence of power-law distributions under different input conditions. As such distributions indicate avalanche criticality, our results stand in agreement with the development of criticality signatures in neural networks in-vitro [6, 18, 32, 57, 71] and with the non-critical dynamics observed in spike avalanches in-vivo [46, 67, 71]. Such input driven adaptation may be favorable for the neural circuits in behaving animals, as it allows them to take advantage of the computational advantages of criticality while avoiding unstable supercritical regimes [46] observed during epileptic seizures [37]. Thus, by keeping neural activity at a healthy, non-epileptic level, biological plasticity mechanisms might play an essential role in tuning the system towards and away from criticality when required by varied input conditions [46, 71].

Appendix

Power-Law Fitting

Although power-laws are very common in nature, their characterization is particularly complex and must be carefully evaluated. In particular, many false positives may appear, ranging from inaccurate exponents to wrong distribution fits [13]. Such problems can be better understood and avoided using maximum likelihood estimators. In this study, we followed the procedure of a previous work [16], employing the *powerlaw* python package [1] in order to estimate the distributions exponents. For all the probability distributions of avalanche durations and sizes shown here, power-laws with cut-offs provided better fits when compared to other single parameter distributions, such as exponential distributions. The cut-off was chosen in a case by case analysis, based on the best power-law fit. For a step-by-step description of power-law fitting via maximum likelihood estimators, see [13].

References

1. Alstott, J., Bullmore, E., Plenz, D.: powerlaw: a python package for analysis of heavy-tailed distributions. PloS One **9**(1), e85,777 (2014)
2. de Arcangelis, L., Perrone-Capano, C., Herrmann, H.J.: Self-organized criticality model for brain plasticity. Phys. Rev. Lett. **96**(2), 028,107 (2006)

3. Arviv, O., Goldstein, A., Shriki, O.: Near-critical dynamics in stimulus-evoked activity of the human brain and its relation to spontaneous resting-state activity. J. Neurosci. **35**(41), 13927–13942 (2015)
4. Bak, P., Tang, C., Wiesenfeld, K.: Self-organized criticality: An explanation of the 1/f noise. Phys. Rev. Lett. **59**(4), 381 (1987)
5. Bedard, C., Kroeger, H., Destexhe, A.: Does the 1/f frequency scaling of brain signals reflect self-organized critical states? Phys. Rev. Lett. **97**(11), 118,102 (2006)
6. Beggs, J.M., Plenz, D.: Neuronal avalanches in neocortical circuits. J. Neurosci. **23**(35), 11167–11177 (2003)
7. Beggs, J.M., Timme, N.: Being critical of criticality in the brain. Front. Physiol. **3** (2012)
8. Bertschinger, N., Natschläger, T.: Real-time computation at the edge of chaos in recurrent neural networks. Neural Comput. **16**(7), 1413–1436 (2004)
9. Bertschinger, N., Rauh, J., Olbrich, E., Jost, J.: Shared information–new insights and problems in decomposing information in complex systems. In: Proceedings of the European Conference on Complex Systems 2012, pp. 251–269. Springer (2013)
10. Bi, G.Q., Poo, M.M.: Synaptic modifications in cultured hippocampal neurons: dependence on spike timing, synaptic strength, and postsynaptic cell type. J. Neurosci. **18**(24), 10,464–10,472 (1998)
11. Boedecker, J., Obst, O., Lizier, J.T., Mayer, N.M., Asada, M.: Information processing in echo state networks at the edge of chaos. Theory Biosci. **131**(3), 205–213 (2012)
12. Chialvo, D.R.: Emergent complex neural dynamics. Nat. Phys. **6**(10), 744–750 (2010)
13. Clauset, A., Shalizi, C.R., Newman, M.E.: Power-law distributions in empirical data. SIAM Rev. **51**(4), 661–703 (2009)
14. Dahmen, D., Diesmann, M., Helias, M.: Distributions of covariances as a window into the operational regime of neuronal networks (2016). arXiv:1605.04153
15. Dahmen, D., Grün, S., Diesmann, M., Helias, M.: Two types of criticality in the brain (2017). arXiv:1711.10930
16. Del Papa, B., Priesemann, V., Triesch, J.: Criticality meets learning: criticality signatures in a self-organizing recurrent neural network. PloS One **12**(5), e0178,683 (2017)
17. Duarte, R., Series, P., Morrison, A.: Self-organized artificial grammar learning in spiking neural networks. In: Proceedings of the 36th Annual Conference of the Cognitive Science Society, pp. 427–432 (2014)
18. Friedman, N., Ito, S., Brinkman, B.A., Shimono, M., DeVille, R.L., Dahmen, K.A., Beggs, J.M., Butler, T.C.: Universal critical dynamics in high resolution neuronal avalanche data. Phys. Rev. Lett. **108**(20), 208,102 (2012)
19. Gautam, S.H., Hoang, T.T., McClanahan, K., Grady, S.K., Shew, W.L.: Maximizing sensory dynamic range by tuning the cortical state to criticality. PLoS Comput. Biol. **11**(12), e1004,576 (2015)
20. Haldeman, C., Beggs, J.M.: Critical branching captures activity in living neural networks and maximizes the number of metastable states. Phys. Rev. Lett. **94**(5), 058,101 (2005)
21. Hartmann, C., Lazar, A., Nessler, B., Triesch, J.: Where's the noise? Key features of spontaneous activity and neural variability arise through learning in a deterministic network. PLoS Comput. Biol. **11**(12), e1004,640–e1004,640 (2015)
22. Hesse, J., Gross, T.: Self-organized criticality as a fundamental property of neural systems. Front. Syst. Neurosci. **8**, (2014)
23. Jaeger, H.: Tutorial on training recurrent neural networks, covering BPPT, RTRL, EKF and the "echo state network" approach, vol. 5. GMD-Forschungszentrum Informationstechnik (2002)
24. Kanders, K., Lorimer, T., Stoop, R.: Avalanche and edge-of-chaos criticality do not necessarily co-occur in neural networks. Chaos Interdiscip. J. Nonlinear Sci. **27**(4), 047,408 (2017)
25. Kanders, K., Lorimer, T., Uwate, Y., Steeb, W.H., Stoop, R.: Robust transformations of firing patterns for neural networks (2017). arXiv:1708.04168
26. Kinouchi, O., Copelli, M.: Optimal dynamical range of excitable networks at criticality. Nat. Phys. **2**(5), 348–351 (2006)

27. Lazar, A., Pipa, G., Triesch, J.: Fading memory and time series prediction in recurrent networks with different forms of plasticity. Neural Netw. **20**(3), 312–322 (2007)
28. Lazar, A., Pipa, G., Triesch, J.: Sorn: a self-organizing recurrent neural network. Front. Comput. Neurosci. **3** (2009)
29. Lazar, A., Pipa, G., Triesch, J.: Emerging bayesian priors in a self-organizing recurrent network. In: Artificial Neural Networks and Machine Learning–ICANN 2011, pp. 127–134. Springer (2011)
30. Legenstein, R., Maass, W.: Edge of chaos and prediction of computational performance for neural circuit models. Neural Netw. **20**(3), 323–334 (2007)
31. Levina, A., Herrmann, J.M., Geisel, T.: Dynamical synapses causing self-organized criticality in neural networks. Nature Phys. **3**(12), 857–860 (2007)
32. Levina, A., Priesemann, V.: Subsampling scaling. Nat. Commun. **8**, 15,140 (2017)
33. Lombardi, F., Herrmann, H., Perrone-Capano, C., Plenz, D., De Arcangelis, L.: Balance between excitation and inhibition controls the temporal organization of neuronal avalanches. Phys. Rev. Lett. **108**(22), 228,703 (2012)
34. Lukoševičius, M.: A practical guide to applying echo state networks. In: Neural Networks: Tricks of the Trade, pp. 659–686. Springer (2012)
35. Maass, W., Natschläger, T., Markram, H.: Real-time computing without stable states: a new framework for neural computation based on perturbations. Neural Comput. **14**(11), 2531–2560 (2002)
36. Massobrio, P., de Arcangelis, L., Pasquale, V., Jensen, H.J., Plenz, D.: Criticality as a signature of healthy neural systems. Front. Syst. Neurosci. **9** (2015)
37. Meisel, C., Gross, T.: Adaptive self-organization in a realistic neural network model. Phys. Rev. E **80**(6), 061,917 (2009)
38. Palva, J.M., Zhigalov, A., Hirvonen, J., Korhonen, O., Linkenkaer-Hansen, K., Palva, S.: Neuronal long-range temporal correlations and avalanche dynamics are correlated with behavioral scaling laws. Proc. Natl. Acad. Sci. **110**(9), 3585–3590 (2013)
39. Palva, S., Palva, J.M.: Roles of brain criticality and multiscale oscillations in temporal predictions for sensorimotor processing. Trends Neurosci. **41**(10), 729–743 (2018)
40. Petermann, T., Thiagarajan, T.C., Lebedev, M.A., Nicolelis, M.A., Chialvo, D.R., Plenz, D.: Spontaneous cortical activity in awake monkeys composed of neuronal avalanches. Proc. Natl. Acad. Sci. **106**(37), 15921–15926 (2009)
41. Plenz, D.: Viewpoint: the critical brain. Physics **6**, 47 (2013)
42. Poil, S.S., Hardstone, R., Mansvelder, H.D., Linkenkaer-Hansen, K.: Critical-state dynamics of avalanches and oscillations jointly emerge from balanced excitation/inhibition in neuronal networks. J. Neurosci. **32**(29), 9817–9823 (2012)
43. Priesemann, V., Munk, M.H., Wibral, M.: Subsampling effects in neuronal avalanche distributions recorded in vivo. BMC Neurosci. **10**(1), 40 (2009)
44. Priesemann, V., Shriki, O.: Can a time varying external drive give rise to apparent criticality in neural systems? PLoS Comput. Biol. **14**(5), e1006,081 (2018)
45. Priesemann, V., Valderrama, M., Wibral, M., Le Van Quyen, M.: Neuronal avalanches differ from wakefulness to deep sleep–evidence from intracranial depth recordings in humans. PLoS Comput. Biol. **9**(3), e1002,985 (2013)
46. Priesemann, V., Wibral, M., Valderrama, M., Pröpper, R., Le Van Quyen, M., Geisel, T., Triesch, J., Nikolić, D., Munk, M.H.: Spike avalanches in vivo suggest a driven, slightly subcritical brain state. Front. Syst. Neurosci. **8** (2014)
47. Ribeiro, T.L., Copelli, M., Caixeta, F., Belchior, H., Chialvo, D.R., Nicolelis, M.A., Ribeiro, S.: Spike avalanches exhibit universal dynamics across the sleep-wake cycle. PloS One **5**(11), e14,129 (2010)
48. Schwab, D.J., Nemenman, I., Mehta, P.: Zipf's law and criticality in multivariate data without fine-tuning. Phys. Rev. Lett. **113**(6), 068,102 (2014)
49. Scott, G., Fagerholm, E.D., Mutoh, H., Leech, R., Sharp, D.J., Shew, W.L., Knöpfel, T.: Voltage imaging of waking mouse cortex reveals emergence of critical neuronal dynamics. J. Neurosci. **34**(50), 16611–16620 (2014)

50. Shew, W.L., Clawson, W.P., Pobst, J., Karimipanah, Y., Wright, N.C., Wessel, R.: Adaptation to sensory input tunes visual cortex to criticality. Nat. Phys. **11**(8), 659–663 (2015)
51. Shew, W.L., Plenz, D.: The functional benefits of criticality in the cortex. Neuroscientist **19**(1), 88–100 (2013)
52. Shew, W.L., Yang, H., Petermann, T., Roy, R., Plenz, D.: Neuronal avalanches imply maximum dynamic range in cortical networks at criticality. J. Neurosci. **29**(49), 15595–15600 (2009)
53. Shew, W.L., Yang, H., Yu, S., Roy, R., Plenz, D.: Information capacity and transmission are maximized in balanced cortical networks with neuronal avalanches. J. Neurosci. **31**(1), 55–63 (2011)
54. Shriki, O., Alstott, J., Carver, F., Holroyd, T., Henson, R.N., Smith, M.L., Coppola, R., Bullmore, E., Plenz, D.: Neuronal avalanches in the resting meg of the human brain. J. Neurosci. **33**(16), 7079–7090 (2013)
55. Stepp, N., Plenz, D., Srinivasa, N.: Synaptic plasticity enables adaptive self-tuning critical networks. PLoS Comput. Biol. **11**(1) (2015)
56. Tagliazucchi, E., Balenzuela, P., Fraiman, D., Chialvo, D.R.: Criticality in large-scale brain fmri dynamics unveiled by a novel point process analysis. Front. Physiol. **3**(15) (2012). https://doi.org/10.3389/fphys.2012.00015
57. Tetzlaff, C., Okujeni, S., Egert, U., Wörgötter, F., Butz, M.: Self-organized criticality in developing neuronal networks. PLoS Comput. Biol. **6**(12), e1001,013 (2010)
58. Tomen, N., Rotermund, D., Ernst, U.: Marginally subcritical dynamics explain enhanced stimulus discriminability under attention. Front. Syst. Neurosci. **8**, 151 (2014)
59. Touboul, J., Destexhe, A.: Can power-law scaling and neuronal avalanches arise from stochastic dynamics. PloS One **5**(2), e8982 (2010)
60. Uhlig, M., Levina, A., Geisel, T., Herrmann, J.M.: Critical dynamics in associative memory networks. Front. Comput. Neurosci. **7** (2013)
61. Wibral, M., Finn, C., Wollstadt, P., Lizier, J.T., Priesemann, V.: Quantifying information modification in developing neural networks via partial information decomposition. Entropy **19**(9), 494 (2017)
62. Wibral, M., Lizier, J.T., Priesemann, V.: Bits from brains for biologically inspired computing. Front. Robot. AI **2**, 5 (2015)
63. Williams, P.L., Beer, R.D.: Nonnegative decomposition of multivariate information (2010). arXiv:1004.2515
64. Williams-García, R.V., Moore, M., Beggs, J.M., Ortiz, G.: Quasicritical brain dynamics on a nonequilibrium widom line. Phys. Rev. E **90**(6), 062,714 (2014)
65. Wilting, J., Dehning, J., Neto, J.P., Rudelt, L., Wibral, M., Zierenberg, J., Priesemann, V.: Dynamic adaptive computation: tuning network states to task requirements (2018). arXiv:1809.07550
66. Wilting, J., Priesemann, V.: Inference, validation and predictions about statistics and propagation of cortical spiking in vivo, p. 363085 (2018)
67. Wilting, J., Priesemann, V.: Inferring collective dynamical states from widely unobserved systems. Nat. Commun. **9**(1), 2325 (2018)
68. Yaghoubi, M., de Graaf, T., Orlandi, J.G., Girotto, F., Colicos, M.A., Davidsen, J.: Neuronal avalanche dynamics indicates different universality classes in neuronal cultures. Sci. Rep. **8**(1), 3417 (2018)
69. Zheng, P., Dimitrakakis, C., Triesch, J.: Network self-organization explains the statistics and dynamics of synaptic connection strengths in cortex. PLoS Comput. Biol. **9**(1), e1002,848 (2013)
70. Zheng, P., Triesch, J.: Robust development of synfire chains from multiple plasticity mechanisms. Front. Comput. Neurosci. **8**, 66 (2014)
71. Zierenberg, J., Wilting, J., Priesemann, V.: Homeostatic plasticity and external input shape neural network dynamics. Phys. Rev. X **8**(3), 031,018 (2018)

Bruno Del Papa studied physics at the University of São Paulo, Brazil and is currently working on his Ph.D. in computational neuroscience at the Frankfurt Institute for Advanced Studies, Germany. He has worked on complex systems and agent-based models during his Master's before moving to his current position and studying self-organization in neural networks. His current research interests include learning algorithms, language processing with recurrent neural networks, and applications of criticality to deep learning.

Viola Priesemann studied physics and neuroscience at TU Darmstadt, the MPI for Brain Research, Caltech and Ecole Normale Superieure. Briefly after obtaining her Ph.D. in 2013, she became independent group leader at the MPI for Dynamics and Self-Organization, investigating the interplay. She employs tools from statistical physics and information theory, and in particular develops approaches that enable inference about the full system when only a tiny fraction is sampled. Solutions to this subsampling problem are crucial to combine results from artificial with those of biological neural networks, where only a tiny fraction of all spiking neurons can be recorded experimentally.

Jochen Triesch studied physics at the University of Bochum, Germany, receiving his Ph.D. in 1999. He spent two years as a post-doctoral fellow at the University of Rochester, USA, before joining the Cognitive Science Dept. at UC San Diego, USA as an Assistant Professor. In 2005 he moved to the newly founded Frankfurt Institute for Advanced Studies (FIAS) in Frankfurt am Main, Germany. Since 2007 he is the Johanna Quandt Chair for Theoretical Life Sciences at FIAS. His research interests include Computational Neuroscience, Machine Learning, Cognitive Science, and Developmental Robotics.

Bruno Del Papa studied [illegible] at the University of São Paulo, Brazil and is currently working on his Ph.D. in computational neuroscience at the Frankfurt Institute for Advanced Studies, Germany. He has worked [illegible] models during his Master's [illegible] [illegible] [illegible] [illegible] and [illegible].

Homeostatic Structural Plasticity Can Build Critical Networks

Arjen van Ooyen and Markus Butz-Ostendorf

Abstract Many neural networks, ranging from in vitro cell cultures to the neocortex in vivo, exhibit bursts of activity ("neuronal avalanches") with size and duration distributions characterized by power laws. The exponents of these power laws point to a critical state in which network connectivity is such that, on average, activity neither dies out nor explodes, a condition that optimizes information processing. Various neural properties, including short- and long-term synaptic plasticity, have been proposed to underlie criticality. Reviewing several model studies, here we show that during development, activity-dependent neurite outgrowth, a form of homeostatic structural plasticity, can build critical networks. In the models, each neuron has a circular neuritic field, which expands when the neuron's average electrical activity is below a homeostatic set-point and shrinks when it is above the set-point. Neurons connect when their neuritic fields overlap. Without any external input, the initially disconnected neurons organize themselves into a connected network, in which all neurons attain the set-point level of activity. Both numerical and analytical results show that in this equilibrium configuration, the network is in a critical state, with avalanche distributions described by precisely the same power laws as observed experimentally. Thus, in building critical networks during development, homeostatic structural plasticity can lay down the basis for optimal network function in adulthood.

Keywords Homeostatic structural plasticity · Activity-dependent neurite outgrowth · Neuronal avalanches · Power laws · Self-organized criticality · Neural networks · Development

A. van Ooyen (✉)
Department of Integrative Neurophysiology, VU University Amsterdam, De Boelelaan 1085, 1081 HV Amsterdam, The Netherlands
e-mail: arjen.van.ooyen@gmail.com

M. Butz-Ostendorf
Biomax Informatics AG, Robert-Koch-Str. 2, 82152 Planegg, Germany
e-mail: butzostendorf@gmail.com

N. Tomen et al. (eds.), *The Functional Role of Critical Dynamics in Neural Systems*, Springer Series on Bio- and Neurosystems 11,
https://doi.org/10.1007/978-3-030-20965-0_7

1 Introduction

Experimental studies have observed an intriguing dynamical state characterized by so-called neuronal avalanches in a variety of neural systems, including acute and cultured cortical slices [5, 6], developing cultures of dissociated cortex cells [48], the developing retina [30], the developing cortex in vivo [25] and the adult neocortex in vivo [49]. Neuronal avalanches are spontaneous bursts of activity that have power-law size and duration distributions [5, 6]. Most studies report that the number of avalanches of a given size (e.g., in terms of number of electrodes on which activity is recorded) decreases proportionally to the size to the power -1.5, and that the number of avalanches of a given duration declines proportionally to the duration to the power -2 [5, 25]. Power laws typically emerge in systems when they are critical, meaning that they are close to a transition in behavior [42]. Simple mathematical models have shown [78] that power laws with exponents -1.5 and -2 can arise if connectivity is such that every neuron that fires an action potential causes, on average and independently of network activity [38], one other neuron to fire. With this connectivity, network activity, on average, neither dies out nor blows up over time.

How do networks develop and maintain such a critical pattern of connectivity? Reviewing several model studies, here we show that activity-dependent outgrowth of neurites (axons and dendrites) can self-organize a network into a critical state. During development, electrical activity controls the elongation, branching and retraction of neurites [34, 44, 60, 77] by modifying the level of intracellular calcium. Calcium, which enters the cell through voltage-gated channels, is the principal regulator of the growth cone, a specialized structure at the tip of outgrowing neurites [24, 34, 37, 40]. A high intracellular calcium concentration, caused by membrane depolarization, a high neuronal firing rate, or stimulation by excitatory neurotransmitters, arrests neurite outgrowth or even causes retraction. Conversely, a low calcium concentration, due to a low firing rate, hyperpolarization, or inhibitory neurotransmitters, promotes neurite elongation [16, 23, 32, 45, 46]. Thus, the way in which electrical activity modulates neurite outgrowth contributes to maintaining neuronal electrical activity at a stable average level (homeostasis). When the electrical activity of a neuron is above a desired value (homeostatic set-point) its neurites retract, breaking-up synaptic connections and so reducing neuronal activity. Conversely, when activity is below this value, neurites grow out, making new synaptic connections and so raising the neuron's activity.

Activity-dependent neurite outgrowth is a form of homeostatic structural plasticity [14, 15, 22], with structural plasticity defined as encompassing all the structural adaptations, such as neurite outgrowth and changes in dendritic spine numbers, that lead to the formation or deletion of synapses [14, 69]. Structural plasticity can connect previously unconnected neurons, disconnect neurons, or change the number of synapses by which neurons are connected. In contrast, synaptic plasticity is defined as a change in the strength of existing synapses. Hebbian synaptic plasticity changes synapse strength depending on the correlation between pre- and postsynaptic activity [8, 28], whereas synaptic scaling (homeostatic synaptic plasticity) modifies the strengths of all the cell's incoming synapses so as to stabilize neuronal activity around some set-point value [63].

One of the first models of homeostatic structural plasticity is the neuritic field model of activity-dependent neurite outgrowth [70–72, 75]. In this model, the neurite extensions of each neuron are represented by a circular neuritic field, which expands when the neuron's electrical activity is below a homeostatic set-point and shrinks when the neuron's activity is above the set-point. Neurons connect synaptically when their neuritic fields overlap.

In this chapter, we give a brief overview of the original neuritic field model, followed by a review of studies [2, 38, 61] that have employed the model to examine the development of criticality. The results show that simple, homeostatic growth rules can construct neural circuits with critical, power-law behavior.

2 The Neuritic Field Model

2.1 Model at a Glance

In constructing the neuritic field model, we were inspired in part by developing cultures of dissociated cortex cells, in which initially disconnected cells assemble themselves, without external input, into a synaptically connected network by neurite outgrowth and synaptogenesis [43, 48, 65, 76]. In the model, growing neurons are described as expanding neuritic fields, representing both axons and dendrites. Neurons become synaptically connected when their neuritic fields overlap, with a connection strength proportional to the area of overlap. The outgrowth of each neuron depends on its own level of electrical activity, as follows. The neuritic field expands when the neuron's electrical activity is below a certain set-point and shrinks when activity is above this set-point. Thus, a reciprocal influence exists between electrical activity (fast dynamics) and outgrowth (slow dynamics): electrical activity determines outgrowth, while in turn outgrowth alters connectivity and consequently activity. Through these interactions, the initially disconnected neurons organize themselves into a synaptically connected network, guided only by the activity generated by the network itself; there is no external input.

2.2 Neuronal Activity

Neuronal electrical activity is described by the shunting model [26]. In this model, excitatory inputs drive the membrane potential towards a maximum (the excitatory saturation potential), while inhibitory inputs drive the membrane potential towards a minimum (the inhibitory saturation potential). For a network containing only excitatory cells, the model becomes [70]:

$$\frac{dX_i}{dt} = -\frac{X_i}{\tau_X} + (1 - X_i)\sum_{j=1}^{N} W_{ij}F(X_j) \tag{1}$$

where X_i is the membrane potential of neuron i, t is time, τ_X is the membrane time constant, $W_{ij} \geq 0$ is the connection strength between presynaptic neuron j and postsynaptic neuron i, $F(X_j)$ is the firing rate of neuron j, and N is the total number of neurons. The term $(1 - X_i)$ implies that inputs from other cells drive the membrane potential towards a saturation potential of 1. The firing rate, with its maximum set to 1, is a sigmoidal function of the membrane potential:

$$F(X_j) = \frac{1}{1 + e^{(\theta - X_j)/\alpha}} \tag{2}$$

where α determines the steepness of the function and θ represents the firing threshold. The low firing rate for sub-threshold membrane potentials reflects spontaneous neuronal activity.

2.3 Outgrowth and Connectivity

Neurons are placed at random positions on a two-dimensional surface. Each neuron has a circular neuritic field, the radius of which is variable. When the fields of neurons i and j overlap, both neurons become connected with a strength $W_{ij} = \sigma A_{ij}$, where $A_{ij} = A_{ji}$ is the area of overlap, representing the total number of synapses formed reciprocally between neurons i and j; and σ is a constant of proportionality, representing the strength of a single synapse.

The change in neuritic field size depends on the neuron's own firing rate:

$$\frac{dR_i}{dt} = \rho G[F(X_i)] \tag{3}$$

where R_i is the radius of the neuritic field of neuron i, and ρ determines the rate of outgrowth. The outgrowth function G is defined as

$$G[F(X_i)] = 1 - \frac{2}{1 + e^{[F_{\text{target}} - F(X_i)]/\beta}} \tag{4}$$

where F_{target} is the homeostatic set-point, i.e., the value of $F(X_i)$ for which $G = 0$; and β determines the steepness of the function. Equation 4 implements that depending on the value of $F(X_i)$, a neuritic field grows out [$G > 0$ if $F(X_i) < F_{\text{target}}$], retracts [$G < 0$ if $F(X_i) > F_{\text{target}}$] or remains constant [$G = 0$ if $F(X_i) = F_{\text{target}}$]. In biological neurons, the effect of electrical activity on neurite outgrowth is mediated by calcium [24, 34, 37, 40], with the concentration of intracellular calcium acting as indicator of the neuron's firing rate [2, 3, 58].

2.4 Network Assembly, Overshoot and Homeostasis

The neurons are initialized with no or small neuritic fields, so most neurons are initially disconnected or organized in small, isolated clusters (Fig. 1a). Consequently, neuronal firing rates $F(X_i)$ are below the homeostatic set-point F_{target}, and neuritic fields start expanding. As the neurons grow out, they begin to form more and stronger connections, linking neurons together and slowly raising the level of activity in the network. At some degree of connectivity, network activity abruptly jumps to a much higher level (Fig. 1d), but activity is then so high that $F(X_i) > F_{\text{target}}$. As a result, neuritic field size and connectivity start decreasing and activity drops. As neurons adjust the size of their neuritic fields, and react to the adjustments of their neighbors, the network eventually reaches a stable equilibrium in which the connectivity between cells is such that for all cells $F(X_i) = F_{\text{target}}$ and neuritic fields and connectivity no longer change (Fig. 1b). The neurons thus self-organize, via a transient phase of high connectivity (overshoot) (Fig. 1c), into a stable network with network-wide homeostasis of activity. They thereby adapt to the local cell density, with neurons acquiring small neuritic fields in areas with a high cell density and large fields in areas with a low cell density (Fig. 1b).

The assembly of initially unconnected model neurons into a connected network strongly resembles development in cultures of dissociated cortex cells, with respect to both activity and connectivity [27, 55, 57, 65, 66]. The first three weeks in vitro show a phase of steady neurite outgrowth and synapse formation [65, 66], with neuron firing and network activity abruptly appearing within a window of a few days [27] and network structure exhibiting a transition from local to global connectivity [57]. In the next week, this is followed by a substantial elimination of synapses until a stable connectivity level is reached [65, 66].

2.5 Analytical Relationship Between Activity and Connectivity

The relationship between activity and connectivity, and the changes in activity and connectivity during development, can be predicted directly from Eq. 1 [70]. For a given connectivity matrix $\mathbf{W}$, the equilibrium points of X_i are solutions of

$$0 = -\frac{X_i}{\tau_X} + (1 - X_i) \sum_{j=1}^{N} W_{ij} F(X_j) \tag{5}$$

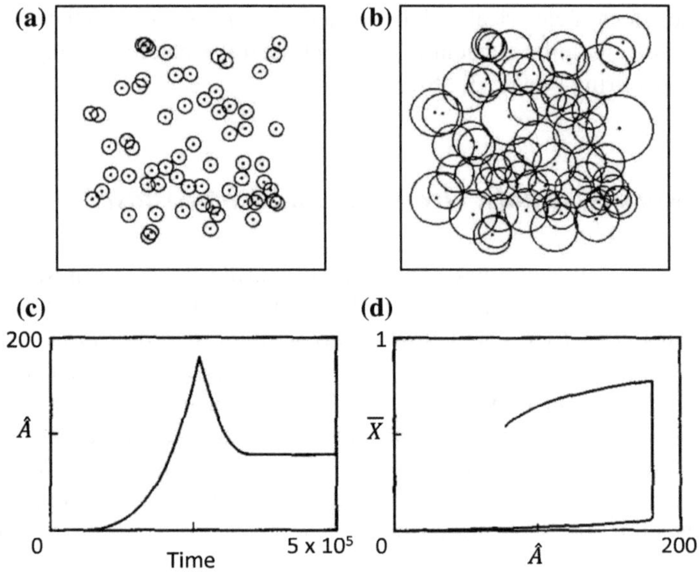

Fig. 1 **Network assembly.** In this example of the original neuritic field model [70, 72], all cells are excitatory. **a** Early stage of network development. Neuritic fields are small, connectivity is low, and cells have a low level of electrical activity. **b** Network at equilibrium. The electrical activity of all cells is at the homeostatic set-point, and the neuritic field sizes remain constant. **c** Development of network connectivity $\hat{A} = \frac{1}{2}\sum_{i=1,j=1}^{N} A_{ij}$ = total area of overlap (see Sect. 2.3) over time. **d** Network-averaged membrane potential $\overline{X}$ against network connectivity $\hat{A}$. Electrical activity is initially low, so connectivity increases. When connectivity is strong enough, activity abruptly jumps to a much higher level. This level exceeds the homeostatic set-point, so connectivity and activity then decrease until activity is at the homeostatic set-point. Parameters of the model: $\tau_X = 8$, $\rho = 2.5 \times 10^{-6}$, $\theta = 0.5$, $\alpha = 0.1$, $\beta = 0.1$, $F_{\text{target}} = 0.6$, $\sigma = 0.4$ (**a** and **b**) or 0.1 (**c** and **d**), $N = 64$. The value of the outgrowth rate ρ is small enough for connectivity to be quasi-stationarity on the time scale of membrane potential dynamics (Figure reproduced, with permission, from [70])

If all cells have the same F_{target} and the variations in X_i are small relative to the average membrane potential $\overline{X}$ of the network, then $0 = -\overline{X}/\tau_X + \left(1 - \overline{X}\right)\overline{W}F\left(\overline{X}\right)$, where $\overline{W}$ is the average connection strength. Rewriting this equation gives

$$\overline{W} = \frac{\overline{X}/\tau_X}{\left(1 - \overline{X}\right)F\left(\overline{X}\right)} \quad 0 \leq \overline{X} < 1 \tag{6}$$

Equation 6, which defines a manifold in $(\overline{W}, \overline{X})$ space (Fig. 2), provides the equilibrium value(s) of $\overline{X}$ for a given, fixed value of $\overline{W}$ (i.e., a bifurcation diagram). Equilibrium states on branch CD of the manifold are unstable with respect to $\overline{X}$; equilibrium states on branches ABC and DEF are stable. Because changes in $\overline{W}$ are slow, being caused by outgrowth and retraction of neuritic fields (Eq. 3), $\overline{W}$ can be considered quasi-stationary on the time scale of membrane potential dynamics. That

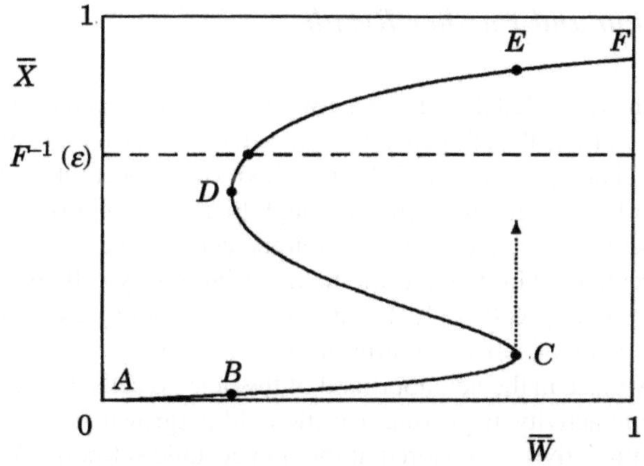

Fig. 2 Relationship between activity and connectivity. In the original neuritic field model [70, 72], the manifold of Eq. 6 defines the equilibrium value(s) of the network-averaged membrane potential $\overline{X}$ for a given, fixed value of the network-averaged connectivity $\overline{W}$ in a purely excitatory network. Equilibrium values on branch CD are unstable with respect to $\overline{X}$; equilibrium values on branches ABC and DEF are stable. The intersection point with the line $\overline{X} = F^{-1}(\varepsilon)$, where F^{-1} is the inverse of the firing rate function and $\varepsilon = F_{\text{target}}$ (see Eqs. 2–4), is the equilibrium state of the whole system, at which $\overline{W}$ remains constant. See further Sect. 2.5 (Figure reproduced, with permission, from [70])

is, in the time that $\overline{X}$ relaxes to its equilibrium value, $\overline{W}$ hardly changes. In other words, at any given value of $\overline{W}$, $\overline{X}$ is at its equilibrium value. Therefore, the slow evolution of $\overline{X}$, i.e., the changes in $\overline{X}$ that are brought about by changes in $\overline{W}$, take place along the manifold.

If for all cells $F(X_i) = F_{\text{target}}$, the neuritic fields, and therefore $\overline{W}$, remain constant. Thus, at the intersection point with the line $\overline{X} = F^{-1}(F_{\text{target}})$ (F^{-1} is the inverse of F), $\overline{W}$ remains constant; above and below that line, it decreases and increases, respectively. Consider, for example, an intersection point on branch DE (Fig. 2). During development, connectivity and activity are initially low, so $\overline{W}$ increases, and $\overline{X}$ follows the branch ABC until it reaches C, at which point it jumps to branch DEF. However, $\overline{X}$ is then so high that the neuritic fields begin to retract and $\overline{W}$ to decrease until $\overline{X}$, moving along branch DEF, reaches the intersection point. Thus, in order to arrive at an intersection point on branch DE, a developing network has to go through a phase in which connectivity is higher than in the final situation (overshoot; see Sect. 2.4). If the intersection point is on branch CD, connectivity and activity will oscillate on the time scale of growth [71]. No overshoot or oscillations occur if the intersection point is on branch ABC or EF.

2.6 Inhibition and Further Results

Simulation studies revealed that also in networks with both excitatory and inhibitory cells (mixed networks), all cells generally achieve homeostasis of activity, just as they do in purely excitatory networks [72]. Overshoot of connectivity can be enhanced in mixed networks [72]. Interestingly, although there are no intrinsic differences in growth rules between excitatory and inhibitory cells in the model, the cells nevertheless differentiate, with the neuritic fields of inhibitory cells becoming smaller than those of excitatory cells [72]. Furthermore, both purely excitatory and mixed networks are capable of self-repair after lesions. Following cell loss, the remaining cells, especially those in the neighborhood of the deleted cells, lose connections and undergo a drop in activity, triggering neuritic field outgrowth and formation of new connections, until activity is restored at the homeostatic set-point [72]. In addition, the model can account for the development of intrinsic firing patterns [1], the development of retinal mosaics [20], developmental changes in network-wide activity bursts [35], and developmental transitions in cognition [51, 52]. For extensive reviews of the model, see [68, 74].

3 Criticality in the Neuritic Field Model

3.1 Model

Abbott and Rohrkemper [2] used a slightly modified version of the original neuritic field model [70, 72]. In their variant of the model, neuronal activity is governed by a Poisson spiking model (rather than being described by a firing rate) and neuritic field outgrowth is dependent on the neuron's internal calcium concentration (rather than directly on the neuron's firing rate). In the purely excitatory network they investigated, neuronal activity is generated by a Poisson spiking model based on a computed firing rate. The firing rate F_i of neuron i is described by

$$\frac{\mathrm{d}F_i}{\mathrm{d}t} = \frac{F_0 - F_i}{\tau_F} \tag{7}$$

where F_0 is a spontaneous background rate and τ_F is the time constant with which F_i relaxes to F_0. At every time step Δt, neuron i fires an action potential with probability $F_i \Delta t$. After a neuron fires, it cannot fire again for a refractory period t_{ref}. Whenever another neuron j fires an action potential, F_i is incremented, $F_i \to F_i + \sigma A_{ij}$, where A_{ij} is the area of overlap between neurons i and j, and the constant σ represents synaptic strength.

The average level of activity of neuron i is monitored by the neuron's internal calcium concentration C_i, which is incremented whenever neuron i fires, $C_i \to C_i + 1$, and decays to zero otherwise,

$$\frac{\mathrm{d}C_i}{\mathrm{d}t} = -\frac{C_i}{\tau_C} \tag{8}$$

with time constant τ_C. The calcium concentration determines the change in the neuritic field radius R_i of neuron i:

$$\frac{\mathrm{d}R_i}{\mathrm{d}t} = \rho\big(C_{\text{target}} - C_i\big) \tag{9}$$

where ρ is the rate of outgrowth. If neuronal activity and thus calcium concentration are low ($C_i < C_{\text{target}}$), neuron i grows out, leading to more excitatory connections and hence higher activity. Conversely, if neuronal activity and calcium concentration are high ($C_i > C_{\text{target}}$), the neuron retracts, reducing connectivity and lowering activity. In this way, each neuron grows out or retracts to try to reach the target level of calcium concentration ($C_i = C_{\text{target}}$).

3.2 Results

In a similar manner to that described for the original model (Sect. 2.4), the neurons grow out and assemble themselves into a synaptically connected network. In the equilibrium state, the calcium concentrations of the neurons remain close to C_{target} and the radii R_i of the neuritic fields are nearly constant, with only small fluctuations over time. In the equilibrium configuration, the pattern of network activity was analyzed in terms of size and duration of networks bursts [2]. A network burst or avalanche was defined as an event in which spiking is observed in at least one neuron for a contiguous sequence of time bins ($t_{\text{bin}} = 10$ ms), bracketed before and after by at least one bin of silence in all neurons. The results of the analysis (Fig. 3) were interpreted to show that burst size and burst duration in the model follow power-law distributions (i.e., linearity in a log-log plot), characteristic of critical dynamics. The occurrence of bursts of a given size (as measured in number of action potentials generated during a burst) was described as following a power law with exponent −1.5 (Fig. 3a), and the number of bursts of a given duration as a power law with exponent −2 (Fig. 3b), similarly to what had been observed in cultures of cortical slices [5, 6] and dissociated cortex cells [48]. The property of the model that neurons grow out when activity is low and withdraw when activity is high forces the network to find a middle ground between all-to-all connectivity (producing excessive activity) and local connectivity (producing insufficient activity). This middle-ground in connectivity, with a stable average level of activity, was believed to underlie the generation of critical dynamics in the model.

The small fluctuations in R_i that are still present in the equilibrium state are not important for the size and duration distributions: shutting off growth completely ($\rho = 0$) once equilibrium is reached did not make any noticeable difference to the results. The distributions do also not crucially depend on the exact values of the

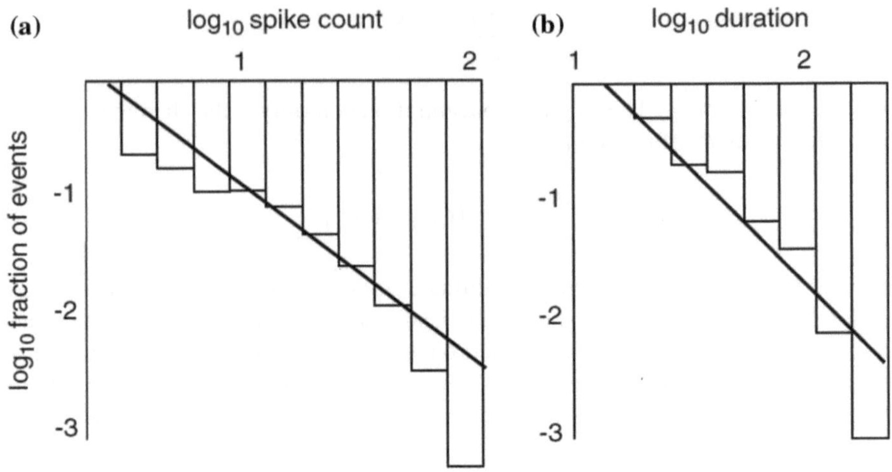

Fig. 3 Burst size and duration. Burst size and duration in the model by Abbott and Rohrkemper [2]. **a** Histogram of the fraction of bursts (events) with different numbers of spikes. The line indicates -1.5 power. **b** Histogram of the fraction of bursts with different durations. The line indicates -2 power. Parameters of the model: $F_0 = 0.1$ Hz, $\tau_F = 5$ ms, $\Delta t = 1$ ms, $t_{\text{ref}} = 20$ ms, $\sigma = 500$ Hz, $\tau_C = 100$ ms, $\rho = 0.002\ \text{s}^{-1}$, $C_{\text{target}} = 0.08$, total number of neurons = 100 (Figure reproduced, with permission, from [2])

model parameters. The value of C_{target} influences the exponents of the power laws with which the distributions are described, but only values much higher or lower than the one used in Fig. 3 lead to essentially different distributions. Much higher values of C_{target} yield flat distributions of burst size and burst duration, whereas much smaller values lead to a shortage of large, long-lasting bursts.

4 Analytical Proof of Criticality in the Neuritic Field Model

Being a relatively small simulation study, the work by Abbott and Rohrkemper [2] could not claim conclusively that the neuritic field model is capable of building critical circuits. Recently, Kossio et al. [38] proved analytically that a slightly different version of the model used by Abbott and Rohrkemper [2] generates activity dynamics characterized by power-law avalanche distributions. In their model, neuronal activity is described by a stochastic, continuous-time spiking model that is very similar to the one used in Abbott and Rohrkemper [2], with an instantaneous firing rate F_i of neuron i and a low spontaneous firing rate F_0 but without a refractory period (but see below). As in Abbott and Rohrkemper [2], a spike from neuron j increments F_i by σA_{ij}, where A_{ij} is the area of overlap between neurons i and j, and the constant σ represents synaptic strength. Without an input spike, F_i decays exponentially to F_0 with time constant τ_F (Eq. 7). A difference from Abbott and Rohrkemper [2] is

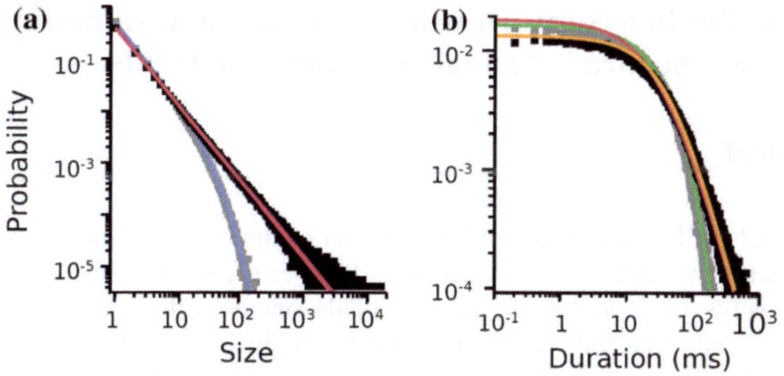

Fig. 4 Avalanche size and duration in the model by Kossio et al. [38]. **a** Analytical size distribution (blue) and simulation results (gray) for a subcritical state ($F_{\text{target}} = 0.04$ Hz), and analytical size distribution (red) and simulation results (black) for a near-critical state ($F_{\text{target}} = 2$ Hz) **b** Analytical duration distribution (green) and simulation results (gray) for the subcritical state, and analytical duration distribution (orange) and simulation results (black) for the near-critical state. Red line shows a closed-form approximation. Parameters of the model: $F_0 = 0.01$ Hz, $\tau_F = 10$ ms, $\sigma = 500$ Hz, $\rho = 10^{-6}\,\text{s}^{-1}$, total number of neurons = 100. For the subcritical state, a time bin t_{bin} of 30 ms was used, and for the near-critical state a t_{bin} of 45 ms (Figure reproduced, with permission, from [38])

that the change in neuritic field radius R_i of neuron i depends directly on its firing rate F_i. In the model, R_i increases linearly with rate ρ between spikes of neuron i and decreases with a constant amount ρ / F_{target} when neuron i fires a spike. Thus, on average, R_i increases if the time-averaged firing rate $\overline{F}_i < F_{\text{target}}$, decreases if $\overline{F}_i > F_{\text{target}}$, and remains constant if $\overline{F}_i = F_{\text{target}}$. The network grows into a stationary state in which all neurons have an average firing rate of F_{target}. Kossio et al. [38] showed mathematically that in this state, provided $F_{\text{target}} \gg F_0$, avalanche size follows a power-law distribution with exponent -1.5, and avalanche duration, for large durations, a power-law distribution with exponent -2 (Fig. 4).

Numerical simulations further demonstrated that halting growth ($\rho = 0$) in the stationary state so that small connectivity fluctuations are eliminated has no effect on the avalanche statistics (as in [2]) and that introducing a biologically plausible refractory period has only a moderate effect on the statistics. However, if the refractory period becomes too long, the power laws begin to break down. This last finding, together with the fact that in Abbott and Rohrkemper [2] F_{target} (based on C_{target}) is not much larger than F_0, may explain the deviations from power law in Fig. 3 (generated with refractory period $t_{\text{ref}} = 4\tau_F$) [38].

5 Criticality in a Network with Excitatory and Inhibitory Cells and Separate Axonal and Dendritic Fields

5.1 Model

In the model by Tetzlaff et al. [61], in contrast to the original neuritic field model [70, 72] and the models by Abbott and Rohrkemper [2] and Kossio et al. [38], each neuron i has two separate circular neuritic fields, one describing the size of its axon (radius R_i^{axo}) and one the size of its dendrites (radius R_i^{den}). The change in R_i^{den} depends in the same way on the internal calcium concentration C_i as in the previous two models:

$$\frac{\mathrm{d}R_i^{\text{den}}}{\mathrm{d}t} = \rho_{\text{den}}\left(C_{\text{target}} - C_i\right) \tag{10}$$

where ρ_{den} is the rate of dendritic outgrowth and C_{target} is the target calcium concentration. However, the change in R_i^{axo} is given by

$$\frac{\mathrm{d}R_i^{\text{axo}}}{\mathrm{d}t} = -\rho_{\text{axo}}\left(C_{\text{target}} - C_i\right) \tag{11}$$

where ρ_{axo} is the rate of axonal outgrowth. Thus, R_i^{axo} increases when $C_i > C_{\text{target}}$ and decreases when $C_i < C_{\text{target}}$, reflecting experimental observations that axons require electrical activity to grow out [53, 79].

The network may contain both excitatory and inhibitory neurons. In the neuron model, which is similar to the one used in Abbott and Rohrkemper [2], the membrane potential X_i (limited by a hard bound to 1) of neuron i is given by

$$\frac{\mathrm{d}X_i}{\mathrm{d}t} = \frac{X_0 - X_i}{\tau_X} \tag{12}$$

where X_0 is the resting potential and τ_X is the time constant with which X_i relaxes to X_0. At every time step, neuron i fires an action potential when $X_i > \varrho_i$, where ϱ_i is a uniformly distributed random number between 0 and 1 (drawn at each time step). After a neuron has fired, it is refractory for four time steps. Whenever another neuron j fires an action potential, X_i is incremented, $X_i \rightarrow X_i + \sigma_j A_{ij}$, where A_{ij} represents the overlap between the axonal field of presynaptic neuron j and the dendritic field of postsynaptic neuron i; and σ_j is a constant representing synaptic strength, defining whether presynaptic neuron j is excitatory ($\sigma_j^{\text{exc}} > 0$) or inhibitory ($\sigma_j^{\text{inh}} < 0$).

As in Abbott and Rohrkemper [2], the calcium concentration C_i of neuron i is incremented whenever neuron i fires an action potential, $C_i \rightarrow C_i + \gamma$, where γ is the increase in calcium concentration. Between action potentials, C_i decays to zero with time constant τ_C (Eq. 8). All the differential equations are solved by the Euler method, with an interval length of one simulated time step.

5.2 Results

During the early stage of development, all cells are taken to be excitatory. Initially, the axonal and dendritic fields of the cells are so small that no connections exist. Consequently, neuronal activity and calcium concentrations are low, triggering dendritic field outgrowth and a slow build-up of connections, together with a gradual rise in neuronal activity (Phase I) (Fig. 5). At a certain point in time, neuronal activity increases rapidly towards a maximum, in parallel with a shrinkage of dendritic fields and an expansion of axonal fields, because of the calcium concentrations rising above C_{target} (Phase II, similar to the overshoot phase described in Sect. 2.4). During Phase II, inhibitory neurons are introduced by changing 20% of all neurons into inhibitory ones (synaptic strength $\sigma < 0$), reflecting the developmental switch of the neurotransmitter GABA from excitatory to inhibitory [7, 33]. Introducing inhibition dampens neuronal activity. In the last stage of development, the system reaches an equilibrium state in which neuronal activity fluctuates around a stable value (homeostasis) and the calcium concentrations remain close to C_{target} (Phase III).

In each developmental phase, the pattern of network activity was analyzed in terms of the number of action potentials contained in networks bursts [61]. As in Sect. 3.2, a network burst or avalanche was defined as a period of network activity between two time bins in which all neurons are silent. In the figures showing frequency of avalanches against number of spikes in an avalanche, the straight dashed lines indicate the best power-law fit (Fig. 6). As before, if an avalanche distribution matches the power-law line, it is called critical. An over-representation of large avalanches is referred to as supercritical, and an under-representation as subcritical [4, 47].

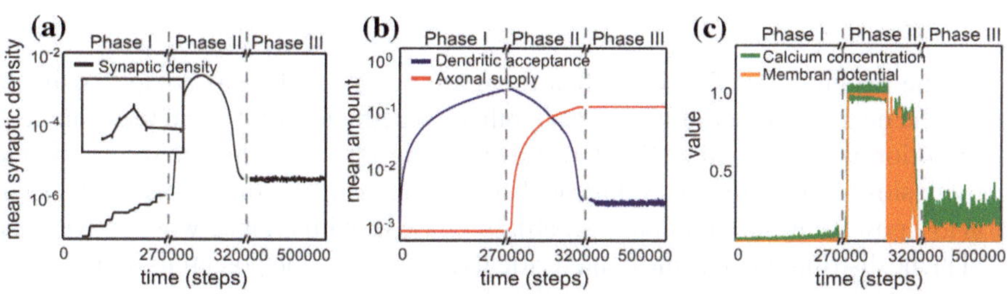

Fig. 5 Developmental phases. Network development in the model by Tetzlaff et al. [61] shows three distinct phases: Phase I, in which synaptic connectivity and neuronal activity gradually increase; Phase II, in which connectivity and activity abruptly rise towards a maximum, followed by pruning of connectivity and a lowering of activity; and Phase III, in which homeostasis of activity is reached. **a** Development of synaptic connectivity (average A_{ij}). Note that the time axis is expanded in the middle. The inset shows the development of synaptic density in cell cultures [65, 66, 70]. **b** Development of axonal extent ("axonal supply"; average R_i^{axo}) and dendritic extent ("dendritic acceptance"; average R_i^{den}). **c** The course of network activity (average X_i) and calcium concentration (average C_i) during network development. Parameters of the model: $\rho_{\text{den}} = 0.02$, $\rho_{\text{axo}} = 0.01$, $C_{\text{target}} = 0.05$, $\tau_X = 5$, $|\sigma^{\text{inh}}| = |\sigma^{\text{exc}}| = 1000$, $\gamma = 0.5$, $\tau_C = 10$, $X_0 = 0.0005$, total number of neurons = 100 (From [61], open access)

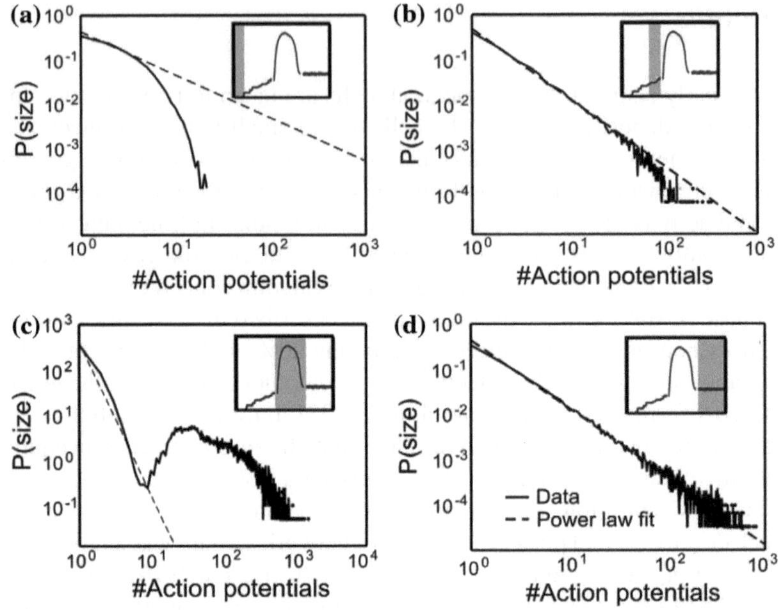

Fig. 6 Avalanche distributions. Avalanche size distributions undergo characteristic changes during network development in the model by Tetzlaff et al. [61]. Gray area in inset indicates stage of development (see Fig. 5). **a** At the beginning of Phase I, when there are hardly any synaptic connections, the distribution is Poisson-like. **b** As more connections are formed, the distribution takes on a power-law form. **c** In Phase II, when connectivity is high, the distribution becomes supercritical. **d** In Phase III (if $|\sigma^{\text{inh}}| = |\sigma^{\text{exc}}|$), when homeostasis is reached, the distribution is critical. The exponent of the power law is close to -1.5 (From [61], open access)

In the beginning of Phase I, when there are no or hardly any synaptic connections, the neurons do not influence each other's electrical activity, and the avalanche distribution is Poisson-like (Fig. 6a). Later during Phase I, when connectivity and activity slowly increase, the avalanche distribution changes from a Poisson distribution to a power-law distribution (Fig. 6b). In Phase II, with high network activity, the avalanche distribution becomes supercritical (Fig. 6c). Action potentials of both excitatory and inhibitory neurons were included in measuring this distribution. Even with much stronger inhibitory synaptic strength ($|\sigma^{\text{inh}}| = 100|\sigma^{\text{exc}}|$, as compared with $|\sigma^{\text{inh}}| = |\sigma^{\text{exc}}|$, as in Fig. 6), the distribution stays supercritical. The system remains supercritical during the whole of Phase II, until shrinkage of dendritic fields has so far pruned connectivity that homeostasis is reached, with calcium concentrations around C_{target} and stable neuronal activity (Phase III). In Phase III, provided $|\sigma^{\text{inh}}| = |\sigma^{\text{exc}}|$, the avalanche distribution becomes critical (Fig. 6d). If inhibition is stronger, the system turns into a subcritical state, whereas without inhibition it remains slightly supercritical (although in all cases homeostasis of activity is reached). The exponent of the power law in Phase III is close to -1.5.

Finding a power law for avalanche distributions is not sufficient to show decisively that the system is in a critical state [47]. Therefore, Tetzlaff et al. [61] performed several additional tests to confirm criticality. They validated that the avalanche distribution remained critical when in the analysis fewer neurons or shorter or longer time bins were used, and that the inter-avalanche distribution and the Fano Factor [21, 41] also provided evidence for criticality.

Developing cultures of dissociated cortical cells show similar transformations in avalanche distribution to those observed in the model [61]. Like the model, dissociated cultures start with an initial stage characterized by Poisson-like avalanche distributions, followed by a supercritical regime as connectivity and neuronal activity sharply increase. As connectivity and activity subsequently decline, the cultures go through a subcritical state before stabilizing in a critical state, a developmental course that can be mimicked in the model by gradually reducing the inhibitory strength in Phase III from $|\sigma^{\mathrm{inh}}| \gg |\sigma^{\mathrm{exc}}|$ to $|\sigma^{\mathrm{inh}}| = |\sigma^{\mathrm{exc}}|$.

6 Discussion

Different variants [2, 38, 61] of the original neuritic field model [70, 72] have shown, as reviewed in this chapter, that homeostatic structural plasticity is a potent developmental mechanism for bringing networks to criticality. In the assembly of a critical network, the developing neurons are guided only by the activity generated by the network itself, and there is no need for any external instructive signal. All model variants employ a spiking neuron model rather than a firing rate neuron model (as used in the original model) so that bursts of activity can arise and avalanches be defined.

Neurons in the neocortex have a broad spectrum of firing rates [54], whereas in the models discussed here all cells have the same average firing rate at equilibrium. However, the relevant firing rate is the time-averaged firing rate on the time scale of structural growth, so cells can have different firing rates on shorter time scales. Moreover, different types of cells may have different homeostatic set-points, with neurons characterized by a high firing rate having their homeostatic set-point at a higher activity level than neurons that fire less frequently [19, 29]. The impact of such variability in set-points on the emergence of criticality could be a topic for future research.

The use of circular neuritic fields in all models is a simple yet powerful way to abstract away from detailed neuronal morphology. A disadvantage is that it puts some constrains on the type of network topologies that can arise, as the strongest connections are usually formed between neighboring cells. Another way to model neuronal morphology, with fewer inherent constrains, is to assign to each neuron a set of axonal synaptic elements (representing axonal boutons) and a set of dendritic synaptic elements (representing dendritic spines), which can combine to form synapses [13, 17]. In this model, which has also been implemented in the neural simulation package NEST [19], neurons generate new elements when neuronal electrical activity is below a target value, and delete elements, including those bound in synapses, when activity is above the target value or below a certain minimum level. The model can account for changes in visual cortex after focal retinal lesions [13], alterations in global network topology following deafferentation and focal stroke [10], the emergence of efficient small-world networks [11], and the inverse relationship between cell proliferation and synaptic rewiring in the adult hippocampus [12], but has not yet been used to study avalanche dynamics.

Future work may also include the analytical analysis of the role of inhibitory cells in the development and maintenance of critical circuits. The variant of the model that was studied analytically contains only excitatory cells [38]. The numerical studies by Tetzlaff et al. [61] predicted that criticality is best reached with 20% inhibitory cells and a synaptic strength of inhibitory connections that equals that of excitatory connections. However, the models by Abbott and Rohrkemper [2] and Kossio et al. [38] proved that inhibition is not required for criticality, thus meriting further investigation into the potential impact of differences in model formulation, especially the use of separate axonal and dendritic neuritic fields in Tetzlaff et al. [61].

In addition to anatomical changes in connectivity, as brought about by homeostatic structural plasticity, two other categories of neural mechanisms have been proposed to explain the emergence of criticality: intrinsic cellular properties [18, 30] and short- and long-term synaptic plasticity [18, 39, 59]. An example of the first category is found in a biophysically realistic model of retinal waves [30]. In the model, starburst amacrine cells are equipped with a slow after-hyperpolarization current, which regulates neuronal excitability. Spontaneous, cell-intrinsic firing activates this current, thereby reducing excitability and desynchronizing the activity sustained by synaptic transmission. The competition between the desynchronizing effect of spontaneous firing and the synchronizing effect of synaptic transmission enables the network to operate at a transition point between purely local and global functional connectedness. These dynamics are somewhat reminiscent of those seen in a simple model for the occurrence of long-lasting periods of activity [73]. For certain parameter settings, the network is in a critical state in which periods of high activity ("long-lasting transients") alternate irregularly with periods of quiescence. Transients are triggered by spontaneous firing but are eventually also terminated by spontaneous firing, as spontaneous firing, by means of inducing refractoriness, renders cells temporarily non-excitable and so interferes with the flow of network-generated activity.

As to the second category of mechanisms for the origin of criticality, various models have shown that short- and long-term synaptic plasticity can tune a neural network into a critical state with power-law avalanche distributions. Levina et al. [39] demonstrated, both analytically and numerically, that synaptic depression—the short-term decrease in synaptic strength due to depletion of neurotransmitter vesicles—can drive the dynamics of a network towards a critical regime (but see [9]). Stepp et al. [59] showed that a combination of short- and long-term synaptic plasticity can produce hallmarks of criticality, with the interplay between Hebbian long-term excitatory and inhibitory plasticity providing a mechanism for self-tuning. Likewise, Del Papa et al. [18] found that a network endowed with firing threshold adaptation and various types of plasticity, including homeostatic synaptic plasticity [62, 63] and a simple form of structural plasticity, can give rise to criticality signatures in network activity.

The power-law exponents -1.5 and -2 for avalanche size and duration, respectively, imply that each firing neuron activates, on average, one other neuron, so activity will on average neither die out nor explode over time [78]. Thus, an important functional advantage of such a critical state is that neural circuits are prevented from becoming hyper- or hypoactive. Although functional properties have not been studied in the models discussed here [2, 38, 61], maintaining a stable average level of activity is in general crucial for processes ranging from memory storage to activity-dependent development [31, 64]. Besides homeostatic structural plasticity, other forms of slow plasticity, such as homeostatic synaptic plasticity or synaptic scaling [63], are directed at stabilizing network activity (and may generate critical dynamics [38]), in order to counter the destabilizing forces of synaptic long-term potentiation (LTP) and long-term depression (LTD) during memory encoding.

Further functional benefits of critical dynamics include the maximization of dynamic range, information transmission and information capacity [56]. A network at criticality is sensitive to external input, exhibiting a wide range of possible response sizes [36]. Activity patterns in critical networks are not biased towards a typical scale or sequence, providing flexibility that may be advantageous during development as connections are established [30]. Avalanches may reflect the transient formation of cell assemblies [50], and the scale-free organization of avalanche size at criticality implies that assemblies of widely different sizes occur in a balanced way [36].

In conclusion, during development, homeostatic structural plasticity can guide the formation of synaptic connections to create a critical network that has optimal functional properties for information processing in adulthood. In this form of plasticity, neurons adapt their axonal and dendritic morphology and, consequently, their connectivity so as to reach and maintain a desired level of neuronal activity. Homeostatic structural plasticity does not require information about pre- and postsynaptic activity, as does Hebbian synaptic plasticity (synapse-centric plasticity), but only needs the local activity state of the neuron itself (neuron-centric plasticity). In general, homeostatic structural plasticity may act as a central organizing principle driving both the formation of networks [11, 61, 67, 70, 72] and the compensatory structural changes following loss of input caused by lesions, stroke or neurodegeneration [10, 13].

References

1. Abbott, L.F., Jensen, O.: Self-organizing circuits of model neurons. In: Bower, J. (ed.) Computational Neuroscience, Trends in Research, pp. 227–230. Plenum, New York (1997)
2. Abbott, L.F., Rohrkemper, R.: A simple growth model constructs critical avalanche networks. Prog. Brain Res. **165**, 13–19 (2007)
3. Aizenman, C.D., Manis, P.B., Linden, D.J.: Polarity of long-term synaptic gain change is related to postsynaptic spike firing at a cerebellar inhibitory synapse. Neuron **21**(4), 827–835 (1998)
4. Bak, P., Tang, C., Wiesenfeld, K.: Self-organized criticality: an explanation of the 1/f noise. Phys. Rev. Lett. **59**(4), 381–384 (1987)

5. Beggs, J.M., Plenz, D.: Neuronal avalanches in neocortical circuits. J. Neurosci. **23**(35), 11167–11177 (2003)
6. Beggs, J.M., Plenz, D.: Neuronal avalanches are diverse and precise activity patterns that are stable for many hours in cortical slice cultures. J. Neurosci. **24**(22), 5216–5229 (2004)
7. Ben-Ari, Y., Khalilov, I., Kahle, K.T., Cherubini, E.: The GABA excitatory/inhibitory shift in brain maturation and neurological disorders. Neuroscientist **18**(5), 467–486 (2012)
8. Bliss, T.V., Lomo, T.: Plasticity in a monosynaptic cortical pathway. J. Physiol. **207**(2), 61P (1970)
9. Bonachela, J.A., de Franciscis, S., Torres, J.J., Muñoz, M.A.: Self-organization without conservation: are neuronal avalanches generically critical? J. Stat. Mech. P02015 (2010)
10. Butz, M., Steenbuck, I.D., van Ooyen, A.: Homeostatic structural plasticity can account for topology changes following deafferentation and focal stroke. Front Neuroanat **8**, 115 (2014)
11. Butz, M., Steenbuck, I.D., van Ooyen, A.: Homeostatic structural plasticity increases the efficiency of small-world networks. Front Synaptic Neurosci. **6**, 7 (2014)
12. Butz, M., Teuchert-Noodt, G., Grafen, K., van Ooyen, A.: Inverse relationship between adult hippocampal cell proliferation and synaptic rewiring in the dentate gyrus. Hippocampus **18**(9), 879–898 (2008)
13. Butz, M., van Ooyen, A.: A simple rule for dendritic spine and axonal bouton formation can account for cortical reorganization after focal retinal lesions. PLoS Comput. Biol. **9**(10), e1003259 (2013)
14. Butz, M., Wörgötter, F., van Ooyen, A.: Activity-dependent structural plasticity. Brain Res. Rev. **60**(2), 287–305 (2009)
15. Butz-Ostendorf, M., Van Ooyen, A.: Is lesion-induced synaptic rewiring induced by activity homeostasis? In: Van Ooyen, A., Butz-Ostendorf, M. (eds.) The Rewiring Brain, pp. 71–92. Academic Press, San Diego (2017)
16. Cohan, C.S., Kater, S.B.: Suppression of neurite elongation and growth cone motility by electrical activity. Science **232**(4758), 1638–1640 (1986)
17. Dammasch, I.E., Wagner, G.P., Wolff, J.R.: Self-stabilization of neuronal networks. I. The compensation algorithm for synaptogenesis. Biol. Cybern. **54**(4–5), 211–222 (1986)
18. Del Papa, B., Priesemann, V., Triesch, J.: Criticality meets learning: criticality signatures in a self-organizing recurrent neural network. PLoS ONE **12**(5), e0178683 (2017)
19. Diaz-Pier, S., Naveau, M., Butz-Ostendorf, M., Morrison, A.: Automatic generation of connectivity for large-scale neuronal network models through structural plasticity. Front Neuroanat **10**, 57 (2016)
20. Eglen, S.J., van Ooyen, A., Willshaw, D.J.: Lateral cell movement driven by dendritic interactions is sufficient to form retinal mosaics. Network **11**(1), 103–118 (2000)
21. Fano, U.: Ionization yield of radiations. II. The fluctuations of the number of ions. Phys. Rev. **72**, 26–29 (1947)
22. Fauth, M., Tetzlaff, C.: Opposing effects of neuronal activity on structural plasticity. Front Neuroanat **10**, 75 (2016)
23. Fields, R.D., Neale, E.A., Nelson, P.G.: Effects of patterned electrical activity on neurite outgrowth from mouse sensory neurons. J. Neurosci. **10**(9), 2950–2964 (1990)
24. Ghiretti, A.E., Moore, A.R., Brenner, R.G., Chen, L.F., West, A.E., Lau, N.C., Van Hooser, S.D., Paradis, S.: Rem2 is an activity-dependent negative regulator of dendritic complexity in vivo. J. Neurosci. **34**(2), 392–407 (2014)
25. Gireesh, E.D., Plenz, D.: Neuronal avalanches organize as nested theta- and beta/gamma-oscillations during development of cortical layer 2/3. Proc. Natl. Acad. Sci. USA **105**(21), 7576–7581 (2008)
26. Grossberg, S.: Nonlinear neural networks: principles, mechanisms and architectures. Neural Netw. **1**, 17–61 (1988)
27. Habets, A.M., van Dongen, A.M., van Huizen, F., Corner, M.A.: Spontaneous neuronal firing patterns in fetal rat cortical networks during development in vitro: a quantitative analysis. Exp. Brain Res. **69**(1), 43–52 (1987)
28. Hebb, D.O.: The Organization of Behavior. Wiley & Sons, New York (1949)

29. Hengen, K.B., Torrado Pacheco, A., McGregor, J.N., Van Hooser, S.D., Turrigiano, G.G.: Neuronal firing rate homeostasis is inhibited by sleep and promoted by wake. Cell **165**(1), 180–191 (2016)
30. Hennig, M.H., Adams, C., Willshaw, D., Sernagor, E.: Early-stage waves in the retinal network emerge close to a critical state transition between local and global functional connectivity. J. Neurosci. **29**(4), 1077–1086 (2009)
31. Houweling, A.R., van Ooyen, A.: Homeostasis at multiple spatial and temporal scales. In: Squire, L. (ed.) New Encyclopedia of Neuroscience. Elsevier Press, Amsterdam (2008)
32. Hui, K., Fei, G.H., Saab, B.J., Su, J., Roder, J.C., Feng, Z.P.: Neuronal calcium sensor-1 modulation of optimal calcium level for neurite outgrowth. Development **134**(24), 4479–4489 (2007)
33. Jiang, B., Huang, Z.J., Morales, B., Kirkwood, A.: Maturation of GABAergic transmission and the timing of plasticity in visual cortex. Brain Res. Brain Res. Rev. **50**(1), 126–133 (2005)
34. Kater, S.B., Mattson, M.P., Cohan, C., Connor, J.: Calcium regulation of the neuronal growth cone. Trends Neurosci. **11**(7), 315–321 (1988)
35. Kawasaki, F., Stiber, M.: A simple model of cortical culture growth: burst property dependence on network composition and activity. Biol. Cybern. **108**(4), 423–443 (2014)
36. Kinouchi, O., Copelli, M.: Optimal dynamical range of excitable networks at criticality. Nat. Phys. **2**, 348–352 (2006)
37. Konur, S., Ghosh, A.: Calcium signaling and the control of dendritic development. Neuron **46**(3), 401–405 (2005)
38. Kossio, F.Y.K., Goedeke, S., Van den Akker, B., Ibarz, B., Memmesheimer, R.-M.: Growing critical: self-organized criticality in a developing neural system. Phys. Rev. Lett. **121**, 058301 (2018)
39. Levina, A., Herrmann, J.M., Geisel, T.: Dynamical synapses causing self-organized criticality in neural networks. Nat. Phys. **3**, 857–860 (2007)
40. Lohmann, C., Wong, R.O.: Regulation of dendritic growth and plasticity by local and global calcium dynamics. Cell Calcium **37**(5), 403–409 (2005)
41. Lowen, S.B., Ozaki, T., Kaplan, E., Saleh, B.E., Teich, M.C.: Fractal features of dark, maintained, and driven neural discharges in the cat visual system. Methods **24**(4), 377–394 (2001)
42. Markovic, D., Gros, C.: Power laws and self-organized criticality in theory and nature. Phys. Rep. **536**, 41–74 (2014)
43. Marom, S., Shahaf, G.: Development, learning and memory in large random networks of cortical neurons: lessons beyond anatomy. Q. Rev. Biophys. **35**(1), 63–87 (2002)
44. Mattson, M.P.: Neurotransmitters in the regulation of neuronal cytoarchitecture. Brain Res. **472**(2), 179–212 (1988)
45. Mattson, M.P., Dou, P., Kater, S.B.: Outgrowth-regulating actions of glutamate in isolated hippocampal pyramidal neurons. J. Neurosci. **8**(6), 2087–2100 (1988)
46. Mattson, M.P., Kater, S.B.: Excitatory and inhibitory neurotransmitters in the generation and degeneration of hippocampal neuroarchitecture. Brain Res. **478**(2), 337–348 (1989)
47. Newman, M.: Power laws, pareto distributions and zipf's law. Contemporary Physics **46**, 323–351 (2005)
48. Pasquale, V., Massobrio, P., Bologna, L.L., Chiappalone, M., Martinoia, S.: Self-organization and neuronal avalanches in networks of dissociated cortical neurons. Neuroscience **153**(4), 1354–1369 (2008)
49. Petermann, T., Thiagarajan, T.C., Lebedev, M.A., Nicolelis, M.A., Chialvo, D.R., Plenz, D.: Spontaneous cortical activity in awake monkeys composed of neuronal avalanches. Proc Natl Acad Sci U S A **106**(37), 15921–15926 (2009)
50. Plenz, D., Thiagarajan, T.C.: The organizing principles of neuronal avalanches: cell assemblies in the cortex? Trends Neurosci. **30**(3), 101–110 (2007)
51. Raijmakers, M.E., Molenaar, P.C.: Modeling developmental transitions in adaptive resonance theory. Dev Sci **7**(2), 149–157 (2004)

52. Raijmakers, M.E.J., Molenaar, P.C.M.: Modelling developmental transitions in neural networks: bifurcations in an adaptive resonance theory model. In: Mareschal, D., Sirois, S., Westermann, G., Johnson, M.H. (eds.) Neuroconstructivism: Perspectives and Prospects, pp. 99–128. Oxford University Press, Oxford (2007)
53. Rekart, J.L., Sandoval, C.J., Routtenberg, A.: Learning-induced axonal remodeling: evolutionary divergence and conservation of two components of the mossy fiber system within Rodentia. Neurobiol. Learn. Mem. **87**(2), 225–235 (2007)
54. Roxin, A., Brunel, N., Hansel, D., Mongillo, G., van Vreeswijk, C.: On the distribution of firing rates in networks of cortical neurons. J. Neurosci. **31**(45), 16217–16226 (2011)
55. Schilling, K., Dickinson, M.H., Connor, J.A., Morgan, J.I.: Electrical activity in cerebellar cultures determines Purkinje cell dendritic growth patterns. Neuron **7**(6), 891–902 (1991)
56. Shew, W.L., Plenz, D.: The functional benefits of criticality in the cortex. Neuroscientist **19**(1), 88–100 (2013)
57. Soriano, J., Rodríguez Martínez, M., Tlusty, T., Moses, E.: Development of input connections in neural cultures. Proc Natl Acad Sci U S A **105**(37), 13758–13763 (2008)
58. Soto-Treviño, C., Thoroughman, K.A., Marder, E., Abbott, L.F.: Activity-dependent modification of inhibitory synapses in models of rhythmic neural networks. Nat. Neurosci. **4**(3), 297–303 (2001)
59. Stepp, N., Plenz, D., Srinivasa, N.: Synaptic plasticity enables adaptive self-tuning critical networks. PLoS Comput. Biol. **11**(1), e1004043 (2015)
60. Tailby, C., Wright, L.L., Metha, A.B., Calford, M.B.: Activity-dependent maintenance and growth of dendrites in adult cortex. Proc Natl Acad Sci U S A **102**(12), 4631–4636 (2005)
61. Tetzlaff, C., Okujeni, S., Egert, U., Wörgötter, F., Butz, M.: Self-organized criticality in developing neuronal networks. PLoS Comput. Biol. **6**(12), e1001013 (2010)
62. Turrigiano, G.G.: The self-tuning neuron: synaptic scaling of excitatory synapses. Cell **135**(3), 422–435 (2008)
63. Turrigiano, G.G., Leslie, K.R., Desai, N.S., Rutherford, L.C., Nelson, S.B.: Activity-dependent scaling of quantal amplitude in neocortical neurons. Nature **391**(6670), 892–896 (1998)
64. Turrigiano, G.G., Nelson, S.B.: Homeostatic plasticity in the developing nervous system. Nat. Rev. Neurosci. **5**(2), 97–107 (2004)
65. van Huizen, F., Romijn, H.J., Habets, A.M.: Synaptogenesis in rat cerebral cortex cultures is affected during chronic blockade of spontaneous bioelectric activity by tetrodotoxin. Brain Res. **351**(1), 67–80 (1985)
66. van Huizen, F., Romijn, H.J., Habets, A.M., van den Hooff, P.: Accelerated neural network formation in rat cerebral cortex cultures chronically disinhibited with picrotoxin. Exp. Neurol. **97**(2), 280–288 (1987)
67. van Ooyen, A.: Activity-dependent neural network development. Netw. Comput. Neural Syst. **5**, 401–423 (1994)
68. Van Ooyen, A.: Network formation through activity-dependent neurite outgrowth: a review of a simple model of homeostatic structural plasticity. In: Van Ooyen, A., Butz-Ostendorf, M. (eds.) The Rewiring Brain, pp. 95–121. Academic Press, San Diego (2017)
69. Van Ooyen, A., Butz-Ostendorf, M.: The Rewiring Brain. Academic Press, San Diego (2017)
70. van Ooyen, A., van Pelt, J.: Activity-dependent outgrowth of neurons and overshoot phenomena in developing neural networks. J. Theor. Biol. **167**, 27–43 (1994)
71. van Ooyen, A., van Pelt, J.: Complex periodic behaviour in a neural network model with activity-dependent neurite outgrowth. J. Theor. Biol. **179**(3), 229–242 (1996)
72. van Ooyen, A., van Pelt, J., Corner, M.A.: Implications of activity dependent neurite outgrowth for neuronal morphology and network development. J. Theor. Biol. **172**(1), 63–82 (1995)
73. van Ooyen, A., van Pelt, J., Corner, M.A., da Silva, F.H., van Ooyen, A.: The emergence of long-lasting transients of activity in simple neural networks. Biol. Cybern. **67**(3), 269–277 (1992)
74. Van Ooyen, A., Van Pelt, J., Corner, M.A., Kater, S.B.: Activity-dependent neurite outgrowth: implications for network development and neuronal morphology. In: Van Ooyen, A. (ed.) Modeling Neural Development, pp. 111–132. The MIT Press, Cambridge, MA (2003)

75. van Oss, C., van Ooyen, A.: Effects of inhibition on neural network development through activity-dependent neurite outgrowth. J. Theor. Biol. **185**(2), 263–280 (1997)
76. Wagenaar, D.A., Pine, J., Potter, S.M.: An extremely rich repertoire of bursting patterns during the development of cortical cultures. BMC Neurosci **7**, 11 (2006)
77. Yamahachi, H., Marik, S.A., McManus, J.N., Denk, W., Gilbert, C.D.: Rapid axonal sprouting and pruning accompany functional reorganization in primary visual cortex. Neuron **64**(5), 719–729 (2009)
78. Zapperi, S., Bækgaard Lauritsen, K., Stanley, H.E.: Self-organized branching processes: mean-field theory for avalanches. Phys. Rev. Lett. **75**(22), 4071–4074 (1995)
79. Zhong, Y., Wu, C.F.: Neuronal activity and adenylyl cyclase in environment-dependent plasticity of axonal outgrowth in Drosophila. J. Neurosci. **24**(6), 1439–1445 (2004)

Arjen van Ooyen studied biology at Utrecht University and received his Ph.D. in computational neuroscience from the University of Amsterdam in 1995. He has worked as postdoctoral researcher at the University of Edinburgh and the Netherlands Institute for Brain Research before moving on to become associate professor at VU University Amsterdam. His research interests include neuronal morphogenesis and formation of synaptic connectivity, with a focus on homeostatic structural plasticity in the development and reorganization of neuronal networks.

Markus Butz-Ostendorf studied informatics and biology and holds a Ph.D. in neuroanatomy. He has worked as postdoctoral researcher at the Bernstein Center for Computational Neuroscience Göttingen, the VU University Amsterdam and the Forschungszentrum Jülich. He now works as Lead Manager Innovations at Biomax Informatics AG, Planegg, Germany. His research focus is on modeling structural plasticity in the healthy and diseased brain.

Linear Stability of Spontaneously Active Local Cortical Circuits: Is There Criticality on Long Time Scales?

Nathan X. Kodama and Roberto F. Galán

Abstract Self-organizing systems acquire their structures and functions without patterned input from the outside world. In the interconnected architectures of the neocortex, spontaneous activity—that is, activity that arises without external sensory or electrical stimulus—predominates over sensory-evoked activity. Thus, spontaneous neuronal activity provides a means to characterizing the structure, function and dynamics of neocortical networks. We have recorded spontaneous, asynchronous network activity from hundreds of neurons constituting local cortical circuits in mice with high-density microelectrode arrays (MEAs) in vitro. The spontaneous activity in the network displayed features of a system at criticality and scale-free structures, such as fluctuation scaling and multiple frequency bands. To investigate dynamical parameters, we have investigated the linear and nonlinear components of the network dynamics. The former allows us not only to define a linear measure of functional connectivity, but also to determine the linear stability of the system through its eigenvalues. Similarly, the latter allows us to define a measure of nonlinear functional connectivity. An important feature revealed by this approach is the large number of eigenvalues with positive real parts and the high density of eigenvalues near the imaginary axis, which demonstrate respectively that this high-dimensional system is linearly unstable and critical on long time scales (>1s). The function of critical dynamics in these networks is discussed with respect to exploratory behavior in rodents.

N. X. Kodama · R. F. Galán (✉)
Department of Electrical Engineering and Computer Science, Case Western Reserve University, Cleveland, OH 44106, USA
e-mail: rfgalan@case.edu

N. X. Kodama
e-mail: nxkodama@case.edu

N. Tomen et al. (eds.), *The Functional Role of Critical Dynamics in Neural Systems*, Springer Series on Bio- and Neurosystems 11,
https://doi.org/10.1007/978-3-030-20965-0_8

1 Introduction

Systems exhibiting self-organized criticality (SOC) reach their critical points without the need to tune order parameters. They display scale-invariant structures characterized by power-laws for many observable quantities [5]. The general class of critical phenomena are well-defined by their instabilities and bifurcations [24, 47, 48]. These instabilities form hierarchies that generate macroscopic patterns which may in turn self-impose constraints on the constituent elements [25]. Stability, both global and local, offer qualitative information about how close a system is to criticality. A SOC system becomes stable when a network of minimally stable states reduces to the point where signals cannot be transmitted through infinite distances [5]. These minimally stable states in SOC systems are a subset of marginally stable states in critical systems. Marginally stable states may be observed directly in linear models, such as generalized linear models [39], by examining the distribution of the real component of the complex eigenvalues derived from the linear coefficient matrix.

In the study of the cerebral cortex, neural network models are mathematically equivalent to SOC models of earthquakes [4, 11, 26]. This has been observed experimentally in neuronal avalanches [7, 18], for which branching ratios of unity are well-defined and represent a directed percolation process. Although power laws may have non-unique origins in the cortex [6, 58], extensive evidence of scale-free phenomena in the graphical topology of neuronal networks [17, 53, 55] points to a critical system [13, 14, 37]. The computational advantages of systems operating at critical states are clear [31], however, the functional role of criticality to information processing, in the cortical substrate and under physiologically relevant conditions, is still unclear.

Activity of the cortical substrate may arise spontaneously without patterned stimuli or external sensory input. The predominance of spontaneous activity over sensory-evoked activity in the neocortex [34] is a consequence of the disproportionately high anatomical connections between neocortical neurons compared to feedforward sensory inputs [1, 16]. Nevertheless, functional relationships only emerge after the underlying networks have adapted to sensory inputs from the body and the environment [12]. The spontaneous activity that occurs after this body- and environment-driven adaptation may be used to infer the functional roles of self-organized criticality. In the local circuits of the cerebral cortex, large networks of neurons are composed of both excitatory synapses [7] and inhibitory synapses [40]. It is important then to examine these networks of excitatory and inhibitory synapses in their intrinsic, unperturbed mode of activity—neuronal spiking—and determine if they self-organize into spatial and temporal scale-free structures.

In the last decade, modern in vitro electrophysiological and imaging technologies have characterized the spontaneous activity of neuronal networks, especially synchronous circuit events that were drug-induced or obtained through manipulations of the solutions bathing these networks [7, 34]. The recent realization of systems displaying asynchronous spiking and other physiologically relevant activity [22, 23, 40], presents a unique opportunity to investigate neuronal network phenomenology pertinent to other domains of network activity. The self-organization of spiking neuronal networks may offer novel insight into behaviorally relevant and functional brain states; asynchronous spiking in particular provides a complementary perspective to that of synchronous discharges or oscillations. A linear stability analysis reveals unintuitive answers to seemingly paradoxical questions such as: is asynchronous firing organized in time? Below we find the effort to answer this question reveals how critical neuronal networks organize without sensory input and minimal spike synchrony. Moreover, interpreting these principles of organization on long time scales (>1 s) suggests several functional advantages that have been conferred onto the animal during exploratory behavior.

Information processing is spatiotemporal and involves the regime of spontaneous neuronal spiking. Not only is it temporal, it is multitemporal—that is, it spans multiple time scales [10]. Do local circuits of the neocortex exhibit features of criticality in the temporal structure of spontaneous neuronal spiking on long time scales? If so, is it possible to measure aspects of criticality, such as linear stability, and also deduce this temporal structure? We propose a method to (1) characterize the scaling of spatiotemporal structure of neuronal spiking, (2) retrieve the linear and nonlinear functional connectivity of spontaneously active networks, (3) infer the overall balance between functional excitation and inhibition, (4) assess the linear stability of the system's dynamics, and (5) interpret the results in a functional and behavioral context. Here, in the study of spontaneous neuronal spiking, we present two features of temporal structure, both of which are characteristic of a neural system operating in a regime near a critical state: a power law in the inter-spike interval statistics, referred to as fluctuation scaling [29], and frequency banding that is present at macroscopic spatial scales, such as the electroencephalogram. We do so at the mesoscopic level of neural networks constituting intact cortical circuits and spanning its anatomical features (columns and layers). We also describe a method for processing binary spike trains, fitting the resulting trajectories to a linear model, and assessing the eigenvalues of the linear coefficients on the complex plane, which simultaneously yields a measure of linear stability and explains for the temporal structure in the neuronal spiking. Finally, we characterize the nonlinear residuals of the model, which display scale-free structure in synchronous network bursts that are not present in the firing rate. We also show that the residuals are centered at zero, non-Gaussian, and non-white. Given our recent work [28], these finding are interpreted in a functional and behavioral context.

2 Methodological Approach

2.1 Experimental Methods

The emergence of spontaneous activity in acute cortical slices, without electrical stimulation or pharmacological manipulation, has been shown to depend on the constituency of the artificial cerebrospinal fluid (aCSF), particularly the solute composition and oxygenation [23]. Neuronal spiking and other cortical activity that was only previously observed in vivo has been observed in vitro only at high oxygen concentrations; this is because of the limited diffusion of oxygen during perfusion in vitro, which is in contrast to the direct delivery of oxygen through capillaries in vivo [22]. Recently, the investigation of spontaneous activity under these conditions has revealed a rich diversity in neuronal firing that is associated with specific cell types [40]. In parallel, synchronous network discharges have been shown to correspond with network-wide 'up' and 'down' states [27, 33, 34, 53, 59].

Slice Preparation. Here we used the artificial cerebrospinal fluid (aCSF) solutions reported by the MacLean lab [54] and high flow/oxygen rates recommended by Hájos et al. which allow us to measure sustained, non-epileptogenic, spiking neuronal activity. Three aCSF solutions were prepared for brain extraction and slicing at 4 °C, slice incubation at 30 °C, and perfusion at 31 °C. All solutions were cooled or warmed to their designated temperature before they were saturated with 95% O_2–5% CO_2. Coronal slices (350 μm thickness) of primary somatosensory cortex (S1) were collected from juvenile (P13–P17) C57BL/6 mice. Animals were anesthetized with vapor isoflurane and decapitated; brains were submerged in ice-cold aCSF. Slices were directly collected and transferred to the oxygenated incubation solution to equilibrate for 30 min. Finally, a single slice was selected and placed into the recording chamber.

Microelectrode Array Recordings. High-density microelectrode arrays (MEAs) with 120 electrodes (100 μm pitch) were used to record from S1. The entire thickness of the somatosensory cortex (~1 mm) was fixed to the 1.2 × 1.2 mm perforated recording field with 15-mbar suction. The slice was perfused with aCSF at 6.5 ml/min and 53.3 kPa O_2 for 30 min before the start of the recording and for the duration of the recording. The activity from acute brain slices was recorded at a sampling frequency of 50 kHz with a resolution of 16 bits in the range of −2 to 2 mV and two sequential hardware filters (2nd order 0.5 Hz high-pass filter; 1st order 10 kHz low-pass filter) were used to eliminate voltage offsets and drifts.

Signal Processing. All processing and other data analyses were carried out in MATLAB. A digital band-pass filter (80–2,000 Hz, 3rd order) was used to pre-process all recordings. Spikes were detected as events based on a time-varying threshold defined for 1 s windows as ±4 × standard deviation. Spike sorting was conducted with a wavelet-based spike sorting algorithm [50]. Spike trains that appeared in more than one electrode with <1 ms delays were considered to come from the same neuron: if the number of sub-millisecond spikes was greater than 30% of the number of spikes in either putative neuron, the neuron with smaller peak amplitude waveforms was omitted from further analyses. Nearly two hundred neurons were resolved in each slice and animal (187 ± 4 neurons, mean ± s.e.m. across 6 mice). Further details on our experimental preparation can be found in Kodama et al. [28].

2.2 Data Analysis

Pre-processing. The spike times of each neuron were represented as binary spike trains with 1 ms resolution. The binary spike-train signals were convolved with a triangular window of unitary area and duration of 1 s to compute the instantaneous firing rate (spikes per second) for each neuron in the recording. We chose a duration of 1 s because we have previously shown that this is the time scale at which the number of significant pairwise correlations between neurons is largest [28]. The derivative of the firing rate was obtained by center differencing the firing rate at the time resolution of 1 ms. From the instantaneous firing rates and their time derivatives, we obtained the linear coefficients and nonlinear residuals introduced in the next paragraph.

Dynamical Model. The state vector $\vec{u}$ of the spiking neuronal network contained the firing rate for each neuron. Thus, the most general expression for the spontaneous network dynamics is

$$\frac{d\vec{u}}{dt} = F(\vec{u}). \tag{1}$$

We are interested in the fluctuations in firing rate around its mean across all neurons $\vec{x} = \vec{u} - \vec{u}_0$ where $\vec{u}_0$ is the mean firing rate. Upon substitution in (1) we obtain

$$\frac{d\vec{x}}{dt} = F(\vec{x} + \vec{u}_0) = L\vec{x} + \vec{G}(\vec{u}_0; \vec{x}),$$

where we have split the right hand side into a linear component (first term, with L being a square matrix) and a nonlinear component which is an implicit function of time via its dependence on $\vec{x}$ (second term). The key idea of our approach to quantifying critical parameters is that we can estimate both L and $\vec{G}$ from empirical data by means of a simple linear regression of the form

$$\vec{y} = L\vec{x} + \vec{\eta}(t), \tag{2}$$

where the dependent variable $\vec{y} = d\vec{x}/dt$ and the independent variable $\vec{x}$ are both known from the experimental recordings. Thus, the regression returns matrix L and the residuals of the regression corresponding to the nonlinear component of the dynamics, $\vec{\eta}(t) = \vec{G}(\vec{u}_0; \vec{x}(t))$. If the network dynamics are linear and stable, then all the eigenvalues of L have negative real parts and the residuals must be white and Gaussian. If the network dynamics are linearly unstable but the firing rates are bounded and therefore, finite, then at least one eigenvalue of L has a positive real part. In general, the more eigenvalues have a positive real part, the more complex the behavior will be. In these cases, the residuals, that is, the nonlinear component of the dynamics cannot be trivial; they cannot be white and must have a non-zero correlation time. In other words, the auto-correlogram of the residuals for each neuron cannot be a Dirac's delta function. From the perspective of time series analysis, the residuals can be thought of as the de-trended time derivative of the firing rate.

Matrix L is not only useful to investigate critical parameters. It can also be considered a general measure of functional connectivity [19, 56]. If the nonlinear component of the dynamics is non-trivial, as it is the case for critical systems, one may define a nonlinear measure of functional connectivity, N, as the correlation matrix of the residuals

$$N_{ij} = \frac{\langle \eta_i(t)\eta_j(t) \rangle}{\sqrt{\langle \eta_i^2(t) \rangle \langle \eta_j^2(t) \rangle}},$$

where the brackets $\langle \ldots \rangle$ denote time averages. By definition, the nonlinear functional connectivity is symmetric, contrary to L, which will be non-symmetric in general.

Surrogate data. To assess the explanatory power of our analytical approach, we compared the results of our analysis of the experimentally observed data with surrogate data obtained by circularly shifting the spike trains: the whole spike train is advanced by the shift value and the portion exceeding the duration of the recording is then wrapped around at the beginning of the recording. For each neuron, the shift in milliseconds was randomly chosen between 1 and the total time in milliseconds.

3 Results

We simultaneously recorded spontaneous, asynchronous network activity from murine primary somatosensory cortex (S1) with high-density microelectrode arrays (MEAs, Fig. 1a–c). Notably, simultaneous neuronal spiking was widely observed across the entire network without pharmacological manipulation or patterned electrical stimulation (Fig. 1d). To assess the spatiotemporal structure of the observed activity, spike timing statistics were evaluated and signal processing methods for binary spike trains were employed (Fig. 1e–g). Temporal structure can be observed at the level of the raster plots, the diversity of spike timing structure in the network obeys a fluctuation scaling law, and the broadband spectral content of neuronal spiking is characteristic of spatiotemporal scaling in cortical networks.

Coronal slices were prepared such that local cortical circuits were intact, but external inputs were severed. The entire thickness of the acute slice was attached to the MEA field with negative pressure (Fig. 1a), with the surface of the slice tangentially aligned to the first row of electrodes. Columns of the barrel field in the somatosensory cortex and layers of the cortical thickness could be observed and selected for, prior to slice placement. The spikes were sorted according to their waveforms. In a single experiment, roughly 200 neurons could be discerned from multi-unit activity (Fig. 1b, 1–7 per electrode). The spikes assigned to multiple neurons were visually validated (Fig. 1c) in a supervised clustering step. Redundant spikes observed in neighboring electrodes were accounted for by identifying sub-millisecond spike

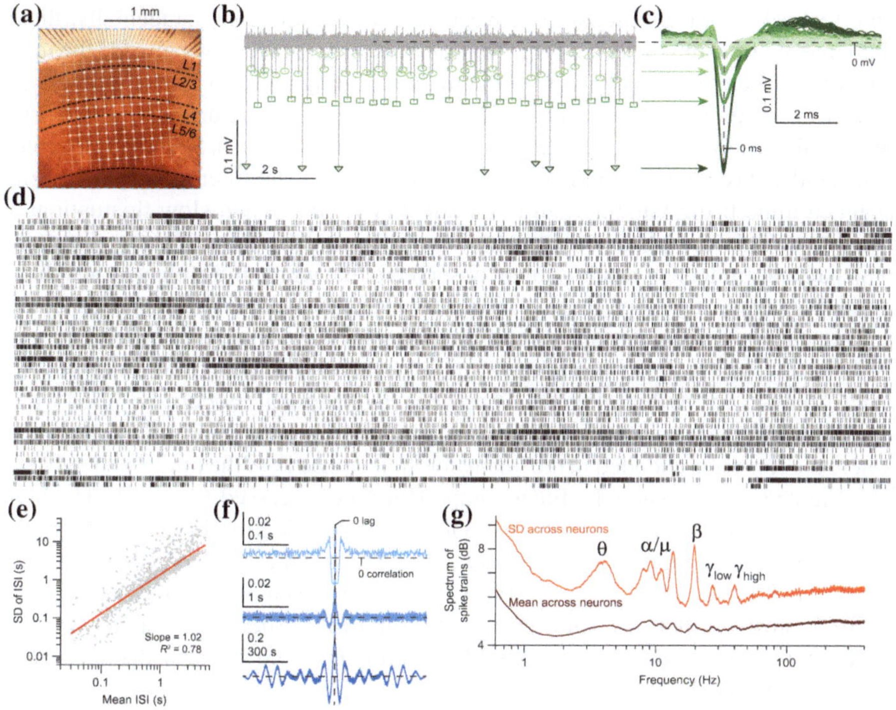

Fig. 1 Experimental preparation, recordings, and scale-free features in time. **a** Microelectrode array on top of primary somatosensory neocortex. There are 120 electrodes (100 μm pitch) sampling the entire cortical thickness across layers. **b** Multiunit recording in a single electrode channel. The electrode captures action potentials from different neurons. **c** These neurons have different waveforms that can be discriminated with standard spike-sorting techniques. **d** Raster plot of action potentials for 46 neurons sorted by vertical location on the array. **e** Fluctuation scaling of neuronal firing. The relationship between the mean and standard deviation of the inter-spike intervals follows a power law of exponent ~ 1, that is, a linear relationship. **f** Auto-correlograms for three representative neurons demonstrate long-term correlations and/or sustained oscillations. **g** Frequency bands of neuronal firing: they are virtually the same as those measured from the whole brain with electroencephalography in rodents and humans

alignments and only the largest one in voltage amplitude was kept. The raster plot of all uniquely identified neurons, recorded simultaneously, was sorted by the neuron's vertical position in the neocortex (Fig. 1d). Temporal structure and large-scale diversity are readily apparent in the raster plot.

The spike times, obtained from spike sorting, were used to compute inter-spike intervals (ISIs) and construct binary spike train signals with 1 ms precision. Both representations of spike timing were used to infer temporal structure in the spontaneous firing of neurons. First, a power law relationship between the mean ISI and its standard deviation was observed (Fig. 1e) with an exponent of 1.02, indicating an effectively linear relationship between the two parameters—incidentally, this is the same relationship expected for a Poisson process. A power law between the mean and standard deviation is known as fluctuation scaling and is a known property of stochastic integrate-and-fire neurons [29]. This scale-free structure of neuronal firing was confirmed by auto-correlograms of the spike trains (Fig. 1f). Several time scales, from tenths of seconds to hundreds of seconds, pointed to multitemporal patterns that spanned tonic firing bursting and low frequency rhythms, and slow fluctuations in firing rate. This complexity is also found in the power spectra of the spike trains, which reveal frequency bands that are observed in the whole brain electroencephalogram (Fig. 1g). Altogether, these results demonstrate that the firing of spontaneously active neurons in local cortical circuits is non-random, diverse, and multiscale.

After the binary spike trains were smoothed to create firing rate estimates (see Data Analysis in Methodological Approach), a dynamical model was fitted to the data. This approach separates the linear and nonlinear components of the dynamics, where the former is encoded in the linear coefficient matrix of the model and the latter is contained in the residuals. Each component was processed separately to obtain measures of functional connectivity, stability, and criticality. Figure 2 shows the firing rate of three representative neurons (Fig. 2a) and the corresponding nonlinear residuals (Fig. 2b) obtained from Eq. (2). The three neurons display, to different degrees, slow oscillations, transients, and tonic firing. There is apparent structure in the nonlinear residuals: most notably, the nonlinear component displays many burst-like events about the mean.

The distribution of the sizes of synchronous network events, obtained from summing firing rate estimates across the network at various bin sizes, points to the absence of neuronal avalanches (Fig. 2c), which is consistent with both the sparse and asynchronous nature of neuronal firing seen in Fig. 1d. Thus, the question of whether the system is critical must be evaluated in the absence of spatial scaling relationships that are seen in feed-forward models [7]. The linear coefficient matrix of our model serves as a starting point for evaluating the stability of the system directly. But first, its interpretation as a measure of functional connectivity is assessed alongside with the correlation coefficient matrix of the nonlinear residuals.

Linear and nonlinear coefficient matrices serve as measures of functional connectivity in the network (Fig. 3). The contributions to the overall connectivity of the network are distinct. The distribution of the coupling coefficients reveals that both matrices are symmetric about zero suggesting that the functional excitation and inhibition are balanced in the network. Interestingly, vertical and horizontal banding structure of the linear coefficient matrix (Fig. 3a) suggests that a substantial portion of the network is not linearly coupled with the rest of the neurons in the network. However, these neurons are coupled in the nonlinear domain (Fig. 3d), though there are neurons in the nonlinear connectivity matrix that are not coupled.

The density about zero in the distribution of linear coefficients, in contrast to the diffuse distribution of nonlinear coefficients, suggests that the network constituents are weakly coupled. However, the nonlinear coupling of the network is, relative to the linear coupling, stronger on the whole.

To determine how critical the spontaneously active system was, the eigenvalues of matrix L were computed. The real components reveal whether the network dynamics are stable ($\lambda < 0$), marginally stable ($\lambda = 0$), or unstable ($\lambda > 0$). Figure 4, displays the eigenvalues for six experiments (Fig. 4a) compared to surrogate data generated as circularly-shifted spike trains (Fig. 4b; see Surrogate Data in Methodological Approach). The latter preserves the intervals between spikes but disrupts the temporal relationships between neurons. Across experiments, the distribution of complex eigenvalues points to marginal stability in the linear regime and thus criticality of the network dynamics. Compared to the eigenvalues derived from the model fitted to the surrogate data, the eigenvalues observed in the experiments were distributed more diffusely (confirmed by a two-sample Kolmogorov–Smirnoff test on the real parts of the eigenvalues, $p = 2.44 \times 10^{-6}$). The observed eigenvalues are also concentrated near the origin which is consistent with the zero and redundant banding structure of L, implying oversampling of the network's dynamics.

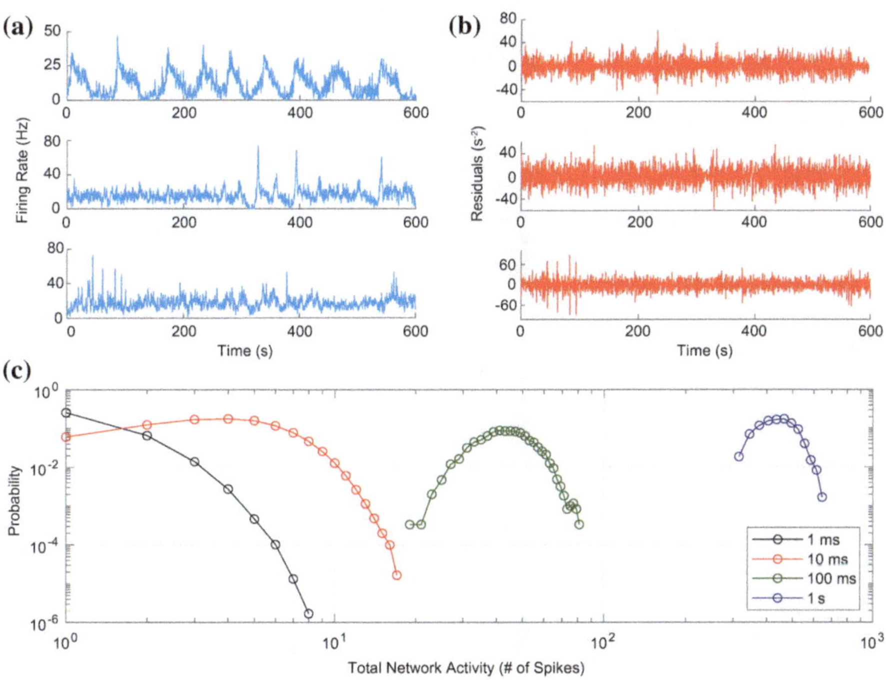

Fig. 2 Neuronal firing rates for several neurons within the same network and their corresponding residuals. **a** Firing rates of three different neurons obtained with a $\Delta t = 1$ s triangular kernel convolution. Several features of neuronal firing dynamics are apparent, including slow oscillations, transient discharges and tonic firing. **b** Residuals obtained through linear regression in Eq. (2). The residuals are centered at zero and contain temporal structure in the form of bursts. **c** The size distribution of synchronized network events, measuring the total number of spikes in the network at various bin sizes, does not follow a power law

Furthermore, the number of positive eigenvalues across experiments points to linear instabilities in the system. All six experiments have 59% or more positive real eigenvalues. This is qualitatively consistent with the asynchronous and complex firing of neurons in these networks. In the distribution of eigenvalues from a representative experiment (Fig. 4c), it is clear that there are more positive real eigenvalues in the observed data and more negative real eigenvalues in the circularly-shifted controls. The distribution of the real components of the eigenvalues across experiments (Fig. 4d) demonstrates that the system is more linearly unstable than the corresponding circular-shifted control. Notably, the distribution of real eigenvalues in the observed system also contains negative real eigenvalues that point to overall stability in the system. Compared to the circular-shifted control, however, there are less negative real eigenvalues, which means the observed system is truly less stable.

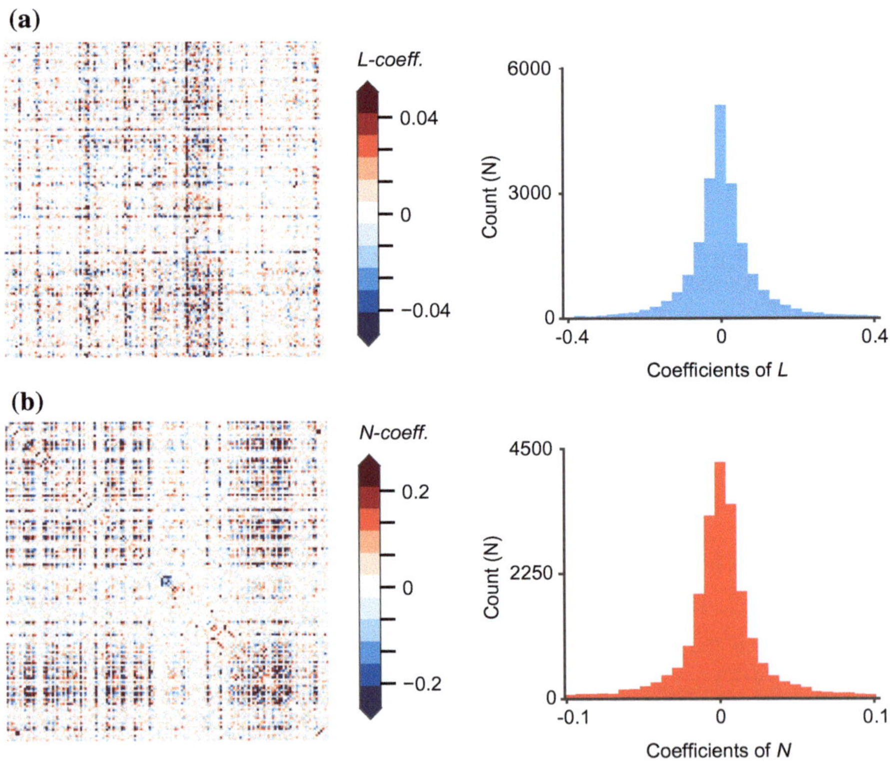

Fig. 3 Linear coupling and nonlinear correlation matrices with their value distributions. **a** Linear connectivity matrix L ($n = 149$ neurons) from a single mouse (left) and the distribution of its entries (right). Note the vertical and horizontal banding structure, which implies that a subset of neurons is not captured in a linear model of the network dynamics. **b** Nonlinear connectivity matrix N as the correlation matrix of the residuals from the linear regression (left) and the distribution of its entries (right). The range of color and x-axis limits are from -3σ to $+3\sigma$, with respect to the corresponding matrices. The diagonal of the matrix N was removed

It is worthwhile to note that the eigenvalues of the system are concentrated on the imaginary axis. The substantial peak near zero imply the existence of neutrally stable modes on a center manifold in the complex plane. This center manifold provides a mechanism for slow fluctuations and suggest that the spontaneously active system is marginally stable and therefore critical in its dynamics over long time scales. Since the system is not displaying avalanches in its firing rate, but is rather in a tonic, multi-rhythmic, and fluctuating state of neuronal spiking, these results point to an alternative approach in assessing the criticality of the system. The eigenvalue distribution of the system occupies a diffuse domain of stability that includes many positive eigenvalues. This suggest that the system is linearly unstable with critical dynamics.

A substantial portion of network dynamics remains in the nonlinear residuals. This is not surprising since only 10% of the variance is captured in L. Across all experiments, 84% or more of the networks had residuals that were not normally distributed

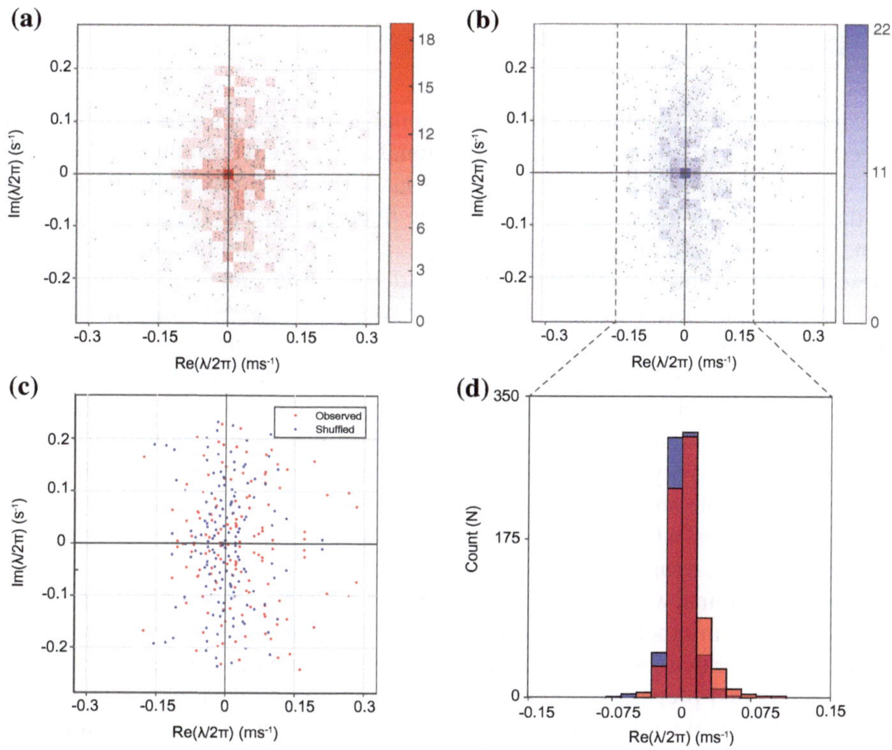

Fig. 4 Eigenvalues of linear coefficient matrices L for several experiments ($n = 6$ mice) on the complex plane plotted with shuffled data. **a** Distribution of complex eigenvalues for 6 mice. **b** Eigenvalues for shuffled data. **c** The eigenvalues of one neuronal network are displayed (red) and after shuffling bins (blue). **d** Distribution of eigenvalues for **c**. Note the number of eigenvalues distributed on the imaginary axis which implies that these networks are linearly unstable and critical

(Kolmogorov–Smirnov (KS) test with a normal distribution at a 5% significance level). These distributions were symmetrically distributed about zero (Fig. 5a) and exhibited temporal structure in their auto-correlograms (Fig. 5b), consistent with the nonlinear component of the dynamics being non-white. Furthermore, the eigenvalues of the correlation matrices of the residuals (Fig. 2b) are distributed non-randomly, which is most clear when these eigenvalues are ranked and overlaid with the eigenvalues corresponding to the circularly-shifted binary spike trains (Fig. 5c). The distribution of eigenvalues is much more spread in its tails and less concentrated near its center (Fig. 5d). Importantly, for the surrogate data, the eigenvalues of the correlation matrix decay linearly with their rank, as expected for random symmetric matrices [36, 38, 51]. In contrast, for the experimental data, there are multiple eigenvalues that significantly deviate from that trend, demonstrating that the nonlinear coupling is not random. Altogether, these features of the residuals suggest there is still dynamical content in the residuals, not captured by the linear model.

4 Discussion

We have recorded spontaneous neuronal spiking from local cortical circuits and investigated the linear and nonlinear functional connectivity, stability, and criticality, of the system's dynamics. The spontaneous activity exhibits fluctuation scaling in the ISI of single neurons, slow firing-rate fluctuations, and the same frequency bands as the whole-brain electroencephalogram in rodents and humans. Altogether, these features point to complexity in spike-derived firing rates of the neurons in the network and spatiotemporal scaling. To investigate linear stability and criticality in the network firing dynamics, we separated the linear and nonlinear coupling components. We found that the linear and nonlinear connectivity matrices were qualitatively different, but were both symmetrically balanced in functional excitation and inhibition. In addition, we found numerous eigenvalues with positive real parts of the linear connectivity matrix, which implies that the spontaneously active system is linearly unstable and critical. We also found that the nonlinear components were non-Gaussian, non-white, non-random, and contained temporal structure, consistent with brain circuit dynamics being strongly nonlinear.

Since the seminal work on neuronal avalanches conducted 15 years ago [7], numerous innovations in experimental tools and numerical methods have been brought forth to understand criticality in the cortex. Although there are qualitative differences in the experimental preparation and mathematical approach taken here, our observations are consistent with the perspective on neuronal avalanches insofar as they both point to cortical networks self-organizing into critical states. However, we do not see a power-law at the level of the firing rate and, as a direct result, take an alternative approach to assessing the criticality of the system.

Recent studies have shown that the absence of observing power laws points to subsampling of a neuronal network and that this subsampling leads to an overestimation of stability in dynamical systems [32, 46, 60]. A limitation of this study is that it

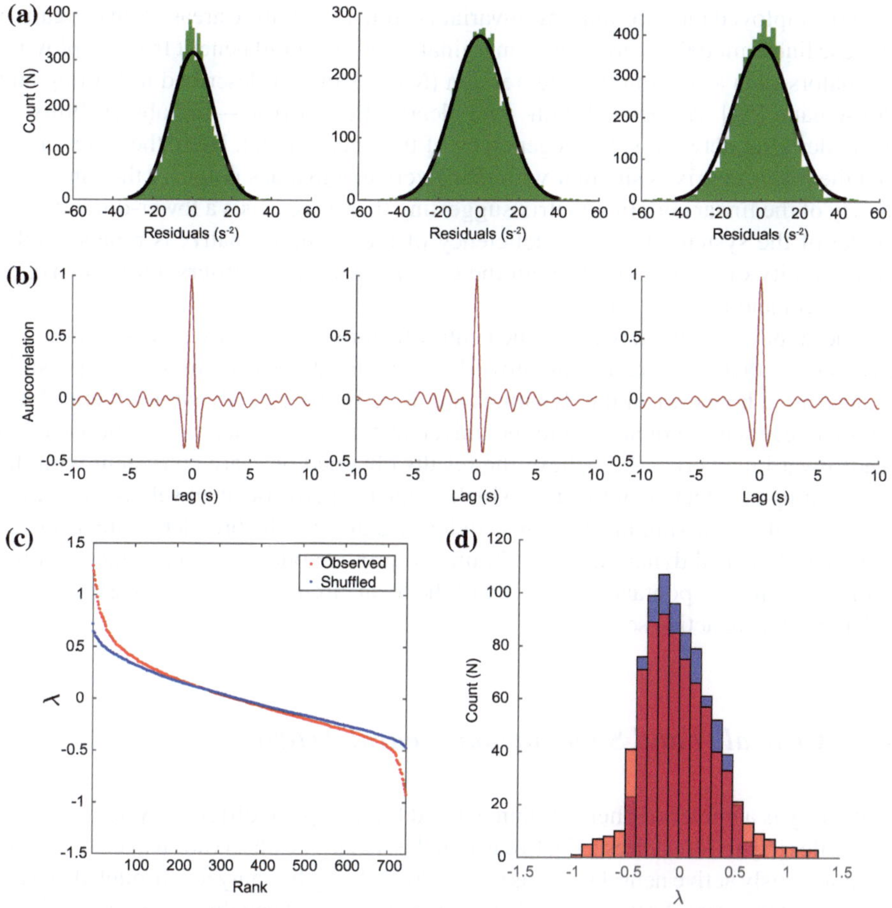

Fig. 5 Distribution and autocorrelation of residuals. **a** Distribution of residuals of the linear stability analysis, with Gaussian fits indicated by the black curve. None are normally distributed ($P \ll 0.01$, Kolmogorov–Smirnov (KS) test with a normal distribution) **b** Auto-correlograms of the nonlinear residuals show that there is still temporal structure (i.e. they are not white). **c** Distribution of the eigenvalues of the correlation matrices of the residuals. The tails of the ranked distribution corresponding to the observed system (red) deviate from those of the circularly-shifted binary spike trains (blue), indicating that the correlation structure is non-random. **d** The distribution of eigenvalues of the correlation matrices of the residuals displays differences in the tails and centers of the distributions

has not employed the subsampling-invariant estimators. Future areas of investigation that use linear models to investigate marginal stability would benefit from employing estimators such as the multiple-regression (MR) estimator described in Wilting and Priesemann [60]. Associated with—yet clearly distinct from—the observability of the underlying network is the degeneracy of the linear model. Here, the observation that the linear matrix is singular with many zero eigenvalues points to the rank deficiency of the linear coupling matrix suggesting that there exists a lower-dimensional model of the system. The rank deficiency of the coupling matrix is related to the observability of the system through the construction of the observability matrix of the system and its row rank.

The stability of the network is the central feature of criticality assessed here. We have shown that in the recurrent networks of the local cortical circuit, it is possible to determine how critical the dynamics are directly from the dynamics themselves, without necessarily connecting to the universal class of directed percolation used in feedforward networks. We believe the results obtained here are consistent with the overarching perspective of the field which is that the networks themselves are critical through self-organizing mechanisms. Below we discuss the broader context for the function of critical dynamics in the brain and provide our own interpretation of the results here in the spontaneous activity of the local circuit as it applies to exploratory behavior during active sensing.

4.1 Criticality and Spontaneous Neural Activity

Criticality is a universal phenomenon defined for a range of different systems—from sand piles [5], to forest fires [35], to earthquakes [21]. The observation of criticality in spontaneously active neural systems [7, 15, 20, 44] may be analyzed through different theoretical frameworks [9] and likely has many functional benefits. Analyzing critical neural phenomena as a branching process [3, 61] leads to fundamental insight into the transmission of information in the brain, by characterizing the localized propagation of activity through cortical networks.

It is worth noting the differences of the spontaneously active networks observed here in comparison to neuronal avalanches. Near-synchronous circuit events appear to be very similar to inter-ictal paroxysmal depolarizing shifts observed in disinhibited network regimes at the single-cell, local circuit, and whole-brain levels [49]. These network-wide events are useful probes into the excitatory synaptic networks of the brain [7, 53, 54]. In vivo, where there are many incoming sensory inputs, networks have been shown to be subcritical in the sense that they do not fully propagate activity through the whole network [46, 57], that is, they operate below the percolation limit associated with the onset avalanches. However, when modeling criticality in these networks as a percolation process, the fine tuning of inhibitory circuits, constituting ~20% of the total neuronal population, appears to be lost. This fine tuning is most apparent during ongoing neuronal spiking and has a characteristic diversity which is well-characterized [40]. Here, we have adopted a methodology to assess criticality in neuronal networks that is based on dynamic systems theory rather than percolation theory. In contrast to the propagating activity of near-synchronous circuit events, for which the theory of branching processes is well-equipped to address, asynchronous spiking activity has recurrent rather than only feed-forward connections (Fig. 3) and criticality is more appropriately assessed by the eigenvalue spectra of this matrix.

The eigenvalues of the linear connectivity matrix govern the qualitative dynamics of the networks, such as bifurcations. Positive real components of the eigenvalue spectrum (1) identifies network elements that destabilize the dynamics from a quiescent to an active state and (2) reveals how unstable or critical the spontaneous activity is [25]. Furthermore, the accumulation of eigenvalues along the imaginary axis with both positive and negative real parts has significant implications on not only how critical the networks are but also how chaotic they might be. Indeed, coherent chaos has been recently described in recurrent neural networks [30].

4.2 *Critical Dynamics, Function, and Exploratory Behavior*

In assessing the spontaneous spiking networks of primary somatosensory cortex in mice, it is fruitful to consider this in vitro work with past and ongoing in vivo studies concerning exploratory whisking behavior of rodents [2, 8, 42, 41]. Critical state transitions might be functionally relevant to the shifting behavioral states and information processing that takes place during sensory-motor integration. There is a rich behavioral repertoire correlated to constructive states of transients discharges, oscillations, and asynchronous spiking. Here, we consider this last phenomenon since asynchronous spiking can lead to the other two phenotypes of neuronal activity.

Spontaneous neuronal spiking is the fundamental mode of activity in neocortical networks [52]; spiking may synchronize to form local field potentials, but there is always a loss of information. This is a feature that becomes strikingly evident during asynchronous firing states which occur during the states of cognition that are most information-rich. Thus, the elucidation of the functional roles critical dynamics serve during asynchronous firing is crucial to understanding the operational organization of local cortical circuits.

Such an attempt has already yielded insight into local cortical circuits: for instance, we have discovered that on behaviorally relevant timescales, the firing rates of asynchronously spiking neurons self-organize into two anti-correlated networks that localize to superficial and deep layers of the cortical anatomy [28]: when one network fires more, the other fires less and vice versa. This interplay manifests itself on longer timescales (>1 s), which is consistent with the localization of multiple eigenvalues to the imaginary axis since these eigenvalues correspond to non-decaying modes of the network dynamics. The anatomical localization of neurons with critical firing dynamics could have profound implications for function and behavior. These competing networks, may be extended to a behavioral context when considering the distinct exploratory modes of the rodent, which shift between feedforward sensation and top-down control configurations. These configurations correspond to active sensing and anticipatory behavioral states that are layer-specific: early activation of deep layers correlates with anticipatory behavior, whereas early activation of superficial layers correlates with active sensing [43]. The rodent's ability to switch between these behavioral modes is crucial since each mode involved the mutually exclusive use of subcortical brain structures and peripheral resources. Criticality in cortical networks could imbue the system with this sort of capacity along with the entire set of possible network configurations that give rise to rich behavior repertoires.

5 Conclusion and Future Directions

The approach presented here allows us to investigate the dynamical parameters that quantify critical phenomena, such as scale-free features, linear and nonlinear functional connectivity, and the linear stability of activity fluctuations. This can be attained efficiently by a straightforward regression analysis of the empirical data to our model. A limitation of our approach, however, is that the dependence of the nonlinear component on the firing rates remains unknown; only its dependence on time is retrieved from our analysis. In other words, the residuals in Eq. (2) are obtained as $\vec{\eta}(t)$ and not as $\vec{\eta}(\vec{x}(t))$. Because of this, the Jacobian of the network dynamics cannot be computed, which prevents us from computing the Lyapunov exponents, and hence, from determining whether the dynamics is chaotic or not. An alternative approach in this direction is to fit well-established firing-rate models to empirical data, and then calculate its Jacobian from those parameters. Along this line, Pikovsky [45] has recently fitted the firing rate model to simulated recordings, and in principle, the same approach could be applied to experimental data. Fitting the firing rates to this model would allow one to determine the functional connectivity, eigenvalues and Lyapunov exponents parametrically. In preliminary studies, we have determined that although this approach is very promising, the model underestimates the firing rate of real neurons significantly. Future work should investigate other more realistic models.

References

1. Ahmed, B., Anderson, J.C., Douglas, R.J., et al.: Polyneuronal innervation of spiny stellate neurons in cat visual cortex. J. Comput. Neurol. **341**, 39–49 (1994). https://doi.org/10.1002/cne.903410105
2. Ahrens, K.F., Kleinfeld, D.: Current flow in vibrissa motor cortex can phase-lock with exploratory rhythmic whisking in rat. J. Neurophysiol. **92**, 1700–1707 (2004). https://doi.org/10.1152/jn.00404.2004
3. Alstrom, P.: Mean-field exponents for self-organized critical phenomena. Phys. Rev. A **38**, 4905–4906 (1988). https://doi.org/10.1103/PhysRevA.38.4905
4. Bak, P., Stassinopoulos, D.: Democratic reinforcement **51** (1995)
5. Bak, P., Tang, C., Wiesenfeld, K.: Self-organized criticality: An explanation of the 1/ f noise. Phys. Rev. Lett. **59**, 381–384 (1987). https://doi.org/10.1103/physrevlett.59.381

6. Bédard, C., Kröger, H., Destexhe, A.: Does the 1/f frequency scaling of brain signals reflect self-organized critical states? Phys. Rev. Lett. **97,** 1–4 (2006). https://doi.org/10.1103/physrevlett.97.118102
7. Beggs, J.M., Plenz, D.: Neuronal avalanches in neocortical circuits. J. Neurosci. **23**, 11167–11177 (2003). doi:23/35/11167 [pii]
8. Berg, R.W., Kleinfeld, D.: Rhythmic whisking by rat: retraction as well as protraction of the vibrissae is under active muscular control. J. Neurophysiol. **89**, 104–117 (2003). https://doi.org/10.1152/jn.00600.2002
9. Buice, M.A., Cowan, J.D.: Statistical mechanics of the neocortex. Prog. Biophys. Mol. Biol. **99**, 53–86 (2009). https://doi.org/10.1016/j.pbiomolbio.2009.07.003
10. Cabral, J., Kringelbach, M., Deco, G.: Functional connectivity dynamically evolves on multiple time-scales over a static structural connectome: models and mechanisms. Neuroimage, 0–1 (2017). https://doi.org/10.1016/j.neuroimage.2017.03.045
11. Chialvo, D., Bak, P.: Commentary: learning from mistakes. Neuroscience **90**, 1137–1148 (1999)
12. Chiel, H.J., Beer, R.D.: The brain has a body: adaptive behavior emerges from interactions of nervous system, body and environment. Trends Neurosci. **20**, 553–557 (1997)
13. Cowan, J.D., Neuman, J., Kiewiet, B., Van Drongelen, W.: Self-organized criticality in a network of interacting neurons. J. Stat. Mech. Theory Exp. 2013:. https://doi.org/10.1088/1742-5468/2013/04/p04030
14. Cowan, J.D., Neuman, J., Van Drongelen, W.: Self-organized criticality and near-criticality in neural networks. In: Criticality in Neural Systems, pp 465–484 (2014)
15. De Arcangelis, L., Perrone-Capano, C., Herrmann, H.J.: Self-organized criticality model for brain plasticity. Phys. Rev. Lett. **96**, 1–4 (2006). https://doi.org/10.1103/physrevlett.96.028107
16. Douglas, R.J., Koch, C., Mahowald, M., et al.: Recurrent excitation in neocortical circuits. Science **269**, 981–985 (1995)
17. Eguíluz, V.M., Chialvo, D.R., Cecchi, G.A., et al.: Scale-free brain functional networks. Phys. Rev. Lett. **94**, 1–4 (2005). https://doi.org/10.1103/physrevlett.94.018102
18. Friedman, N., Ito, S., Brinkman, B.A.W., et al.: Universal critical dynamics in high resolution neuronal avalanche data. Phys. Rev. Lett. **108**, 1–5 (2012). https://doi.org/10.1103/physrevlett.108.208102
19. Galán, R.F.: On how network architecture determines the dominant patterns of spontaneous neural activity. PLoS One **3**, (2008). https://doi.org/10.1371/journal.pone.0002148
20. Gireesh, E.D., Plenz, D.: Neuronal avalanches organize as nested theta- and beta/gamma-oscillations during development of cortical layer 2/3. Proc. Natl. Acad. Sci. **105**, 7576–7581 (2008). https://doi.org/10.1073/pnas.0800537105
21. Gutenberg B, Richter C (1954) Seismicity of the earth: Princeton, NJ
22. Hájos, N., Ellender, T.J., Zemankovics, R., et al.: Maintaining network activity in submerged hippocampal slices: importance of oxygen supply. Eur. J. Neurosci. **29**, 319–327 (2009). https://doi.org/10.1111/j.1460-9568.2008.06577.x
23. Hájos, N., Mody, I.: Establishing a physiological environment for visualized in vitro brain slice recordings by increasing oxygen supply and modifying aCSF content. J. Neurosci. Methods **183**, 107–113 (2009). https://doi.org/10.1016/j.jneumeth.2009.06.005
24. Haken, H.: Synergetics: An Introduction. Springer, Berlin (1978)
25. Haken, H.: Advanced Synergetics: Instability Hierarchies of Self-organizing Systems. Springer, Berlin (1983)
26. Herz, A.V.M., Hopfield, J.J.: Earthquake cycles and neural reverberations: collective oscillations in systems with pulse-coupled threshold elements andreas. Phys. Rev. Lett. **75**, 4–7 (1995)
27. Hoffman, K.L., Battaglia, F.P., Harris, K., et al.: The upshot of up states in the neocortex: from slow oscillations to memory formation. J. Neurosci. **27**, 11838–11841 (2007). https://doi.org/10.1523/JNEUROSCI.3501-07.2007
28. Kodama, N.X., Feng, T., Ullett, J.J., et al.: Anti-correlated cortical networks arise from spontaneous neuronal dynamics at slow timescales. Sci. Rep. (2017)

29. Koyama, S.: On the spike train variability characterized by variance-to-mean power relationship. Neural Comput. **27**, 1530–1548 (2015). https://doi.org/10.1162/NECO_a_00748
30. Landau, I.D., Sompolinsky, H.: Coherent chaos in a recurrent neural network with structured connectivity, 1–29 (2018). https://doi.org/10.1101/350801
31. Langton, C.G.: Computation at the edge of chaos: phase transitions and emergent computation. Phys. D Nonlinear Phenom. **42**, 12–37 (1990). https://doi.org/10.1016/0167-2789(90)90064-V
32. Levina, A., Priesemann, V.: Subsampling scaling. Nat. Commun. **8**, 1–9 (2017). https://doi.org/10.1038/ncomms15140
33. Luczak, A., Maclean, J.N.: Default activity patterns at the neocortical microcircuit level. Front. Integr. Neurosci. **6**, 30 (2012). https://doi.org/10.3389/fnint.2012.00030
34. MacLean, J.N., Watson, B.O., Aaron, G.B., Yuste, R.: Internal dynamics determine the cortical response to thalamic stimulation. Neuron **48**, 811–823 (2005). https://doi.org/10.1016/j.neuron.2005.09.035
35. Malamud, B.D.: Forest fires: an example of self-organized critical behavior. Science **1840**, 1998–2001 (2008). https://doi.org/10.1126/science.281.5384.1840
36. Mehta, M.L.: Random Matrices (2005)
37. Millman, D., Mihalas, S., Kirkwood, A., Niebur, E.: Self-organized criticality occurs in non-conservative neuronal networks during "up" states. Nat. Phys. **6**, 801–805 (2010). https://doi.org/10.1038/nphys1757
38. Mitra, P., Pesaran, B.: Analysis of dynamic brain imaging data. Biophys. J. **76**, 691–708 (1999)
39. Nelder, J., Wedderburn, R. Generalized linear models. J. R. Stat. Soc. (1972). https://doi.org/10.2307/2344614
40. Neske, G.T., Patrick, S.L., Connors, B.W.: Contributions of diverse excitatory and inhibitory neurons to recurrent network activity in cerebral cortex. J. Neurosci. **35**, 1089–1105 (2015). https://doi.org/10.1523/JNEUROSCI.2279-14.2015
41. Nicolelis, M.A.L., Baccala, L.A., Lin, R.C.S., Chapin, J.K.: Sensorimotor encoding by synchronous neural ensemble activity at multiple levels of the somatosensory system. Science **268**, 1353–1358 (1995)
42. Nicolelis, M.A.L., Fanselow, E.E.: Thalamocortical optimization of tactile processing according to behavioral state. Nat. Neurosci. **5**, 517–523 (2002). https://doi.org/10.1038/nn0602-517
43. Pais-Vieira, M., Kunicki, C., Tseng, P.-H., et al.: Cortical and thalamic contributions to response dynamics across layers of the primary somatosensory cortex during tactile discrimination. J. Neurophysiol. **114**, 1652–1676 (2015). https://doi.org/10.1152/jn.00108.2015
44. Petermann, T., Thiagarajan, T.C., Lebedev, M.A., et al.: Spontaneous cortical activity in awake monkeys composed of neuronal avalanches. Proc. Natl. Acad. Sci. **106**, 15921–15926 (2009). https://doi.org/10.1073/pnas.0904089106
45. Pikovsky, A.: Reconstruction of a neural network from a time series of firing rates. Phys. Rev. E **93**, 062313 (2016)
46. Priesemann, V.: Spike avalanches in vivo suggest a driven, slightly subcritical brain state **8**, 1–17 (2014). https://doi.org/10.3389/fnsys.2014.00108
47. Prigogine, I., Lefever, R.: Symmetry breaking instabilities in dissipative systems. II. J Chem Phys **48**, 1695–1700 (1968). https://doi.org/10.1063/1.1668896
48. Prigogine, I., Nicolis, G.: On symmetry-breaking instabilities in dissipative systems. J. Chem. Phys. **46**, 3542–3550 (1967). https://doi.org/10.1063/1.1841255
49. Puzerey, P.A., Kodama, N.X., Galán, R.F.: Abnormal cell-intrinsic and network excitability in the neocortex of serotonin-deficient *Pet-1* knockout mice. J. Neurophysiol. (2016). https://doi.org/10.1152/jn.00996.2014
50. Quiroga, R.Q., Nadasdy, Z., Ben-Shaul, Y.: Unsupervised spike detection and sorting with wavelets and superparamagnetic clustering. Neural Comput. **16**, 1661–1687 (2004). https://doi.org/10.1162/089976604774201631
51. Rajan, K., Abbott, L.F.: Eigenvalue spectra of random matrices for neural networks. Phys. Rev. Lett. **97**, 2–5 (2006). https://doi.org/10.1103/physrevlett.97.188104

52. Renart, A., de la Rocha, J., Bartho, P., et al.: The asynchronous state in cortical circuits. Science **327**, 587–590 (2010). https://doi.org/10.1126/science.1179850
53. Sadovsky, A.J., MacLean, J.N.: Scaling of topologically similar functional modules defines mouse primary auditory and somatosensory microcircuitry. J. Neurosci. **33**, 14048–14060 (2013). https://doi.org/10.1523/JNEUROSCI.1977-13.2013
54. Sederberg, A.J., Palmer, S.E., MacLean, J.N.: Decoding thalamic afferent input using microcircuit spiking activity. J. Neurophysiol. **113**, 2921–2933 (2015). https://doi.org/10.1152/jn.00885.2014
55. Sporns, O., Zwi, J.D.: The small world of the cerebral cortex. Neuroinformatics **2**, 145–162 (2004). https://doi.org/10.1385/NI:2:2:145
56. Steinke, G.K., Galán, R.F.: Brain rhythms reveal a hierarchical network organization. PLoS Comput. Biol. **7**, e1002207 (2011). https://doi.org/10.1371/journal.pcbi.1002207
57. Tomen, N., Rotermund, D., Ernst, U.: Marginally subcritical dynamics explain enhanced stimulus discriminability under attention. Front. Syst. Neurosci. **8**, 1–15 (2014). https://doi.org/10.3389/fnsys.2014.00151
58. Touboul, J., Destexhe, A.: Power-law statistics and universal scaling in the absence of criticality. Phys. Rev. E **012413**, 1–15 (2017). https://doi.org/10.1103/PhysRevE.95.012413
59. Watson, B.O., MacLean, J.N., Yuste, R.: UP States protect ongoing cortical activity from thalamic inputs. PLoS One **3**, (2008). https://doi.org/10.1371/journal.pone.0003971
60. Wilting, J., Priesemann, V.: Inferring collective dynamical states from widely unobserved systems (2018)
61. Zapperi, S., Lauritsen, K.B., Stanley, H.E.: Self-organized branching processes: Mean-field theory for avalanches. Phys. Rev. Lett. **75**, 4071–4074 (1995). https://doi.org/10.1103/physrevlett.75.4071

Nathan X. Kodama studied physics at Case Western Reserve University in Cleveland, United States. He has worked on computational condensed matter physics at Georgetown University before moving on to studying biological neural networks as a Graduate Student in systems and control engineering. His research interests include generative models, neural coding, and graphical topology.

Roberto Fernández Galán studied physics at the Autonomous University of Madrid, Spain, and obtained a Ph.D. in theoretical biology (emphasis on Computational Neuroscience) from the Humboldt University in Berlin, Germany. He worked as a research associate at Carnegie Mellon University in Pittsburgh, USA, before becoming a faculty member and research group leader at Case Western Reserve University in Cleveland, USA. His research interests include neural dynamics, biophysical mechanisms of noise in brain circuits and its functional role.

Optimal Fisher Decoding of Neural Activity Near Criticality

Eric S. Kuebler, Matias Calderini, Philippe Lambert and Jean-Philippe Thivierge

Abstract Studies on the functional role of criticality in the brain have thus far mainly examined the role of neural dynamics on stimulus encoding, with scant attention devoted to the impact of these dynamics on downstream decoding. Here, we consider the question of how a linear decoder may classify spontaneous cortical activity both near and away from a critical state. We show that accurate performance of the decoder is obtained only when network activity is near criticality. Simulations of a branching process capture these results and argue for a potential role of the critical state in providing a format for neural activity that can be adequately processed by downstream brain structures.

Neuronal activity both in vitro and in vivo has increasingly been shown to behave like a physical system operating near a critical state [30]. The proposed advantages of critical activity are manyfold. In the critical state, neural activity is suggested to display a maximal range of responses to sensory input [15]. Further, the critical state may optimize information capacity and transmission [33]. However, investigations on the role of neural criticality have primarily focused on neural encoding, with relatively little attention paid to how criticality may impact neural decoding by downstream brain structures [12, 38]. A priori, there is no reason to assume that a state that is optimal for encoding is also optimal for decoding. Indeed, in a simplified scenario, we can segregate information coding and decoding into two distinct steps along an information processing pathway leading from "lower" to "higher" sensory areas (Fig. 1a).

To address the role of criticality in neural decoding, this chapter examines three approaches. First, we apply a linear decoder to network activity recorded with multielectrode arrays. Second, we compare experimental results with numerical simulations where we decode the activity of a simple branching model. Finally, we provide analytical results relating critical branching to linear decoding. Together,

E. S. Kuebler · M. Calderini · P. Lambert · J.-P. Thivierge (✉)
School of Psychology, University of Ottawa, Ottawa K1N 6N5, Canada
e-mail: jthivier@uottawa.ca

N. Tomen et al. (eds.), *The Functional Role of Critical Dynamics in Neural Systems*, Springer Series on Bio- and Neurosystems 11,
https://doi.org/10.1007/978-3-030-20965-0_9

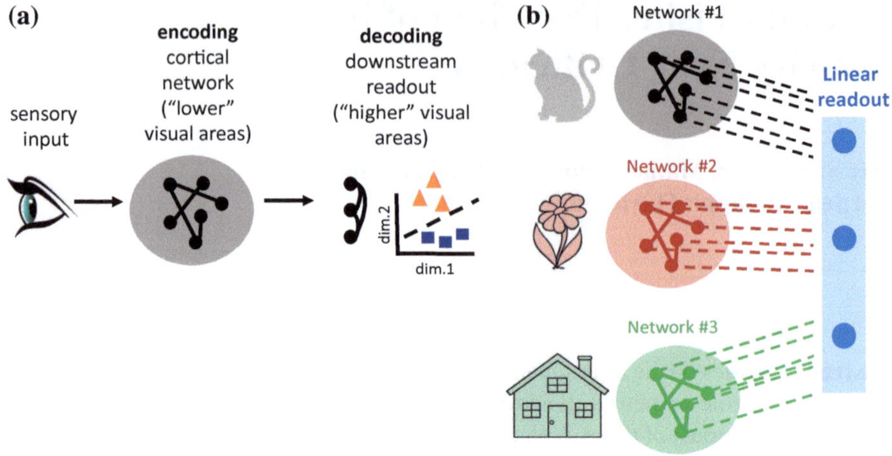

Fig. 1 Setting up a linear decoder to read out a neuronal population. a Schematic description of sensory encoding versus decoding along an information processing pathway from "lower" to "higher" brain areas. A sensory stimulus activates neural pathways leading to cortex. The response of cortical networks is read out by downstream structures in order to classify and discriminate different stimuli. **b** Schema illustrating several networks (#1–3) projecting to a readout layer of units that identify the active network at a given time. For illustration purposes, each network represents an abstracted concept of "cat", "flower", and "house"

findings highlight the role of critical dynamics in providing a format for accurate linear decoding of network activity.

1 Thought Experiment

To introduce the problem of neural decoding examined here, we begin with a "thought experiment" in which several cortical networks feed into a set of downstream readout units. One could think of these networks as representing stimulus-specific cell assemblies [17] that code for concepts such as "cat", "flower", or "house" (Fig. 1b). When these different networks are activated, the readout layer must identify the concept that is encoded. Although highly simplified, this is a reasonable scenario in cortical circuits where different networks perform distinct functions despite a degree of shared interconnectivity [24, 28]. This thought experiment serves as the basis for the analyses presented here.

2 Optimal Linear Readout of Neural Activity

We recorded spontaneous population activity from 7 cultured cortical networks for a 20 min duration over days in vitro 17–23. Culturing of cortical neurons and recording of multi-unit spiking activity was performed on a 64-electrode array as described previously [20, 32, 36, 39]. Neural activity was characterized by network bursts that extended over several hundreds of milliseconds (Fig. 2a). Bursts under control conditions were compared to those obtained with the addition of either a $GABA_A$ receptor antagonist (5 μM picrotoxin, PTX) or an NMDA receptor antagonist (2R)-amino-5-phosphonovaleric acid (20 μM APV) combined with an AMPA receptor antagonist 6,7dinitroquinoxaline-2,3-dione (2 μM DNQX). As shown previously [39], PTX resulted in increased bursting activity (specifically, in the number of electrodes active during bursts), while APV/DNQX resulted in decreased bursting (Fig. 2a).

In control recordings, the number of electrodes active during avalanches followed a power-law (Fig. 2b). The range of this distribution is limited to ~1.5 orders of magnitude due to the limited size of the electrode array. We computed a maximum likelihood estimate for the slope of this distribution [11], assuming a continuous power-law function,

$$\hat{\tau} = 1 + n\left[\sum_{i=1}^{n} \log \frac{x_i}{x_{\min}}\right]^{-1}, \tag{1}$$

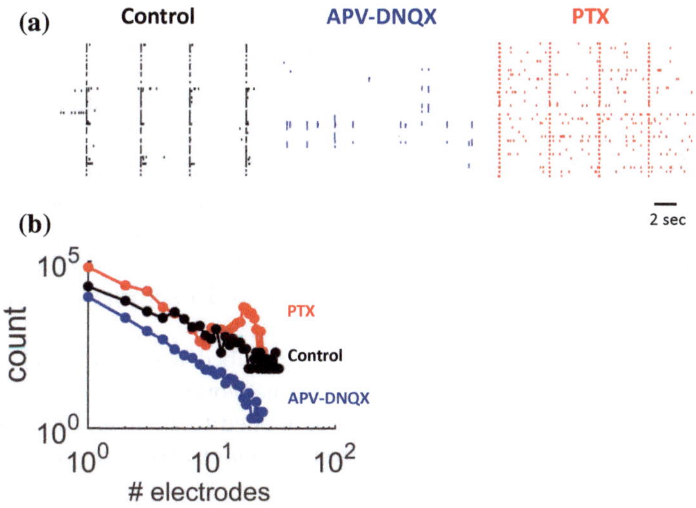

Fig. 2 **Neuronal avalanches.** **a** Examples of population rasters under control and pharmacological conditions. **b** Number of electrodes active per avalanche for control, APV-DNQX, and PTX conditions. Panel b is taken from [39]

where $i = 1 \ldots n$ are the observed values of the data (x) corresponding to avalanche sizes. We set $x_{\min}$ to the minimum value of the distribution. Statistical error in the estimation of τ is given by

$$error = \frac{\hat{\tau} - 1}{\sqrt{n}}, \tag{2}$$

yielding $\hat{\tau} = 1.52$ (statistical error: 0.02), a value close to the expected mean field exponent in the critical state [22]. As previously reported, pharmacological alterations in neural activity shifted this slope (PTX, $\hat{\tau} = 1.47$, statistical error: 0.01; APV/DNQX, $\hat{\tau} = 1.63$, statistical error: 0.04) [39]. Related work reports a power-law of avalanche amplitudes [21] and duration [32] using the same dataset. A full treatment of the goodness-of-fit of these data is considered elsewhere [21, 39, 40], along with a comparison between function fitting with power-law and exponential distributions [21]. The effect of binning the data using windows of various durations (between 1 and 16 ms) has been shown to have a minimal impact on estimates of power-law slopes [39].

One key feature of criticality is the presence of a systematic relation between different measures of avalanches, namely their duration, the number of cells activated, and the number of spikes generated [32]. If we denote the power-law exponent of these measures as α (duration), τ (number of active cells), and $1/\sigma\nu z$ (relation between duration and number of active cells), the mean-field exponent relation for critical systems [26] is given by

$$\frac{\alpha - 1}{\tau - 1} = \frac{1}{\sigma \nu z}. \tag{3}$$

This relation is approximated in cortical recordings under control conditions, but disrupted with pharmacology [32].

Aside from the presence of power-law distributed activity, a key feature of neural activity in the critical state is the presence of avalanches whose duration can be rescaled to show a similar temporal profile of activity [14]. In previous work, we developed a statistical tool to test for this phenomenon, and reported reliable rescaling of avalanches under control conditions [32]. This effect was not found with pharmacologically-induced activity.

Next, following the thought experiment described above (Fig. 1), we asked whether network bursts belonging to different networks could be accurately classified by a linear readout. This readout was applied to spontaneous activity. It is unknown whether results would generalize to stimulus-driven activity, a point that we return to in the Discussion.

To classify network bursts, we began by applying a threshold to the sum of network activity to identify large network bursts. This threshold was adjusted such that the total number of threshold-crossing events would be similar across experimental conditions. For the control condition, the threshold was set to 15 concurrently active electrodes within non-overlapping 10 ms windows. This threshold was lowered to 12 electrodes for the PTX condition and increased to 20 electrodes for the APV/DNQX condition.

An example of a single network burst is shown in Fig. 3a. These bursts are characterized by a sharp rise in population activity followed by a slightly longer timescale of decay, as detailed in Thivierge and Cisek [37]. Note that the network activity shown in this figure would provide multiple training samples if the burst extends over several non-overlapping windows of 10 ms each.

Given N total electrodes, we summarized each threshold-crossing event by a binary vector of length N where each element is set to "1" if the corresponding unit spiked at least once (and "0" otherwise). We then merged activity from all 7 different cortical networks into one dataset. We trained a Fisher linear decoder [6, 9] on one half of this dataset, tested the performance of the readout on the other half, and measured the percentage of errors in identifying the network of origin.

To understand the workings of this decoder, let's consider a simplified scenario where it receives input from two networks (generalizing to more networks is straightforward). Assume a dataset $\mathbf{X} = \{\mathbf{x}^{(1)}, \mathbf{x}^{(2)}, \ldots \mathbf{x}^{(M)}\}$, where each sample $\mathbf{x}$ is an N-dimensional vector corresponding to a single threshold-crossing event, and M is the total number of events recorded. Of these events, we have $C = 2$ subsamples

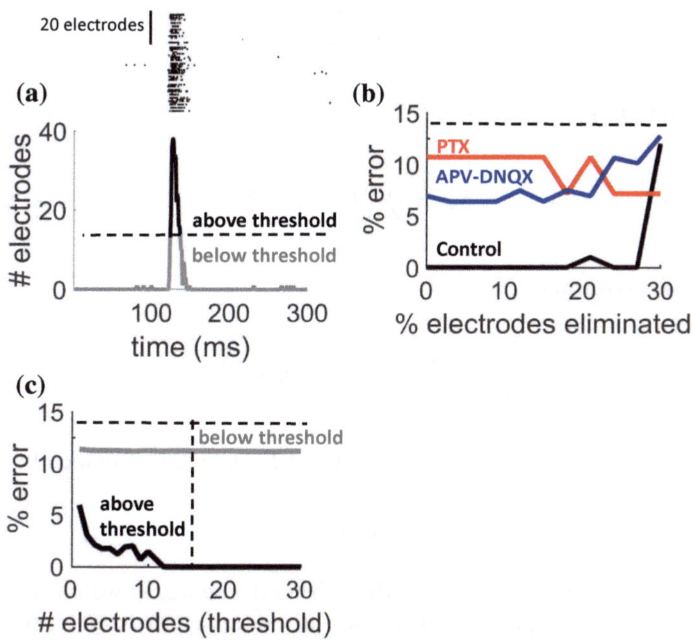

Fig. 3 Linear classification of neural activity. a Example of a single network burst showing a spike raster (top panel) and number of concurrently active electrodes (bottom panel). Dashed line, example of threshold set to 15 active electrodes. **b** Readout accuracy of population activity based on linear discriminant analysis. Dashed line, chance level. **c** Performance of the linear readout when altering the threshold to detect network bursts. Vertical dashed line, default threshold value

(or classes) M_1 and M_2 each belonging to a given network. The goal of the linear decoder is to compute a scalar $y = \mathbf{w}^{\mathrm{T}}\mathbf{x}$ such that the separation between the two networks M_1 and M_2 is maximal. Formally, this is the line that maximizes the Fisher criterion

$$J(\mathbf{w}) = \frac{\mathbf{w}^T S_B \mathbf{w}}{\mathbf{w}^T S_W \mathbf{w}}, \tag{4}$$

where S_B and S_W are matrices of the between- and within-class covariance, respectively. The between-class variance is obtained as

$$S_B = \sum_{i=1}^{C} M_i (\boldsymbol{\mu}_i - \bar{\mathbf{x}})(\boldsymbol{\mu}_i - \bar{\mathbf{x}})^T. \tag{5}$$

where $\bar{\mathbf{x}}$ is the grand mean of the dataset. In the above equation, patterns for all classes are sequentially indexed from 1 to M. In order to reference all patterns within one class, we introduce $k_{im} \in [1, M]$, which is the index for the m-th pattern of class i. With this notation, we define the mean for each class,

$$\boldsymbol{\mu}_i = \frac{1}{M_i} \sum_{m=1}^{M_i} \mathbf{x}^{(k_{im})}. \tag{6}$$

The within-class variance is

$$S_W = \sum_{i=1}^{C} (M_i - 1)\Sigma_i, \tag{7}$$

given the class covariance,

$$\Sigma_i = \frac{1}{M_i - 1} \sum_{m=1}^{M_i} \left(\mathbf{x}^{(k_{im})} - \boldsymbol{\mu}_i\right)\left(\mathbf{x}^{(k_{im})} - \boldsymbol{\mu}_i\right)^T. \tag{8}$$

A closed-form solution to maximizing Eq. 4 for $C = 2$ is provided by

$$\mathbf{w} = S_W^{-1}(\boldsymbol{\mu}_1 - \boldsymbol{\mu}_2). \tag{9}$$

The statistical assumptions underlying Fisher decoding, as well as the impact of their violation in neural population activity, are addressed in detail elsewhere [6, 9].

We trained a separate classifier for each experimental condition (controls and pharmacologically altered networks). The linear readout we employed provides a lower bound on classification capacity and has shown good agreement with behavioral performance on decision making tasks when employed to decode cortical activity [6,

31]. While this readout is not meant as a biologically grounded decoding mechanism, it can be implemented biologically, for instance through a Hebbian learning rule [8].

In the control condition, the linear readout accurately classified bursts with nearly 0% testing error until approximately 90% of electrodes were randomly discarded (after applying the burst detection threshold), at which point error rose sharply (Fig. 3b). Thus, the linear decoder was successful at classifying network activity. By comparison, pharmacological alterations in neural activity markedly disrupted performance, regardless of the percentage of electrodes eliminated (Fig. 3b). In other words, decoding was more accurate in networks near the critical state (in the control condition) compared to pharmacologically-altered conditions.

Next, we asked whether accurate classification of network activity was limited to network bursts or whether it could also be achieved with lower levels of activity. As before, we employed a fixed threshold applied to summed population activity, which we separated into above- and below-threshold components (Fig. 3a). We then computed classification error of the linear decoder across different threshold values (Fig. 3c). As the threshold was increased beyond ~15 electrodes, above-threshold activity attained near-zero error while below-threshold activity remained near chance. Thus, accurate classification of network activity was limited to large network bursts where several electrodes were active concurrently.

To illustrate the performance of the linear readout, we generated a raster containing all network bursts of the 7 arrays. We color-coded this raster according to either the network of origin, readout classification using network bursts, or randomly assigned labels. Classification using randomized data showed that most of the exemplars were misclassified (Fig. 4); by comparison, classification using network bursts was nearly identical to the original data.

While network activity near the critical state seems to convey an advantage for decoding, the basic underlying mechanism responsible for this finding is thus far unclear. To examine this question, the next section of this chapter turns to a simple branching model that allows us to explore the role of criticality in linear classification.

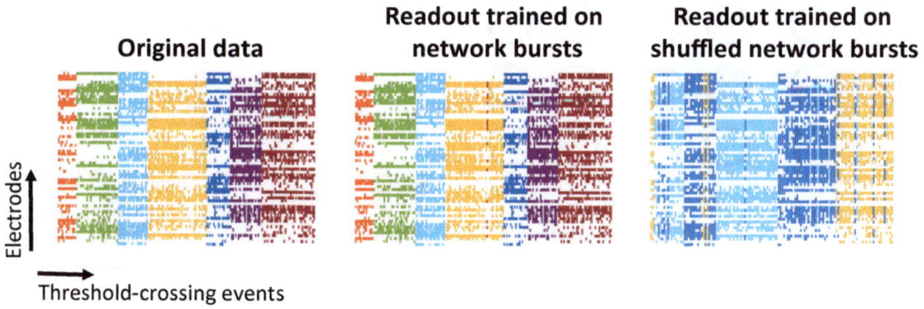

Fig. 4 Linear readout of network bursts. Rasters showing consecutive threshold-crossing events when the linear readout was presented with either original or shuffled spikes. Different networks of origin are labeled by color

3 Neuronal Decoding in a Critical Branching Model

To examine why in vitro networks near the critical state can be accurately classified by a linear readout, we turned to a simple phenomenological model of critical branching that has been employed to describe the power-law distribution of neural avalanches [5] and $1/f$ scaling dynamics [18]. Branching processes are fundamental for capturing the mean-field behavior of critical systems as detailed in classic work [2, 10, 41]. While it is possible to produce both power-law avalanches and $1/f$ scaling without criticality [3, 4], critical dynamics can exist at critical branching [23].

In the branching model, each spike causes a given number of spikes in downstream units (Fig. 5a). This number is determined by a branching ratio (σ) expressed as the average number of descendant to ancestor units,

$$\sigma = \frac{\#descendants}{\#ancestors}. \tag{10}$$

When the branching ratio parameter (σ) is less than 1, activation fades over time, while $\sigma = 1$ leads to short-term sustained activation and $\sigma > 1$ leads to amplification. A value of $\sigma = 1$ is associated with critical branching [41]. This value closely matches our control data (mean of $\sigma = 0.974$ with standard error of 0.06 across all control networks; the average is taken over all consecutively active time bins). Pharmacologically altering network activity is associated with a shift in branching parameter away from $\sigma = 1$. Specifically, APV/DNQX networks yielded a mean branching ratio of 0.82 (standard error, 0.15), compared to 1.05 (standard error, 0.13) for PTX networks. These shifts are captured by the branching model, as described below (Fig. 5b).

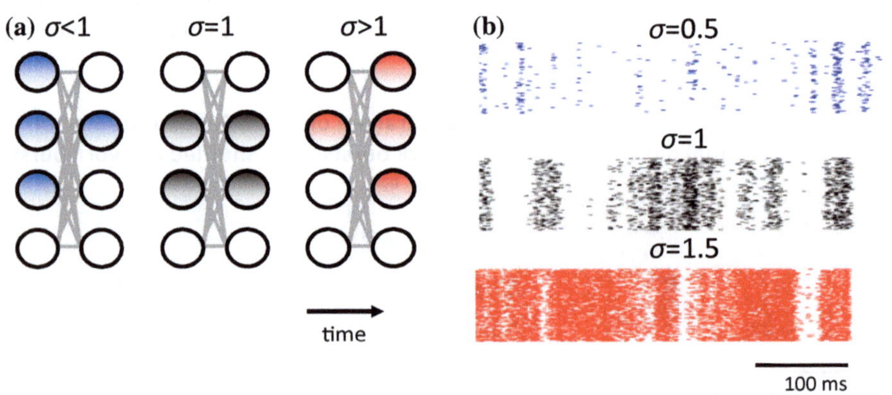

Fig. 5 Neural activity generated by a branching model. a Propagation of activity in a branching model with synapses (grey lines) connecting four neurons (circles) arranged in two layers representing different time points. **b** Spiking activity generated by the branching model with different branching ratios

A dynamical model of branching is described as follows. A population of $N = 100$ units are randomly connected following a uniform distribution with a connection probability of $p_{conn} = 0.5$ between pairs of units [7]. Each unit is either active or inactive and may become spontaneously active with probability $g = 0.001$. The connectivity of the whole circuit is given by a matrix $W \in [0, 1]^{N \times N}$ with elements $0 \leq w_{ij} \leq 1$ describing the probability of a unit i propagating its active state to a unit j within two consecutive time points,

$$W = \begin{bmatrix} w_{11} & \dots & w_{1j} & \dots & w_{1N} \\ \vdots & \dots & \vdots & \dots & \vdots \\ w_{i1} & \dots & w_{ij} & \dots & w_{iN} \\ \vdots & \dots & \vdots & \dots & \vdots \\ w_{N1} & \dots & w_{Nj} & \dots & w_{NN} \end{bmatrix}. \tag{11}$$

Connections are constrained such that

$$\sum_{j=1}^{N} w_{ij} = \sigma. \tag{12}$$

Thus defined, the connectivity matrix does not allow for inhibitory connections. Further, no refractory period was incorporated in the model.

We simulated 5 networks for 10^6 time-steps with various branching ratios. As with in vitro networks, we identified network bursts using a threshold of 15 concurrently active units within time windows of 10 ms duration. Then, we stored network bursts and trained a linear readout to classify them according to their network of origin. Classification accuracy of the readout was highest near $\sigma = 1$. This result was robust to changing the discharge probability for spontaneous activity (Fig. 6a) as well as changing the density of connections (Fig. 6b).

Focusing on the critical branching condition $\sigma = 1$, we examined the performance of the linear readout subject to input from an increasing number of networks. In a model with $N = 100$ units, 10 networks were accurately classified before error approached chance-level performance (Fig. 6c). By adding more units to each network, this performance was markedly improved. Indeed, classification error was lower with $N = 500$ units than with $N = 100$ units. Regardless of the number of units per network, error increased non-linearly with the number of networks to classify. In sum, the branching model captured in vitro results showing that a regime of $\sigma \sim 1$ leads to accurate linear readout of population activity. Next, we take an analytical approach to examine the origin of optimal readout in critical branching.

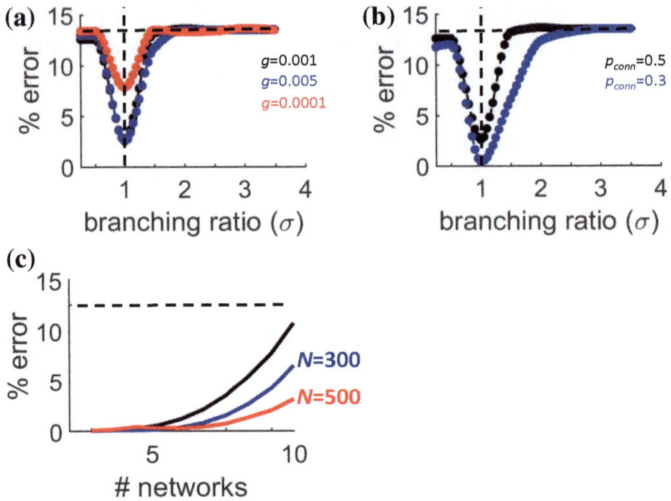

Fig. 6 Linear decoding of activity in a branching model. **a** Optimal linear discrimination of network bursts from networks with various probabilities of spontaneous activity (g). Dashed vertical line corresponds to $\sigma = 1$. Dashed horizontal line: chance level. **b** Optimal linear discrimination with different synaptic densities obtained by varying the probability of pairwise connections in the branching model from sparser ($p_{conn} = 0.3$) to denser ($p_{conn} = 0.5$). **c** Classification performance relative to the number of networks entered in the decoder and the number of units (N) per network

4 Analysis of the Critical Branching Model

To study the dynamics of the branching model, we developed the following formalization. We began with the assumption that all units are independent. This assumption does not hold in practice given the connectivity amongst units (Eq. 11) but provides a useful approximation, as shown below. We define a "network state" according to a vector $\vec{X} \in \{0, 1\}^{N\times 1}$ with elements x_i representing whether unit i is active ($x_i = 1$) or silent ($x_i = 0$). Second, we define the probability of spontaneous spiking $\vec{G} \in [0, 1]^{N\times 1}$ with elements g_i representing the probability of spontaneous activity for unit i.

Given the above definitions, we aim to calculate the probability of a neuron j spiking at time-step $t = 0, 1, 2, \ldots, T$. This problem can be approached by first considering the probability of *not* spiking,

$$P(x_j = 1|t) = 1 - P(x_j = 0|t), \tag{13}$$

where t denotes the time since the last point when the sum of activity across units was zero. In a large network, this can be assumed to be $t = 0$.

A given neuron can spike for two reasons: either it becomes active spontaneously, or it is activated by a neuron to which it is connected. Hence, we can reformulate the above as

$$P\left(x_j = 1|t\right) = 1 - P\left(x_j = 0|t, g_j, \vec{X}_{t-1}, W\right). \tag{14}$$

Assuming independence between spontaneous activity and network connectivity,

$$\begin{aligned} P\left(x_j = 0|t\right) &= P\left(x_j = 0|g_j\right)P\left(x_j = 0|t, \vec{X}_{t-1}, W\right) \\ &= \left(1 - g_j\right)P\left(x_j = 0|t, \vec{X}_{t-1}, W\right). \end{aligned} \tag{15}$$

By the law of total probability, expanding Eq. 15 for a single presynaptic unit yields

$$P\left(x_j = 0|t\right) = \left(1 - g_j\right)\left[1 - w_{ij}P(x_i = 1|t-1)\right]. \tag{16}$$

Converting this result back to the probability of spiking in the case of one presynaptic unit,

$$P\left(x_j = 1|t\right) = 1 - \left(1 - g_j\right)\left[1 - w_{ij}P(x_i = 1|t-1)\right]. \tag{17}$$

For N independent units, this becomes

$$P\left(x_j = 1|t\right) = 1 - \left(1 - g_j\right)\prod_{i=1}^{N}\left[1 - w_{ij}P(x_i = 1|t-1)\right]. \tag{18}$$

To find a distribution $P(x)$ which is invariant under the temporal dynamics, the above equation can be iterated from $t = 0$ until convergence into a steady-state solution.

A special case of Eq. 18 considers that all elements of each row in W are the same, resulting in an equal probability of transmission across all outgoing connections of a given unit, but allowing for different transmission probabilities across units. In this case, Eq. 18 can be formulated as an explicit function of the branching parameter,

$$P\left(x_j = 1|t\right) = 1 - \left(1 - g_j\right)\prod_{i=1}^{N}\left[1 - \frac{\sigma}{o_i}P(x_i = 1|t-1)\right], \tag{19}$$

where o_i is the out-degree of unit i. We verified that $P\left(x_j = 1|t\right)$ obtained from Eq. 19 agrees with mean spike probabilities obtained from numerical simulations of the branching model (Fig. 7).

The expected value for the state of unit j at generation t is given by

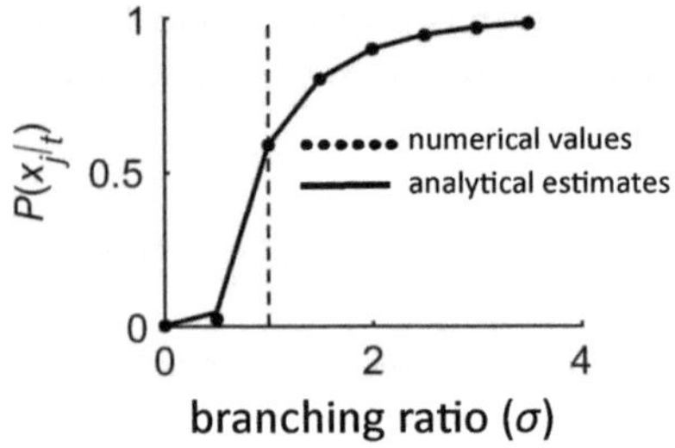

Fig. 7 Spike probability ($P(x_j|t)$) obtained from numerical simulations and analytical estimate (Eq. 19). Vertical dashed line, $\sigma = 1$

$$E\left[x_j|t\right] = 1 \times P\left(x_j = 1|t\right) + 0 \times P\left(x_j = 0|t\right)$$
$$= P\left(x_j = 1|t\right), \tag{20}$$

with variance

$$Var\left(x_j|t\right) = 1 \times P\left(x_j = 1|t\right) + 0 \times P\left(x_j = 0|t\right) - E\left[x_j|t\right]^2$$
$$= P\left(x_j = 1|t\right) - P\left(x_j = 1|t\right)^2. \tag{21}$$

To perform an optimal linear readout of network activity using a Fisher discriminant criterion [25], we assume that two networks each with a different connectivity matrix W (Eq. 11) generate spontaneous activity. This activity is then summed across neurons and compared to a threshold θ to focus on time-steps when a given proportion of neurons are active simultaneously. This scenario is equivalent to the analysis performed previously on experimental data (Fig. 3a). The probability of observing a network burst at generation t,

$$P\left(\sum_{j=1}^{N} x_j \geq \theta|t\right),$$

is equivalent to the probability of observing at least θ successful trials out of N coin flips, each with probability $p_i = P(x_i = 1|t)$ of success. With large N, the probability of observing at least θ successful trials can be approximated by a normal distribution ξ,

$$\xi \sim \mathrm{N}\left(\mu_\xi, s_\xi^2\right), \tag{22}$$

with mean μ_ξ and variance s_ξ^2 given by

$$\mu_\xi = \sum_{i=1}^{N} p_i, \tag{23}$$

and

$$s_\xi^2 = \sum_{i=1}^{N} p_i(1 - p_i). \tag{24}$$

One then calculates the probability of observing at least θ successful trials by integrating the Gaussian probability density function of ξ,

$$P\left(\sum_{j=1}^{N} x_j \geq \theta | t\right) = \int_{\theta}^{\infty} f_\xi \, dz = \int_{\theta}^{N} f_\xi \left(z | \mu_\xi, s_\xi^2\right) dz. \tag{25}$$

Solving the above integral yields

$$P\left(\sum_{j=1}^{N} x_j \geq \theta | t\right) = \frac{1}{2}\left[erf\left(\frac{\mu_\xi - \theta}{\sqrt{2s_\xi^2}}\right) - erf\left(\frac{\mu_\xi - N}{\sqrt{2s_\xi^2}}\right)\right], \tag{26}$$

where $erf(\cdot)$ is the error function. We verified that the estimate provided by Eq. 26 matched numerical simulations (Fig. 8).

By combining Eqs. 17 and 19, we computed the probability that a unit j is active given the occurrence of a network burst at a given time t,

$$P\left(x_j = 1 \middle| \sum_{j=1}^{N} x_j \geq \theta, t\right) = P\left(\sum_{j=1}^{N} x_j \geq \theta | t\right) P\left(x_j = 1 | t\right), \tag{27}$$

where independence of the two terms on the right-hand side arises in a network of large N. We examined the impact of the branching ratio (σ) on the probability of a given unit being active during a network burst (Eq. 27) (Fig. 9). We found that this probability was near zero for low values of σ (below 0.5) and near one for higher values (above 1.5). Given that it would be impossible for a readout to classify network

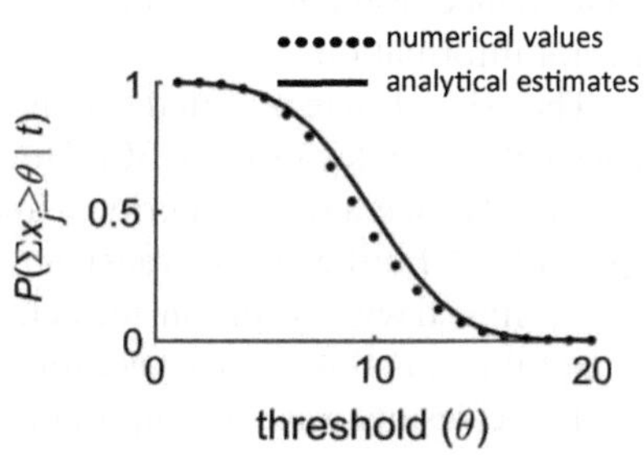

Fig. 8 Estimating the probability that the sum of spikes exceeds a given threshold Θ at time t (Eq. 26). The branching ratio was set to $\sigma = 1$

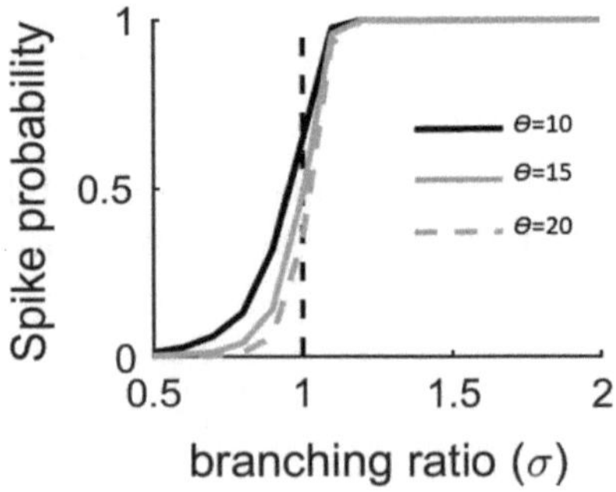

Fig. 9 Impact of branching ratio on the spiking probability of individual units during network bursts (Eq. 27). Vertical dashed line, $\sigma = 1$. Numerical values are averaged over units and time-steps

bursts when either no units are active or all units are active, a necessary condition for accurate classification is that this value is somewhere in between these two extremes. This is the case with networks around critical branching ($\sigma = 1$).

In sum, this analysis provides insights into the role of criticality in terms of "formatting" neural activity for downstream decoding. Through a set of probabilistic equations that capture the statistics of network bursts, we show that criticality yields network activity that is poised between overly active and inactive bursts. This form of activity is well suited for an accurate linear decoder.

5 Discussion

In this work, we considered whether a linear readout could accurately classify spontaneous neural activity both near and far from the critical state. Our results show that networks near the critical state were classified by a linear readout with near-perfect accuracy. Classification error increased in pharmacologically altered networks that shifted activity away from a critical state.

We captured these results using a branching process and showed that critical branching near $\sigma = 1$ was associated with highest readout accuracy. Together, our findings show that an accurate readout of neural activity by downstream populations depends upon the dynamical regime of this activity. Further, in both the experimental data and the model, readout accuracy was highest with dynamics near the critical state, hence arguing for a potential role of this state in facilitating the readout of neural information.

The critical state reported here is in line with previous in vitro literature as well as anesthetized states in vivo [30]. However, recent work suggests that awake animals transiently visit a range of critical and subcritical states on a timescale of seconds or longer [16]. Further, these transient visits may correlate with behavior: while drowsiness is linked with synchronized activity reminiscent of critical dynamics, heightened wakefulness is linked to asynchronous activity akin to subcritical dynamics [34].

These results raise an important question: if critical dynamics are not typically observed during active sensing, what is their functional role? One possibility is that critical dynamics arise during periods of low sensory awareness when the cortex spontaneously explores various network states [16]. During low sensory awareness, resting-state cortical activity may explore and reshape its dynamical repertoire, thus promoting memory consolidation [1], reinforcing key synaptic pathways [19], and regulating activity-dependent homeostatic plasticity [27]. This proposal is a departure from the thought experiment described in Fig. 1, where active sensing activates cortical representations.

The results provided in this chapter do not directly address the relative role of spontaneous versus task-evoked activity, as we do not record from awake behaving animals. We can, however, speculate on the following question: given that we record from spontaneously active in vitro networks, what does it mean that decoding is optimal near the critical state? Given the lack of reference to an external stimulus, the type of neuronal decoding described here is performed with respect to *intrinsic* cortical activity. We thus propose that criticality may allow the cortex to accurately process intrinsic brain activity during periods of low sensory awareness.

What would be the impact of *not* producing critical dynamics during periods of low sensory awareness? Speculatively, this might lead to poor communication between cortical regions. As a consequence, distinct cortical networks may fail to adequately reinforce distributed pathways representing complex, multi-modal concepts such as "cat", "house", or "flower" (Fig. 1). This proposal would benefit from experimental support where critical dynamics in vivo are altered during periods of low sensory awareness.

One prerequisite for the adequate decoding of network activity is that individual networks generate a restricted range of possible burst configurations. If every network explored the full range of possible spike combinations during bursts, the overlap across networks would prohibit classification. A restricted number of configurations makes it possible for downstream structures to accurately classify network dynamics. While it is difficult to estimate the full range of burst configurations that a network can generate, given the limited neurons and bursts recorded here, it is clear that these burst configurations are not drawn randomly and uniformly across different networks.

The classification strategy employed here—making use of network bursts—represents only one of the many potential routes for investigating neural decoding. Other approaches could attempt to decode neural information by applying dimensionality reduction [13] or by focusing on pairwise interactions as estimated by maximum entropy [35]. These approaches, however, come with their own set of caveats and are beyond the scope of this chapter.

Despite emerging evidence for the presence of critical-state brain dynamics, it remains unclear whether this form of neural activity reflects a mere by-product of biological organization (for instance, recurrent network activity) or whether it carries important functional consequences for cognition and behavior. Our results on neural decoding of in vitro activity suggest that criticality provides a useful format for brain dynamics to be accurately decoded by downstream structures. These findings are part of a larger effort to link critical brain dynamics to cognitive processing, memory formation, sensory discrimination, and motor control. Optimal linear decoding as described here may be instrumental to understanding and formalizing these links.

In conclusion, our work aims to fill a gap in the literature on neural criticality, which has been largely focused on encoding and information transmission, with relatively little attention paid to the role of critical dynamics on decoding. Ultimately, our goal is to seek neural codes that provide advantages for both sensory coding and cognitive/motor decoding [29]. A complete characterization of such codes constitutes a major pillar of research in modern neuroscience.

References

1. Albert, N.B., Robertson, E.M., Mehta, P., Miall, R.C.: Resting state networks and memory consolidation. Commun. Integr. Biol. **2**, 530–532 (2009)
2. Alstrøm, P.: Mean-field exponents for self-organized critical phenomena. Phys. Rev. A **38**, 4905–4906 (1988)
3. de Arcangelis, L., Herrmann, H.J.: Learning as a phenomenon occurring in a critical state. Proc. Natl. Acad. Sci. U. S. A. **107**, 3977–3981 (2010)
4. de Arcangelis, L., Perrone-Capano, C., Herrmann, H.J.: Self-organized criticality model for brain plasticity. Phys. Rev. Lett. **96**, 28107 (2006)
5. Benayoun, M., Cowan, J.D., van Drongelen, W., Wallace, E.: Avalanches in a stochastic model of spiking neurons. PLoS Comput. Biol. **6**, e1000846 (2010)
6. Berberian, N., MacPherson, A., Giraud, E., Richardson, L., Thivierge, J.-P.: Neuronal pattern separation of motion-relevant input in LIP activity. J. Neurophysiol. **117**, 738–755 (2017)
7. Braitenberg, V., Schüz, A.: *Cortex: statistics and geometry of neuronal connectivity* (Springer, 2013)
8. Buonomano, D.V., Maass, W.: State-dependent computations: spatiotemporal processing in cortical networks. Nat. Rev. Neurosci. **10**, 113–125 (2009)
9. Calderini, M., Zhang, S., Berberian, N., Thivierge, J.-P.: Optimal readout of correlated neural activity in a decision-making circuit. Neural Comput. **30**, 1573–1611 (2018)
10. Christensen, K., Olami, Z.: Sandpile models with and without an underlying spatial structure. Phys. Rev. E **48**, 3361–3372 (1993)
11. Clauset, A., Shalizi, C.R., Newman, M.E.: Power-law distributions in empirical data. SIAM Rev. **51**, 661–703 (2009)
12. Clawson, W.P., Wright, N.C., Wessel, R., Shew, W.L.: Adaptation towards scale-free dynamics improves cortical stimulus discrimination at the cost of reduced detection. PLoS Comput. Biol. **13**, e1005574 (2017)
13. Cunningham, J.P., Yu, B.M.: Dimensionality reduction for large-scale neural recordings. Nat. Neurosci. **17**, 1500–1509 (2014)
14. Friedman, N., Ito, S., Brinkman, B.A.W., Shimono, M., DeVille, R.E.L., Dahmen, K.A., Beggs, J.M., Butler, T.C.: Universal critical dynamics in high resolution neuronal avalanche data. Phys. Rev. Lett. **108**, 208102 (2012)
15. Gautam, S.H., Hoang, T.T., McClanahan, K., Grady, S.K., Shew, W.L.: Maximizing sensory dynamic range by tuning the cortical state to criticality. PLoS Comput. Biol. **11**, e1004576 (2015)
16. Hahn, G., Ponce-Alvarez, A., Monier, C., Benvenuti, G., Kumar, A., Chavane, F., Deco, G., Frégnac, Y.: Spontaneous cortical activity is transiently poised close to criticality. PLoS Comput. Biol. **13**, e1005543 (2017)
17. Harris, K.D.: Neural signatures of cell assembly organization. Nat. Rev. Neurosci. **6**, 399–407 (2005)

18. Ihlen, E.A.F., Vereijken, B.: Interaction-dominant dynamics in human cognition: beyond 1/f(alpha) fluctuation. J. Exp. Psychol. Gen. **139**, 436–463 (2010)
19. Kelly, C., Castellanos, F.X.: Strengthening connections: functional connectivity and brain plasticity. Neuropsychol. Rev. **24**, 63–76 (2014)
20. Kuebler, E.S., Tauskela, J.S., Aylsworth, A., Zhao, X., Thivierge, J.-P.: Burst predicting neurons survive an in vitro glutamate injury model of cerebral ischemia. Sci. Rep. **5**, 17718 (2015)
21. Langlois, D., Cousineau, D., Thivierge, J.P.: Maximum likelihood estimators for truncated and censored power-law distributions show how neuronal avalanches may be misevaluated. Phys. Rev. E **89**, 12709 (2014)
22. LeBlanc, M., Angheluta, L., Dahmen, K., Goldenfeld, N.: Universal fluctuations and extreme statistics of avalanches near the depinning transition. Phys. Rev. E **87**, 22126 (2013)
23. Lübeck, S.: Universal scaling behavior of non-equilibrium phase transitions. Int. J. Mod. Phys. B **18**, 3977–4118 (2004)
24. Martin, K.A.C., Schröder, S.: Functional heterogeneity in neighboring neurons of cat primary visual cortex in response to both artificial and natural stimuli. J. Neurosci. **33**, 7325–7344 (2013)
25. McLachlan, G.: *Discriminant Analysis and Statistical Pattern Recognition* (Wiley, 2004)
26. Mehta, A.P., Mills, A.C., Dahmen, K.A., Sethna, J.P.: Universal pulse shape scaling function and exponents: critical test for avalanche models applied to Barkhausen noise. Phys. Rev. E **65**, 46139 (2002)
27. Ngodup, T., Goetz, J.A., McGuire, B.C., Sun, W., Lauer, A.M., Xu-Friedman, M.A.: Activity-dependent, homeostatic regulation of neurotransmitter release from auditory nerve fibers. Proc. Natl. Acad. Sci. U. S. A. **112**, 6479–6484 (2015)
28. Ohki, K., Chung, S., Ch'ng, Y.H., Kara, P., Reid, R.C.: Functional imaging with cellular resolution reveals precise micro-architecture in visual cortex. Nature **433**, 597–603 (2005)
29. Panzeri, S., Harvey, C.D., Piasini, E., Latham, P.E., Fellin, T.: Cracking the neural code for sensory perception by combining statistics, intervention, and behavior. Neuron **93**, 491–507 (2017)
30. Plenz, D., Thiagarajan, T.C.: The organizing principles of neuronal avalanches: cell assemblies in the cortex? Trends Neurosci. **30**, 101–110 (2007)
31. Rich, E.L., Wallis, J.D.: Decoding subjective decisions from orbitofrontal cortex. Nat. Neurosci. **19**, 973–980 (2016)
32. Shaukat, A., Thivierge, J.-P.: Statistical evaluation of waveform collapse reveals scale-free properties of neuronal avalanches. Front. Comput. Neurosci. **10**, 29 (2016)
33. Shew, W.L., Yang, H., Yu, S., Roy, R., Plenz, D.: Information capacity and transmission are maximized in balanced cortical networks with neuronal avalanches. J. Neurosci. **31**, 55–63 (2011)
34. Tan, A.Y.Y., Chen, Y., Scholl, B., Seidemann, E., Priebe, N.J.: Sensory stimulation shifts visual cortex from synchronous to asynchronous states. Nature **509**, 226–229 (2014)
35. Tang, A., Jackson, D., Hobbs, J., Chen, W., Smith, J.L., Patel, H., Prieto, A., Petrusca, D., Grivich, M.I., Sher, A., et al.: A maximum entropy model applied to spatial and temporal correlations from cortical networks in vitro. J. Neurosci. **28**, 505–518 (2008)
36. Thivierge, J.-P.: Scale-free and economical features of functional connectivity in neuronal networks. Phys. Rev. E **90**, 22721 (2014)
37. Thivierge, J.-P., Cisek, P.: Nonperiodic synchronization in heterogeneous networks of spiking neurons. J. Neurosci. **28**, 7968–7978 (2008)
38. Tomen, N., Rotermund, D., Ernst, U.: Marginally subcritical dynamics explain enhanced stimulus discriminability under attention. Front. Syst. Neurosci. **8**, 151 (2014)
39. Vincent, K., Tauskela, J.S., Thivierge, J.-P.: Extracting functionally feedforward networks from a population of spiking neurons. Front. Comput. Neurosci. **6**, 86 (2012)
40. Vincent, K., Tauskela, J.S., Mealing, G.A., Thivierge, J.-P.: Altered network communication following a neuroprotective drug treatment. PLoS One **8**, e54478 (2013)
41. Zapperi, S., Bækgaard, Lauritsen K., Stanley, H.E.: Self-organized branching processes: mean-field theory for avalanches. Phys. Rev. Lett. **75**, 4071–4074 (1995)

Eric S. Kuebler studied psychology at Laurentian University, Canada and received his Ph.D. in experimental psychology from the University of Ottawa in 2018. He has worked on characterizing single neurons within neural circuitry as a Postdoctoral researcher at Robarts Research Institute of Western University, Canada.

Matias Calderini finished his Bachelor's degree in physics and psychology at the University of Ottawa, Canada in 2017. He combined his two fields of training during his Master's degree at the University of Ottawa in experimental psychology, specifically on behavioral neuroscience and computational neuroscience. His research interests include neural population dynamics and decision-making systems.

Philippe Lambert studied psychology at the University of Ottawa in Ottawa, Canada; receiving an Honours Bachelor of Arts in Psychology with a minor in Criminology in 2018. During this time, he has worked as a research assistant to Dr. Jean-Philippe Thivierge at the University of Ottawa where he primarily explored topics surrounding the critical dynamics found within neural systems.

Jean-Philippe Thivierge received his Ph.D. from McGill University in 2006. He worked as a postdoctoral fellow at Indiana University before moving to the University of Ottawa where he is now an Associate Professor of Psychology. He employs both experimental and computational approaches to study neural dynamics in cortical circuits of the brain.

Critical Behavior and Memory Function in a Model of Spiking Neurons with a Reservoir of Spatio-Temporal Patterns

Silvia Scarpetta

Abstract Two intriguing phenomena characterize cortical dynamics both in-vitro and in-vivo: (1) the memory function, with the cue-induced and spontaneous reactivation of precise dynamical spatio-temporal patterns of spikes, and (2) criticality and scale-free neuronal avalanches, characterized by power law distributions of bursts of spikes both in-vitro and in the resting spontaneous activity. In this paper we review recent results which link together both these features, memory function and critical behavior, in a model with leaky neurons whose structured connectivity comes from learning multiple spatio-temporal phase-coded patterns using a rule based on Spike-Timing-Dependent Plasticity. We first study the stimulus-driven replay of stored patterns, measuring the storage capacity, i.e. the number of spatio-temporal patterns of spikes that can be stored and selectively reactivated. Then we focus on the noise-evoked spontaneous dynamics, in absence of stimulation. Collective patterns, which replay some of the patterns used to build the connectivity, spontaneously emerge. Near the phase transition between persistent replay and no-replay regimes of the spontaneous dynamics, critical phenomena and neural avalanches are observed, with critical exponents close to the ones experimentally measured. Previous studies have separately addressed the topics of phase-coded memory storage and neuronal avalanches, and this is one of the few works which show how these ideas converge in a single cortical model. This work therefore helps to link the bridge between criticality and the need to have a reservoir of spatio-temporal metastable memories.

1 Introduction

Many experimental results support the idea that cortical resting activity has the signature of a system posed near a critical point [1–7], as reflected by scale-free distributions of neural avalanches. The functional role of being near the critical

S. Scarpetta (✉)
Department of Physics, University of Salerno, Fisciano, Italy
e-mail: sscarpetta@unisa.it

INFN unit. Napoli grupp.coll. di Salerno, Fisciano, Italy

N. Tomen et al. (eds.), *The Functional Role of Critical Dynamics in Neural Systems*, Springer Series on Bio- and Neurosystems 11,
https://doi.org/10.1007/978-3-030-20965-0_10

instability has been investigated from the point of view of the ability of the brain to respond to a wide range of inputs, to process the information in an optimal way [8–10], and to enhance stimulus discriminability [11]. In this paper we review our recent results which link together the associative memory function and the critical scale-free neural avalanches observed in the spontaneous dynamics of cortex.

Notably, an important feature of cortical activity, reported in a variety of in vivo studies, is the similarity of spontaneous and sensory-evoked activity patterns [12–14]. Recently the spontaneous and evoked activity similarity has been reliably observed also in dissociated cortical networks [15], reinforcing the idea that the emergence of cortical recurring patterns, both during spontaneous and evoked activity, is the results of the cortical connectivity (with its micro-circuits and dynamical attractors) [15, 16], which result in the similarity of the patterns.

Many in-vivo results shows that repeatable precise spatiotemporal patterns of spikes play a crucial role in coding and storage of information. Temporally structured replay of spatiotemporal patterns have been observed to occur in the awake state [17] and during sleep [18], both in the cortex and hippocampus [19], and it has been hypothesized that this replay may subserve memory consolidation. Among repeating patterns of spikes a central role is played by phase-coded patterns [20–22], i.e. periodic patterns with precise relative phases of the spikes of neurons participating to a collective oscillation, or precise phases of spikes relatively to the ongoing oscillation.

We model the cortical activity over the course of a learning phase of periodic spatiotemporal patterns and after the learning phase with connections fixed, comparing stimulus-evoked activity from noise-evoked spontaneous dynamics. The model undergoes a discontinuous phase transition between replay and no-replay of stored patterns, mediated by excitability threshold. The model have the following features (1) a learning rule inspired by spike-time-dependent-plasticity, which, as opposed to symmetric Hebbian rules, allows for dynamical attractors to arise in the network, (2) two firing thresholds to account for the heterogeneity of neuron excitability and (3) poissonian background noise.

Multiple periodic spatio-temporal patterns (periodic spike sequences) are store in the connectivity as dynamical metastable attractors. The learning rule inspired by the spike-time dependent plasticity (STDP) observed in neocortical and hippocampal pyramidal cells [23–26] is particularly suitable to store spatiotemporal patterns. While within the class of symmetric hebbian rules the associative memory dynamics converge towards a static patterns, the plasticity which depends on the relative timing of pre- and post-synaptic firing underscores the importance of timing and dynamics, and it's able to stores dynamical attractors, as opposed to static ones.

The proposed associative memory approach, with replay of the stored sequences, can be a method for recognize an item, by activating the same memorized pattern in response of a similar input, or could be a method to transfer the memorized item to another area of the brain (such as for memory consolidation during sleep).

As studied before [27–31], the network is able to selectively replay each of the stored patterns when a partial cue (a short stimulus similar to the patterns) is presented. We review the results on the storage capacity of the system, comparing the capacity of this network (for dynamical patterns) with the storage capacity of the Hopfield model (for static patterns).

Even in absence of cue stimulation [32–34] the noise (focused by leader neurons and disordered structure of connections) may generate a spreading replay of a spatio-temporal pattern (randomly chosen between the encoded ones).

At the critical value of the threshold, that marks the transition from the metastable self-sustained replay regime to the no-replay regime, and viceversa, we observe critical phenomena precursors of the discontinuous transition [35], with scaling invariance and large fluctuations, and power laws in the size and duration of the avalanches, with critical exponents similar to the one experimentally observed [6]. Interestingly, the scale-free neural avalanches observed near this critical threshold are composed of short replays of some of the stored patterns. Notably this is in agreement with the observation that experimentally neuronal avalanches are highly repeatable [36–38] and can be clustered into statistically significant families of activity patterns that satisfy several requirements of a memory substrate.

The systems is therefore both able to work as a stimulus-activated reservoir of spatio-temporal attractors, in the region where dynamic attractors are metastables with long lifetimes, and as a more flexible device when used at the border of the critical instability. Indeed at the critical threshold the normalized variance of the spontaneous dynamics has a maximum, and the susceptibility, which is the response to a stimulus, is also maximal.

2 Emerging of Collective Dynamics

The model, described in the Appendix, is composted of N coupled leaky integrate-and-firing (LIF) units with connectivity dictated by a STDP-based learning of P periodic spatio-temporal patterns of spikes. The emerging complex collective dynamics which makes the net able to work as associative memory is the result of the connectivity built in the learning stage, while the single units are simply LIF units. To take into account the heterogeneity of the neurons we use two values for the firing threshold, a value θ_{th} for the majority of the units, and a lower value θ_{th1} for few, more excitable, units. To model the spontaneous neurotransmitter release at individual synapses, as well as other sources of irregular background synaptic noise, we include a noise term to the membrane potential, whose times and strengths are extracted randomly and independently for each unit. The times of noise contribution are poissonian, while noise strengths are extracted from a Gaussian distribution with zero mean and standard deviation σ_{noise}.

We distinguish two stages: a learning stage in which plasticity rule in Eqs. (9, 10), inspired to the STDP, is used to store P phase-coded patterns into the network connectivity, from a dynamic stage (or retrieval stage) in which connections J_{ij} are fixed to the value found after learning, and the collective dynamics of the neurons is studied. This distinction in two stages (learning and retrieval), even though is not well assessed in real neural dynamics, is useful to simplify the analysis and also finds some neurophysiological motivations [39, 40]. Even with fixed connections J_{ij} the emerging dynamics is not trivial, showing associative memory functioning and an interesting interplay between criticality and memory replay when spiking threshold is close to critical value corresponding the the phase transition between sustained replay and no replay. The dynamics when connections J_{ij} are not fixed but plastic will be even richer, and will be studied in a future work.

In the following we study both the stimulus-evoked dynamics (Sect. 2.1) and the spontaneous dynamics (Sect. 2.2).

2.1 Cue-Evoked Replay and Storage Capacity

The LIF network, with the connectivity that comes from the learning procedure described in the appendix, is able to work as an associative memory for periodic spatio-temporal patterns of spikes. Each pattern is a periodic train of spikes, with one spike per neuron and per cycle, with the neuron i firing at times $t_i^\mu + nT^\mu$, with t_i^μ randomly and uniformly extracted in the interval $[0; T^\mu]$; T^μ is the period, and $\phi_i^\mu = 2\pi t_i^\mu / T^\mu$ are the phases encoded. As studied in [27], a short cue-stimulation, with M spikes with the proper phase relationship, is able to induce the selective replay of the stored pattern.

We first review the cue-evoked dynamics when P patterns with period $T^\mu = 333$ ms are stored, and we give a short cue stimulation with M spikes, with $M = N/10$, at times $t_i^\mu = T_{stim}\phi_i^\mu/2\pi$, $i = 1, \ldots, M$, with $T_{stim} = 50$ ms, where $\phi_1^\mu, \ldots, \phi_M^\mu$ are M consecutive phases in pattern μ. Then, in Sect. 2.2 the behavior without any cue stimulation (M = 0) is studied, in presence of noise.

Figure 1 shows the network dynamics at different spiking threshold, when $M = N/10$, in absence of noise ($\sigma_{stim} = 0$). In the example shown in Fig. 1a and b, the short stimulation (which lasts less then 5 ms, shown in green in all the figures) has the effect to selectively trigger the sustained replay of pattern μ. At high threshold, such as in Fig. 1c, the cue stimulation only triggers a short transient replay, and at even higher threshold no spontaneous activity is triggered by the stimulation.

Depending on the partial cue presented to the network, a different collective activity emerges with the phases of the firing neurons which resemble one or another of the stored patterns. Indeed, in the regime of correct replay, shown in Fig. 1a and b, we check that if the partial cue is taken from pattern $\mu = 2$, the neurons phase relationships recall the phase of pattern $\mu = 2$ (uncorrelated with pattern $\mu = 1$) (not shown).

Note that, after the short cue presentation, the retrieval dynamics has the same phase relationship among units than the stored pattern, but the reactivation may happen on a time scale different from the scale used to store the pattern (and also different from the period T_{stim} of the cue), and the collective dynamics is a time compressed (or dilated) replay of the stored pattern. The time scale of replay changes slightly with spiking threshold. In the example of Fig. 1a and b, the time scale of the

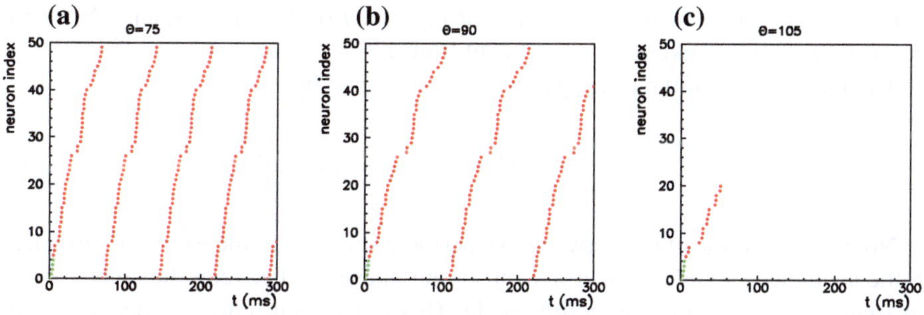

Fig. 1 Cue-induced network dynamics at different values of spiking threshold θ_{th} in absence of noise ($\sigma_{noise} = 0$). In these simulations a cue external stimulation (M spikes, shown in green) is used, and a single value of spiking threshold is used for all units of the network. The network has $N = 3000$ IF neurons, $M = N/10$ spike of stimulation, $\theta_{th} = 75, 90, 105$ respectively in **a, b** and **c**. **a, b** At proper value of the spiking threshold the cue initiates a persistent retrieval of the spatio-temporal pattern, while at too high threshold (**c**) the network is silent or with only a short transient retrieval. Example of selective successful retrieval (**a, b**) and example of failure (**c**) are shown. The raster plot of 50 units (randomly chosen) are shown sorted on the vertical axis according to increasing values of phase ϕ_i^1 of the first stored pattern $\mu = 1$. Connections are given by Eqs. (9, 10) with $P = 2$ stored patterns at $\nu^{\mu} = 3$ Hz

retrieval dynamics is faster at lower threshold, and in both cases it is different from the time scale used to learn the patterns.

The storage capacity, defined as the maximum number of encoded and successfully retrieved patterns, as a function of the spiking threshold of the units, is studied.

To measure quantitatively the success of the cue-induced self-sustained retrieval, we introduce an order parameter, in analogy with the Hopfield model, which estimates the overlap between the network emerging activity and the stored phase-coded pattern. This quantity is 1 when the phases ϕ_j of spike timings coincide with the stored phases ϕ_j^{μ}, and is close to zero when the phases are uncorrelated with the stored ones. Therefore, we consider the following time-dependent dot product $|M^{\mu}(t)| =< \xi(t)|\xi^{\mu} >$ where ξ^{μ} is the vector having components $e^{i\phi_j^{\mu}}$, namely:

$$|M^{\mu}(T^w, t)| = \left| \frac{1}{N} \sum_{\substack{j=1,\dots,N \\ t<t_j^*<t+T^w}} e^{-i2\pi t_j^*/T^w} e^{i\phi_j^{\mu}} \right| \tag{1}$$

where t_j^* is the spike timing of neuron j during the emerging dynamics, and T^w is an estimation of the period of the emerging dynamics.

Then, we consider the following order parameter:

$$m^{\mu}(T^w) = \frac{1}{\langle N_s \rangle} \langle |M(t, T^w)| \rangle \tag{2}$$

where the average $\langle \cdots \rangle$ is done on the starting time t of the window, and $\langle N_s \rangle$ is the average number of spikes on a window of time T^w.

The fluctuations of the order parameter are defined by

$$\sigma(m^\mu)^2 = \frac{1}{\langle N_s \rangle^2} \left[\left\langle \left| M(t, T^w) \right|^2 \right\rangle - \left\langle \left| M(t, T^w) \right| \right\rangle^2 \right]. \tag{3}$$

Note that the value of m^μ between two periodic spike trains measures the similarity in the sequence of spiking neurons and in the phase lag between the spikes, being invariant by a simple change in time scale. This is a suitable choice especially when the replay of a spatio temporal pattern has to be detected independently from the compression of the time scale. Note that if we have a spike train that is not periodic, we cannot define the period, however we can define the order parameter (2) looking at the time-window T^w which maximize the order parameter. This can be useful in the case when one looks for a short replay hidden in a not-periodic spike train, such as here and in many experimental situations. The overlap in Eq. (2) is equal to 1 when the phase-coded pattern is perfectly retrieved (i.e. same sequence and phase relationships among spikes, even though on a different time scale), while is of order $\simeq 1/\sqrt{N}$ when phases of spikes are uncorrelated to the stored phases.

The order parameter $M^\mu(t, T^w)$ at large times, and its average $m^\mu(T^w)$, for the optimal T^w, has been used to measure the network storage capacity in the space of parameters θ_{th} and $\nu^\mu = 1/T^\mu$ in [27]. The same order parameter $m^\mu(T^w)$ and its fluctuations have been used to study the phase transition between spontaneous replay and no-replay in [32].

The storage capacity, as defined as usual as the ratio P_{max}/N, is shown as a function of the spiking threshold θ_{th}, in Fig. 2, when $M = N/10$ spikes are used as cue stimulation. P_{max} is the maximum value of P, for which the cue is able to selectively

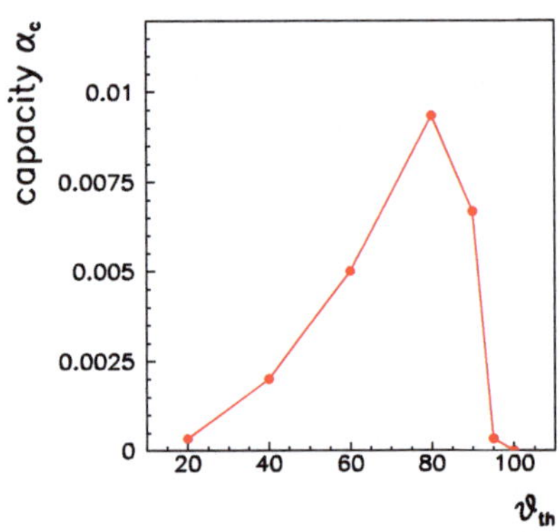

Fig. 2 Storage capacity $\alpha^c = P_{max}/N$ is shown as a function of spiking threshold θ_{th}, when $M = N/10$ spikes are used as cue stimulation. The number of units is $N = 3000$, $T^\mu = 333$ ms, $T_{stim} = 50$ ms, noise strength $\sigma_{noise} = 0$. The figure shows that near $\theta_{th} \simeq 95$ there's a transition from a region of cue-evoked persistent replay to a region of no cue-evoked replay

activate the persistent replay of the stored pattern with an overlap $m^\mu > 0.5$. We fixed the desired similarity value to 0.5, however the whole storage capacity analysis is very robust with respect to this parameter, since in our numerical simulations m^μ has mostly two possible values, close to one (success) or close to $1/\sqrt{N}$ (failure).

The storage capacity, evaluated as P_{max}/N, has a maximum value close to 0.01, which is lower than the storage capacity of the Hopfield model. However, the information content of a single pattern in our dynamical model with N units is higher than the information content of a pattern in the Hopfield model with N-units. Indeed, an Hopfield pattern is only a set of N binary values, while our phase-coded spatio-temporal pattern is a set real number $\phi_j^\mu \in [0 : 2\pi]$ setting the spike timing differences. Moreover coding through spike phase-relationship leaves the possibility to encode ancillary informations in the population rate. Preliminary results shows that a mixed phase- and rate-coding, where only part of the units participates to each of the pattern and the other part of the units is silent in that pattern, is able to reach much higher value of storage capacity.

Notably for $P < P_{max}$ the replay is robust wrt noise (see [27] for details). Indeed stored pattern is an attractor of the network dynamics, that is the dynamics spontaneously goes back to the retrieved phase-coded pattern for all the perturbations which leave the system within the basin of attraction. Therefore phase errors do not sum up, and the phase relationships may be transferred and kept stable over long time scales.

The storage capacity in Fig. 2 is maximal at a value of the threshold around $\theta_{th}^{max} = 80$. At too low threshold, the excitability of the network is too high, and when $P > P_{max}(\theta_{th})$ the network responds with a persistent activity that is not correlated with any of the stored patterns, i.e. a spurious state. At threshold around $\theta_{th} = 95$ the capacity drops off and the network has a transition to a regime where no self-sustained activity is possible for any value of P in response to the cue stimulation. Near this value the cue-induced responses are short transient (non-persistent) replay responses [27], while at even larger threshold the network is unresponsive.

This critical regime, near the phase transition between self-sustained replay and no replay regimes, is further investigated in the following, in the absence of any external cue stimulation ($M = 0$).

2.2 *Spontaneous Collective Dynamics*

While in previous subsection we studied the dynamics triggered by a short stimulation cue, in the following we focus on the spontaneous dynamics without any cue stimulation, in the presence of Poissonian noise coupled with strength of standard deviation σ_{noise}.

While in the region of persistent replay the evoked dynamics is robust wrt noise [27], in the sense that once the dynamics falls in the basin of attraction of one pattern the replay is persistent and robust wrt noise, near the critical point between persistent-replay and no-replay, the system is more sensible to noise. Therefore it is important

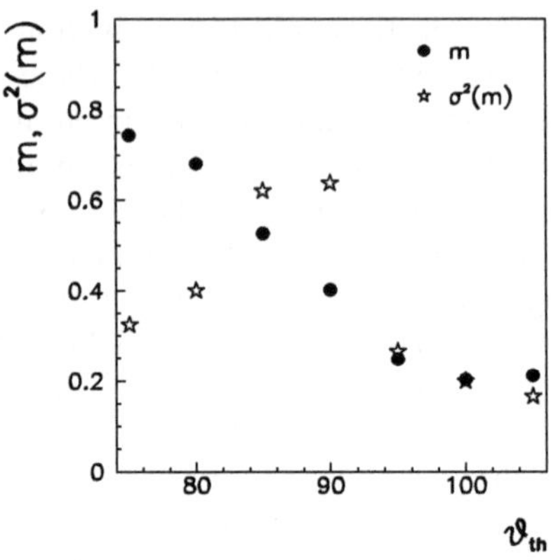

Fig. 3 Order parameter $m(T^W)$, at optimal T^w, and its fluctuations as a function of θ_{th}, of the noise-evoked activity for $N = 3000$, $P = 2$, $\theta_{th1} = 26$, $\nu^\mu = 3$ Hz and noise $\sigma_{noise} = 6$. There is no cue-stimulation ($M = 0$). Note that at the critical point ($\theta_{th} = 90$) between the two regimes, the fluctuations of the order parameter are maximized, as expected near a second order phase transition or near the spinodal line of a first order transition

to study the effects of noise in this transition region, and we therefore will focus on the spontaneous dynamics in the absence of cue stimulation, in presence of only poissonian noise.

Notably we find that, in absence of any external stimulation, it is the noise itself which may initiate a alternation of short replays of stored patterns and quite states (up/down alternation) at the critical value of the threshold, while at lower threshold (higher excitability) the noise is able to induce a persistent replay of the stored patterns (as in the cue-induced case). This is in agreement with the fact that the spontaneous and evoked activity similarity has been reliably observed experimentally [12–15].

In Fig. 3 the mean value of order parameter m^μ and its fluctuations are shown as a function of spiking threshold θ_{th}, when $\theta_{th1} = 26$ and for the optimal T^w, when spontaneous dynamics is studied ($M = 0$, $\sigma_{noise} = 6$). At low spiking threshold the order parameter is high and fluctuations are low, indicating that the noise is able to initiate a successful retrieval (persistent replay) of one of the patterns which are stored into the connections. At high threshold both order parameter and its fluctuations are low. At the critical point ($\theta_{th} = 90$) between the two regimes, the fluctuations of the order parameter are maximized, as expected near the instability of a phase transition.

Notably recent results [41] shows, using large scale fMRI measures, that the brain spent most of the time near a critical point, with maximization of fluctuations. Interestingly, in a recent paper [35] we observe in our model the presence of metastability and hysteresis, which indicate that the transition in Fig. 3 is a non-equilibrium first order one. Maximization of fluctuations, scale-free and critical behaviour in cortical dynamics are frequently associated with second-order (continuous) phase transitions. However power law and critical phenomena also emerge in first-order (discontinuous) phase transitions as one enters the metastability region and approaches the

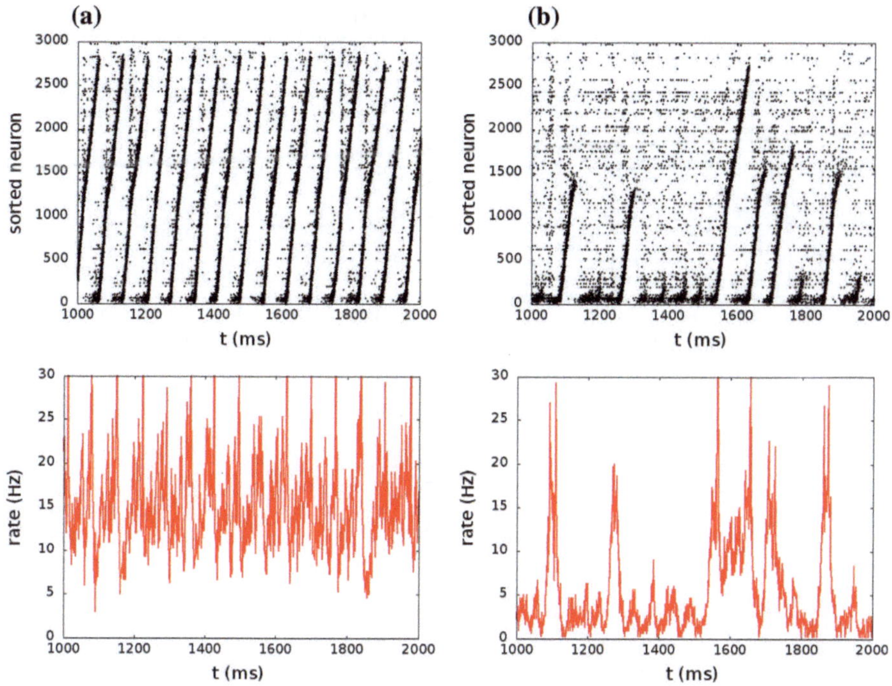

Fig. 4 Raster plot (up) and corresponding firing rate (bottom) as a function of time, of the spontaneous dynamics of the network at spiking threshold $\theta_{th} = 80, 90$, respectively, in panels **a**, **b**. As in previous figure, $N = 3000$, $P = 2$, $\theta_{th1} = 26$, $M = 0$, and noise $\sigma_{noise} = 6$. The raster plots are shown on the upper panels, with neurons sorted according to increasing values of the phases in pattern $\mu = 1$, and the corresponding firing rates (measures on a time window of 1 ms), as a function of time, are shown on the bottom panels. **a** At low threshold (high excitability), a persistent reactivation of one of the patterns (randomly chosen by the noise realization) emerges, in this case, pattern $\mu = 1$, as shown by the regular behavior in raster plot in (**a**). **b** At the critical value of the threshold $\theta_{th} = 90$, the noise is able to initiate an intermittent short collective replay of a pattern (in this case pattern $\mu = 1$). UP/DOWN alternation is shown in (**b**), with alternation of states of quiescence with states of replay activity is shown. During the states of short collective replay, the firing rate is high, while the firing rate is low in the quite periods

spinodal line, in systems with long range interactions [42, 43]. We show in [35] that the maximization of fluctuations in Fig. 3, and the up-down alternation with scale-free features described in Sect. 2.2.1, are indeed associated with the spinodal line of an underlying discontinuous transition, when one enters the metastable region from below.

In Figs. 4 and 5 it is shown the raster plot (and the firing rate) of the spontaneous dynamics of the network used in Fig. 3, at spiking threshold $\theta_{th} = 80, 90, 110$ respectively, in Figs. 4a, b, 5a. In the low threshold regime, the noise-evoked activity triggers the sustained replay of one of the patterns encoded in the network (Fig. 4a). The intermediate regime, corresponding to the critical value of spiking threshold $\theta_{th} = 90$, is shown in Fig. 4b. In this case the noise is able to start the replay of

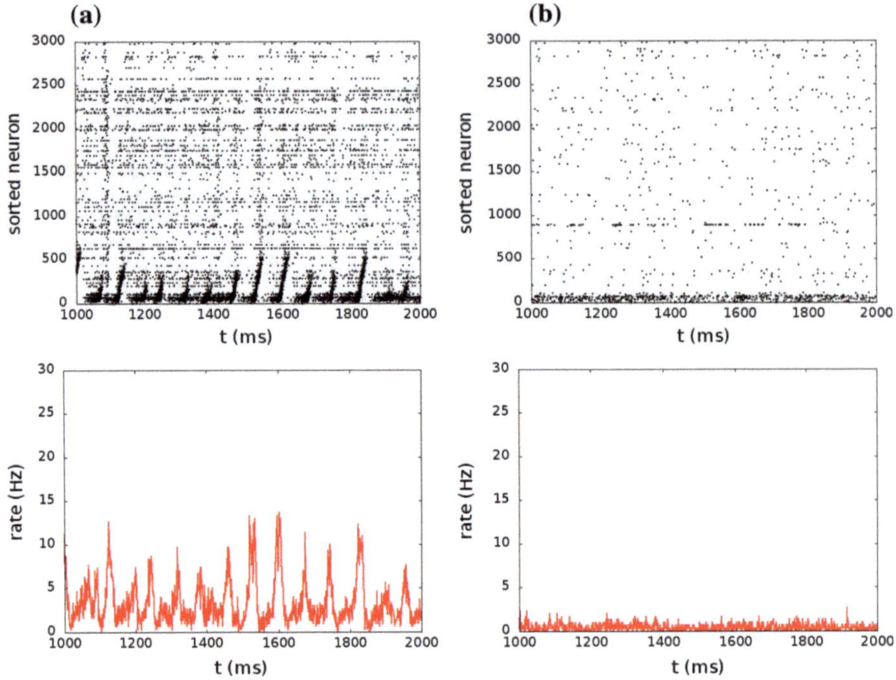

Fig. 5 The spontaneous dynamics as in previous figure, but at high values of spiking thresholds. $\theta_{th1} = 26$, and $\theta_{th} = 110$ in (**a**), while we use a higher value of leaders spiking threshold $\theta_{th1} = 40$, and $\theta_{th} = 120$ in (**b**). The potential of neurons is governed by a Ornstein-Uhlenbeck process, and with some probability crosses the threshold giving rise to a spike, that is not able, however, to generate a spreading activity in the network

a pattern, but also to stop it, so that the activity is intermittent, and resembles the experimentally observed alternation of up and down states [44]. The firing rate is high in the UP state (during short replays) and is low during DOWN states, therefore the distribution of the rates is bimodal [33]. At high threshold (Fig. 5a) the interplay between noise and the leader units generated only very short replay. When also the leader's spiking threshold θ_{th1} is higher (Fig. 5b), the dynamical behavior is dominated by noise, and the rate distribution is nearly exponential with a maximum at zero rate.

The maximization of the order parameter's fluctuations at the critical threshold $\theta_{th} = 90$ in Fig. 3 is in agreement with the UP/DOWN alternation, with intermittent replay, that we observe at $\theta_{th} = 90$ in Fig. 4b. To further investigate the dynamical features near the critical point, in the next subsection we measure the neural avalanche distribution to look for scale-free behaviour.

2.2.1 Scale-Free Behavior

To characterize the spontaneous dynamics near the critical point, corresponding to maximization of fluctuations (Fig. 3) and the UP/DOWN alternation shown in Fig. 4b, we measure the neural avalanches distribution. According with original work of Beggs and Plenz [1] we define an avalanche as a sequence of spikes preceded and succeeded by a time interval of length at least τ^{min} without any spikes. The value of τ^{min} has been chosen looking at the network Inter Spike Interval (ISI) distributions, as a value greater than the short timescale of ISIs within an event but shorter than the timescale of the longer quiescent periods, which are not distributed exponentially. Therefore, we take a value of $\tau^{min} = 0.03$ ms as the time at which the ISI distribution at the critical point deviates from the initial exponential behavior. Figure 6 shows the distribution of the sizes (number of spikes) and durations of the avalanches, $P(S)$ and $P(T)$, as well as the mean size as a function of duration $<S>(T)$, for $N = 20000$, $\theta_{th1} = 26$, $\nu^{\mu} = 3$ Hz and noise $\sigma_{noise} = 6$, at thresholds $\theta_{th1} = 75, 90, 110$. At the critical value of the threshold, the distributions approach a power law with critical exponents $\beta = 1.55$ for size and $\alpha = 1.63$ for duration, and the average size $<S>(T)$ as a function of duration approaches a power law with exponent $k = 1.14$, which notably satisfies the relation, initially derived in Ref. [45] in relation to crackling noise,

$$\frac{\alpha - 1}{\beta - 1} = k \tag{4}$$

It is interesting that relation (4) between the critical exponents is verified in our model and also experimentally, both in-vivo [46] and in-vitro [5], while it is not verified in models where power law is not a manifestation of a critical point [47].

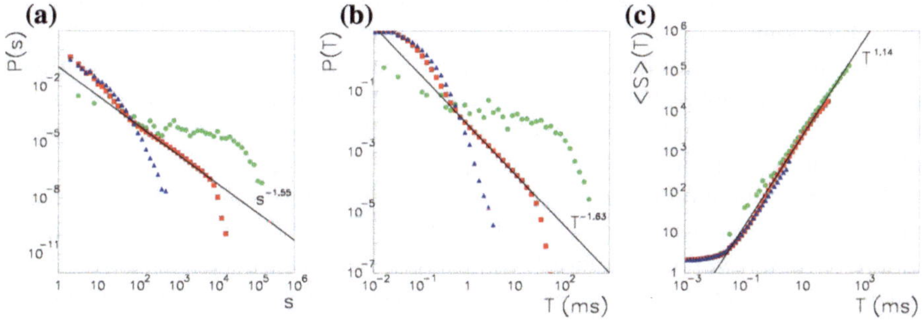

Fig. 6 Avalanches distributions at three values of the spiking threshold: $\theta_{th} = 110$ (blue circles), $\theta_{th} = 90$ (red squares), $\theta_{th} = 75$ (green triangles). The distribution of avalanche's size (**a**) and duration (**b**), and the mean size as a function of duration (**c**) are shown for a network with $N = 20000$, $\sigma_{noise} = 6$, $M = 0$, $\theta_{th1} = 26$. At the critical point the size and duration distribution approach a power law, with exponents respectively $\beta = 1.55$ and $\alpha = 1.63$. The average size $<S>(T)$ as a function of duration approach a power law with exponent $k = 1.14$. (See [32] for details)

Notably recent results [35] shows that the phase transition between the self-sustained replay regime and the no-replay regime (shown in Fig. 3) is a non-equilibrium first order transition. Interestingly first-order transitions are usually associated with memory, as in the Hopfield model. The scale-free behavior observed at the critical value of the threshold could be read, therefore, as the critical behavior that one expects near the spinodal line of a first order transition [48]. As shown in [35] the cut off of the power law in Fig. 6a, b scale with the system size N, indicating a scale-free behavior.

3 Conclusion

In this paper we review a simple Leaky Integrate and Fire model where stimulus-evoked patterns and spontaneous noise-evoked patterns are similar, since cortical connectivity (with its stored dynamical attractors) imposes common constraints on spontaneous and evoked activity flow. The learning rule, based on the STDP plasticity, makes the network able to work as associative memory, replaying each of the periodic spatio-temporal patterns that have been stored in a cue-selective manner. Near the critical point, i.e. near the edge of instability of spontaneous self-sustained replay regime, the spontaneous network activity wanders through the various activity patterns which are encoded in the connectivity, with alternation of up and down states. Near the edge of this instability, neural avalanches with power law distribution are observed. Notably the exponents of size and duration distribution, and the exponent of the average size as a function of duration, are in agreement with the experimental observations [5, 6, 46]. In the framework analyzed in this paper, the neural avalanches and the UP/DOWN alternations are both manifestation of a network with many dynamical attractors stored in the connectivity, posed near the instability point.

One may speculate that indeed to stay in the regime where the reactivation is not a persistent replay, but it's only a transient short replay, may be convenient, in terms of flexibility, and therefore a feedback mechanisms may regulates the degree of excitability of the network to put it near this critical line.

Near this criticality line indeed the susceptibility is maximal, and the response of the network is more various and flexible, and sensitive to novel experience, while the region of persistent replay at threshold lower then the critical value is more suitable when persistent activation of previous stored experience is needed. It is plausible that the brain is able to change its state, by changing the value of excitability (threshold) or with other mechanisms, going out of the critical region toward a supercritical state (more suited for either spontaneous or cue-triggered persistent reactivation of previous experience) or subcritical state (which favors faithful representation of sensory inputs) depending on the different behavioural state.

The coexistence of cue-evoked persistent replay and spontaneous critical dynamics in our model reminds the behaviour observed experimentally in Ref. [46], where spontaneous critical state is observed, while a transient state characterized by large non-critical avalanches is observed in response to an external stimulus.

Another model that study criticality together with associative memory was proposed in [49]. In their model, a Hebbian learning rule is used to store static patterns. However, they found that Hebbian learning alone destroys criticality even when the synaptic strength is properly scaled. Applying an optimization procedure that drives the synaptic couplings either toward the critical regime, or toward the memory state in an alternating fashion, they finally arrive at a configuration both critical and that retains an associative memory. The reason why in our model the learning procedure does not destroy criticality may be due to the difference in the learning rule, that in our case is based on STDP and stores dynamical attractors, as opposed to static ones.

The result of learning spatio-temporal patterns (each with quenched randomly-chosen phase ordering) gives rise to quenched disorder in the network connections structure, and the distribution of connection weight that comes out from our learning rule is very skewed and long-tailed [35]. Note however that in our model the distribution of the weights is not a sufficient condition to determine the dynamical phase transition. Indeed, when we shuffle [50] the connections leaving their distribution unchanged, this kind of transition disappears. It seems therefore that the topology of the network, with its relative abundance of motifs and its corresponding stored dynamical attractors, is crucial [50] for the manifestation of these kind of phase transition between sustained memory reactivation and no-reactivation and its critical features such as scale-free avalanches.

Appendix: The Model and the Learning Procedure

The model and the learning rule have been introduced in [27, 28, 30, 32, 34]. Here we briefly review them. We have N coupled Leaky Integrate-and-Fire (LIF) units with connectivity dictated by a STDP-based learning of P spatio-temporal patterns of spikes.

The post-synaptic membrane potential of neuron i, represented by the Spike-Response-Model [51], in presence of Poisson noise, is simply given by:

$$h_i(t) = \sum_j J_{ij} \sum_{t>t_j>t_i} \varepsilon(t - t_j) + \eta_i(t), \tag{5}$$

$$\eta_i(t) = \sum_{t>t_{i_{noise}}>t_i} J_{noise}(t_{i_{noise}})\varepsilon(t - t_{i_{noise}}). \tag{6}$$

where $\eta_i(t)$ is the Poissonian noise term, J_{ij} are the synaptic connections, $\varepsilon(t)$ describes the response kernel to incoming spikes, and the sum over t_j runs over all pre-synaptic firing times following the last spike of neuron i. The sign of the synaptic connection J_{ij} sets the sign of the post-synaptic potentials change. The noise times $t_{i_{noise}}$ are random and independent times, extracted from a Poissonian distribution $P(\delta t) \propto e^{-\delta t/\tau_{noise}}$, with $\tau_{noise} = 1$ ms, while the strength J_{noise} is extracted for each time $t_{i_{noise}}$ from a Gaussian distribution with mean $\bar{J}_{noise} = 0$ and standard deviation σ_{noise}. The response kernel $\varepsilon(t)$ is given by

$$\varepsilon(t) = K\left[\exp\left(-\frac{t}{\tau_m}\right) - \exp\left(-\frac{t}{\tau_s}\right)\right]\Theta(t) \tag{7}$$

where τ_m is the membrane time constant (here 10 ms), τ_s is the synapse time constant (here 5 ms), $\Theta(t)$ is the Heaviside step function, and K is a multiplicative constant chosen so that the maximum value of $\varepsilon(t)$ is one.

When the membrane potential $h_i(t)$ exceeds the spiking threshold θ^i_{th}, a spike is scheduled, and the membrane potential is reset to the resting value zero. To take account of the heterogeneity of the neurons, we use two values of θ^i_{th}: a low threshold θ_{th1} for $N_1 \ll N$ units, to model units more sensible to noise ("leader units"), and a higher threshold θ_{th} for the others $N_2 = N - N_1$ units. The leaders units (modeled here with lower threshold), are neurons that fire more than other ones in response to noise, and give rise to a cue able to initiate the short collective replay. They are chosen as a fraction of 3% of neurons with consecutive phases, for each pattern. The presence of a few highly active sites, driving cortical neural activity (leaders), has been reported experimentally [52–54].

The emerging network dynamics is determined by the connectivity. Connections J_{ij} are determined via a learning rule inspired by STDP [23, 26, 55], where long-term potentiation and depression of the synaptic efficacies depends on the relative timing of the pre- and post-synaptic activities. In the learning stage, we force the network to store P spatio-temporal patterns of spikes.

According to the learning model [29–31, 56, 57] the connection change J_{ij} that occurs in the time interval [−T, 0] to learn activity pattern $y^{\mu}(t)$ can be modelled as follows:

$$\delta J^{\mu}_{ij} = 1/N \int_{-T}^{0}\int_{-T}^{0} y^{\mu}_i(t)A(t-t')y^{\mu}_j(t')dtdt' \tag{8}$$

where $y^{\mu}_j(t)$ is the activity of the pre-synaptic neuron at time t, and $y^{\mu}_i(t)$ the activity of the post-synaptic one. The learning window $A(\tau)$ is the measure of the strength of synaptic change when a time delay τ occurs between pre and post-synaptic train. To model the experimental results of STDP, the learning window $A(\tau)$ should be an asymmetric function of τ, mainly positive (LTP) for $\tau > 0$ and mainly negative (LTD) for $\tau < 0$. The learning window $A(\tau)$ used in our model is the one introduced and motivated by [55]

$A(\tau) = a_p e^{-\tau/T_p} - a_D e^{-\eta\tau/T_p}$ if $\tau > 0$,

$A(\tau) = a_p e^{\eta\tau/T_D} - a_D e^{\tau/T_D}$ if $\tau < 0$,

with the same parameters used in [55] to fit the experimental data of [23], $a_p = \gamma\,[1/T_p + \eta/T_D]^{-1}$, $a_D = \gamma\,[\eta/T_p + 1/T_D]^{-1}$, with $T_p = 10.2$ ms, $T_D = 28.6$ ms,

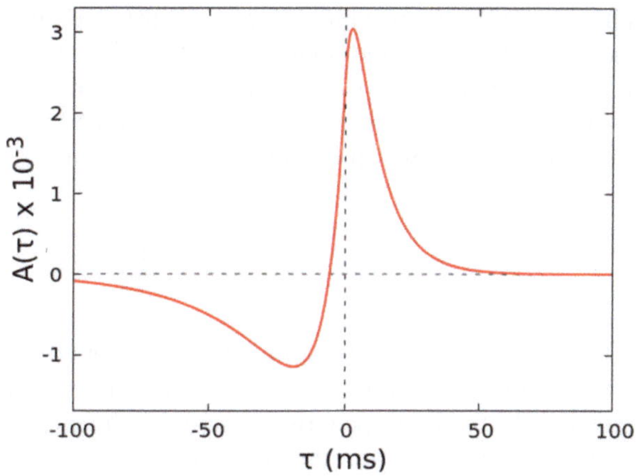

Fig. 7 Plot of the learning window $A(\tau)$ used in the learning rule (8, 9) to model STDP effects. Potentiation $A(\tau) > 0$ or depression $A(\tau) < 0$ depend on the interspike interval τ between pre-synaptic and post-synaptic spikes. The parameters of the function $A(\tau)$ are determined by fitting the experimental data reported in [23]

$\eta = 4$, $\gamma = 1260$. Notably, this function $A(\tau)$, shown in Fig. 7, satisfies the balance condition $\int_{-\infty}^{\infty} A(\tau)d\tau = 0$.

While Eq. 8 holds for activity pattern $y(t)$ which represents instantaneous firing rate and it has been studied in associative memory models with analog rate [30, 31, 56], including a model of theta phase precession [57] and in a spin network model [28], here we want to study the case of spiking neurons. Therefore, here the patterns to be stored are defined as phase-coded patterns of spikes, i.e. periodic spatio-temporal patterns of spikes made up of one spike per cycle. Each unit in each stored pattern has a phase ϕ_j^μ randomly chosen from a uniform distribution in $[0, 2\pi)$. The set of timing of spikes of unit j can be defined as

$$t_j^\mu + nT^\mu = \left(\frac{\phi_j^\mu}{2\pi} + n\right) T^\mu.$$

Thus, the activity of the neuron j in pattern μ is a spike train at times t_j^μ with period T^μ, namely $y_j^\mu(t) = \sum_n \delta(t - (t_j^\mu + nT^\mu))$.

The synaptic strength of the connection J_{ij}, due to pattern $y_i^\mu(t) = \sum_n \delta(t - (t_i^\mu + nT^\mu))$, when the learning time is longer than the period T^μ of the learned pattern, is therefore formulated as follows:

$$\delta J_{ij}^\mu = 1/N \sum_{n=-\infty}^{\infty} A(t_j^\mu - t_i^\mu + nT^\mu) \tag{9}$$

where t_j^μ are the spike times of the neuron j in the pattern μ. At the end of the learning procedure, part of the connections are positive (excitatory) and part are negative (inhibitory). Inhibitory neurons are not explicitly simulated, but negative negative connections can be considered as connections mediated by fast inhibitory interneurons. When the STDP kernel $A(\tau)$, shown in Fig. 7, is used in Eq. (9) to learn phase-coded patterns with uniformly distributed phases, then the balance condition assures that the averaged connections $(1/N)\sum_j J_{ij}$ are of order $1/\sqrt{N}$, and therefore it assures a balance between excitation and inhibition.

When multiple patterns are stored, the learned connections are simply the sum of the contributions from individual patterns, namely

$$J_{ij} = \sum_{\mu=1}^{P} \delta J_{ij}^{\mu}. \tag{10}$$

where P is the number of stored patterns. Note that ring-like topology with strong unidirectional connections is formed only in the case $P = 1$, when a single pattern is stored. When multiple patterns are stored in the same connectivity, with phases of one pattern uncorrelated with the others, bidirectional connections are possible, and the more the stored patterns, the more disordered and less ring-like is the connectivity. The connectivity that comes out from this learning rule, even when P is much larger then one, is still able to support the emerging of the stored patterns as discussed in next section.

References

1. Beggs, J., Plenz, D.: Neuronal avalanches in neocortical circuits. J. Neurosci. **23**(35), 11167–11177 (2003)
2. Chialvo, D.R.: Emergent complex neural dynamics. Nat. Phys. (2010)
3. Petermann, T., Thiagarajan, T., Lebedev, M., Nicolelis, M., Chialvo, D., Plenz, D.: PNAS **106**, 15921 (2009)
4. Pasquale, V., Massobrio, P., Bologna, L., Chiappalone, M., Martinoia, S.: Neuroscience **153**, 1354 (2008)
5. Friedman, N., Ito, S., Brinkman, B.A.W., Shimono, M., Lee DeVille, R.E., Dahmen, K.A., Beggs, J.M., Butler, T.C.: Universal critical dynamics in high resolution neuronal avalanche data. Phys. Rev. Lett. **108**, 208102 (2012)
6. Plenz, D., Niebur, E. (eds.): Criticality in Neural Systems. Wiley (2014)
7. Massobrio, P., de Arcangelis, L., Pasquale, V., Jensen, H., Plenz, D.: Front. Syst. Neurosci. **9**, 22 (2015)
8. Kinouchi, O., Copelli, M.: Nat. Phys. **2**, 348 (2006)
9. Gautam, S., Hoang, T., Clanahan, K.M., Grady, S., Shew, W.: PLoS Comp. Bio. **11**, e1004576 (2015)
10. Yang, H., Shew, W., Roy, R., Plenz, D.: J. Neurosci. **32**, 1061 (2012)
11. Tomen, N., Rotermund, D., Ernst, U.: Front. Syst. Neurosci. **8**, 151 (2014)
12. Luczak, A., Bartho, P., Harris, K.D.: Spontaneous events outline the realm of possible sensory responses in neocortical populations. Neuron **62**, 413–425 (2009). https://doi.org/10.1016/j.neuron.2009.03.014
13. Tsodyks, M1., Kenet, T., Grinvald, A., Arieli, A.: Linking spontaneous activity of single cortical neurons and the underlying functional architecture. Science **286**, 1943–1946 (1999)
14. Miller, J.E., Ayzenshtat, I., Carrillo-Reid, L., Yuste, R.: Visual stimuli recruit intrinsically generated cortical ensembles. Proc. Nat. Acad. Sci. U.S.A. **111**, E4053–4061 (2014). https://doi.org/10.1073/pnas.1406077111

15. Pasquale, V., Martinoia, S., Chiappalone, M.: Stimulation triggers endogenous activity patterns in cultured cortical networks. Scientific Report 7 (2017)
16. Luczak, A., MacLean, J.N.: Default activity patterns at the neocortical microcircuit level. Front. Integr. Neurosci., 12 June (2012)
17. Carr, M.F., Jadhav, S.P., Frank, L.M.: Hippocampal replay in the awake state: a potential substrate for memory consolidation and retrieval. Nat. Neurosci. **14**, 14753 (2011)
18. Euston, D.R., Tatsuno, M., McNaughton, B.L.: Fastforward playback of recent memory sequence in prefrontal cortex during sleep. Science **318**, 1147–1150 (2007)
19. Ji, D., Wilson, M.A.: Coordinated memory replay in the visual cortex and hippocampus during sleep. Nat. Neurosci. **10**, 100–107 (2007)
20. Kayser, C., Montemurro, M.A., Nikos, K., Logothetis, N.K., Panzeri, S.: Spike-phase coding boosts and stabilizes information carried by spatial and temporal spike patterns. Neuron **61**, 597–608 (2009)
21. Siegel, M., Warden, M.R., Miller, E.K.: Phase-dependent neuronal coding of objects in short-term memory. Proc. Nat. Acad. Sci. U.S.A. **106**, 21341–21346 (2009)
22. Montemurro, M.A., Rasch, M.J., Murayama, Y., Logothetis, N.K., Panzeri, S.: Phase-of-firing coding of natural visual stimuli in primary visual cortex. Curr. Biol. **18**, 375–380 (2008)
23. Bi, G.Q., Poo, M.M.: Precise spike timing determines the direction and extent of synaptic modifications in cultured hippocampal neurons. J. Neurosci. **18**, 10464–10472 (1998)
24. Debanne, D., Gahwiler, B.H., Thompson, S.M.: Long-term synaptic plasticity between pairs of individual CA3 pyramidal cells in rat hippocampal slice cultures. J. Physiol. **507**, 237–247 (1998)
25. Feldman, D.E.: Timing-based LTP and LTD and vertical inputs to layer II/III pyramidal cells in rat barrel cortex. Neuron **27**, 45–56 (2000)
26. Markram, H., Lubke, J., Frotscher, M., Sakmann, B.: Regulation of synaptic efficacy by coincidence of postsynaptic APs and EPSPs. Science **275**, 213 (1997)
27. Scarpetta, S., Giacco, F.: Associative memory of phase-coded spatiotemporal patterns in leaky integrate and fire networks. J. Comput. Neurosci. (2012). https://doi.org/10.1007/s10827-012-0423-7
28. Scarpetta, S., Giacco, F., De Candia, A.: Storage capacity of phase-coded patterns in sparse neural networks. EPL **95**(2) (2011)
29. Scarpetta, S., de Candia, A., Giacco, F.: Storage of phase-coded patterns via STDP in fully-connected and sparse network: a study of the network capacity. Front. Synaptic Neurosci. **2**, 1–12 (2010)
30. Scarpetta, S., Zhaoping, L., Hertz, J.: Hebbian imprinting and retrieval in oscillatory neural networks. Neural Comput. **14**, 2371–2396 (2002)
31. Yoshioka, M., Scarpetta, S., Marinaro, M.: Spatiotemporal learning in analog neural networks using spike-timing-dependent synaptic plasticity. Phys. Rev. E **75**, 051917 (2007)
32. Scarpetta, S., de Candia, A.: Neural avalanches at the critical point between replay and non-replay of spatiotemporal patterns. PLoS One **8**, e64162 (2013)
33. Scarpetta, S., de Candia, A.: Alternation of up and down states at a dynamical phase-transition of a neural network with spatiotemporal attractors. Front. Syst. Neurosci. **8**, 88 (2014)
34. Scarpetta, S., Giacco, F., de Candia, A., Lombardi, F.: Effects of Poisson noise in a IF model with STDP and spontaneous replay of periodic spatiotemporal patterns, in absence of cue stimulation. BIOSYSTEMS. Vol. https://doi.org/10.1016/j.biosystems.2013.03.017. Pag.1-7. ISSN:0303-2647 (2013)
35. Scarpetta, S., Minati, L., Apicella, I., de Candia, A.: Hysteresis, neural avalanches and critical behaviour near a first-order transition of a spiking neural network. Phys. Rev. E **97**, 062305 (2018)
36. Beggs, J.M., Plenz, D.: Neuronal avalanches are diverse and precise activity patterns that are stable for many hours in cortical slice cultures. J. Neurosci. **24**, 5216–5229 (2004)
37. Stewart, C.V., Plenz, D.: Inverted-u profile of dopamine-nmda-mediated spontaneous avalanche recurrence in superficial layers of rat prefrontal cortex. J. Neurosci. **26**, 8148–8159 (2006)

38. Ribeiro, T.L., Ribeiro, S., Copelli, M.: Repertoires of spike avalanches are modulated by behavior and novelty. Front. Neural Circuits **10**, 16 (2016)
39. Hasselmo, M.E.: Acetylcholine and learning in a cortical associative memory. Neural Comput. **5**, 32–44 (1993)
40. Hasselmo, M.E.: Neuromodulation: acetylcholine and memory consolidation. Trend Cogn. Sci. **3**(9), 351 (1999)
41. Tagliazucchi, E., Balenzuela, P., Fraiman, D., Chialvo, D.R.: Criticality in large-scale brain fMRI dynamics unveiled by a novel point process analysis. Front. Physiol. (2012)
42. Heermann, D.W., Klein, W., Stauffer, D.: Spinodals in a long-range interaction system. Phys. Rev. Lett. **49**, 1682 (1982)
43. Heermann, D.W., Coniglio, A., Klein, W., Stauffer, D.: Nucleation and metastability in three-dimensional ising models. J. Stat. Phys. **36**, 447 (1984)
44. Cossart, R., Aronov, D., Yuste, R.: Attractor dynamics of network UP states in the neocortex. Nature **423**, 283–288 (2003). https://doi.org/10.1038/nature01614
45. Sethna, J.P., Dahmen, K.A., Myers, C.R.: Crackling noise. Nature **410**, 242 (2001)
46. Shew, W.L., Clawson, W.P., Pobst, J., Karimipanah, Y., Wright, N.C., Wessel, R.: Adaptation to sensory input tunes visual cortex to criticality. Nat. Phys. **11**, 659 (2015)
47. Touboul, J., Destexhe, A.: Power-law statistics and universal scaling in the absence of criticality. Phys. Rev. E **95**, 012413 (2017)
48. Minati, L., de Candia, A., Scarpetta, S.: Critical phenomena at a first-order phase transition in a lattice of glow lamps: experimental findings and analogy to neural activity. Chaos **26**, 073103 (2016). https://doi.org/10.1063/1.4954879
49. Uhlig, M., Levina, A., Geisel, T., Herrmann, J.M.: Critical dynamics in associative memory networks. Front. Comput. Neurosci. **7**, 87 (2013)
50. Apicella I., Scarpetta S., de Candia A.: Cortical phase transitions as an effect of topology of neural network. In: Esposito, A., Faudez-Zanuy, M., Morabito, F., Pasero, E. (eds.) Multidisciplinary Approaches to Neural Computing. Smart Innovation, Systems and Technologies, vol. 69. Springer (2018)
51. Gerstner, W., Kistler, W.: Spiking Neuron Models: Single Neurons, Populations, Plasticity. Cambridge University Press, NY (2002)
52. Luczak, A., MacLean, J.: Front. Integr. Neurosci. **6**, 30 (2012)
53. Orlandi, J., Soriano, J., Alvarez-Lacalle, E., Teller, S., Casademunt, J.: Nat. Phys. **9**, 582 (2013)
54. Pasquale, V., Martinoia, S., Chiappalone, M.: Sci. Rep. **7**, 9080 (2017)
55. Abarbanel, H., Huerta, R., Rabinovich, M.I.: Dynamical model of long-term synaptic plasticity. PNAS **99**, 10132–10137 (2002)
56. Scarpetta, S., Zhaoping, L., Hertz, J.: Spike-timing-dependent learning for oscillatory networks. Advances in Neural Information Processing Systems 13. MIT Press (2001)
57. Scarpetta, S., Marinaro, M.: A learning rule for place fields in a cortical model: theta phase precession as a network effect. Hippocampus **15**(7), 979–89 (2005)

Silvia Scarpetta studied physics in University of Naples, Italy and received her Ph.D. in physics from Salerno University in 1999. She works on Computational Neuroscience as a researcher in Salerno University from 2002. Her research interests include statistical mechanics, critical phenomena in living matter, cortical dynamics and artificial intelligence.

Assessing Criticality in Experiments

Viola Priesemann, Anna Levina and Jens Wilting

Abstract Criticality is considered an attractive candidate state for the dynamics and function of the brain, because in models criticality maximizes a number of properties related to information transmission and computations. These include the dynamic range, the susceptibility, the sensitivity to input, the correlation length, and the pattern diversity. And indeed, numerous studies accumulated evidence that supports the criticality hypothesis. However, some observations are also contradictory. The latter might in part be explained by the fact that criticality is a "full system" property, whereas experimental neural recordings can only assess a tiny part of the full network ("subsampling problem"). Here, we first recapitulate the basic properties of a dynamical system at and around the critical point in a pedagogic manner. We show how subsampling can bias inference about the underling network dynamics, and then present two recent analytical approaches to overcome the subsampling bias, based either on neural avalanches, or on the time-series of neural activity proper. The novel approach typically allows *quantify the distance to criticality* from less then ten recorded neurons and a few minutes of recording only. Thereby, it offers a novel and very powerful approach to assess criticality and task-related deviations thereof.

V. Priesemann (✉) · J. Wilting
Max-Planck-Institute for Dynamics and Self-Organziation, Am Fassberg 17, 37075 Göttingen, Germany
e-mail: viola.priesemann@ds.mpg.de

J. Wilting
e-mail: jens.wilting@ds.mpg.de

A. Levina
Computer Science Department, University of Tübingen, Max-Planck-Ring 8, 72076 Tübingen, Germany
e-mail: anna.levina@uni-tuebingen.de

N. Tomen et al. (eds.), *The Functional Role of Critical Dynamics in Neural Systems*, Springer Series on Bio- and Neurosystems 11,
https://doi.org/10.1007/978-3-030-20965-0_11

1 Introduction

Criticality is considered an attractive candidate state for the dynamics and function of the brain, because in models criticality maximizes a number of properties related to information transmission and computations. These include the dynamic range, the susceptibility, the sensitivity to input, the correlation length, and the pattern diversity [10, 11, 44, 46, 77, 79]. And indeed, numerous studies accumulated evidence that supports the criticality hypothesis [7, 33, 66, 80, 83]. However, some observations are also contradictory [6, 67, 70, 75, 76, 92]. The latter might in part be explained by the fact that criticality is a "full system" property, whereas experimental neural recordings can only assess a tiny part of the full network.

Testing the criticality hypothesis experimentally is a challenge, because criticality is a property that inherently engages the entire system. However, when sampling neural activity in experiments, only a tiny fraction of all neurons can be assessed. More precisely, of the millions or billions of neurons in a mammalian brain, at most a few hundred neurons can be sampled with millisecond precision. Such spatial subsampling can lead to strong bias of the inferred dynamical properties of neural activity [50, 67, 69, 70, 75, 76, 92]. Alternatively, criticality is assessed based on "coarse" measures of neural activity, e.g. in the form of local field potentials (LFP), electroencephalography (EEG), electrocorticography (ECoG), magnetoencephalography (MEG) and functional magnetic resonance imaging (fMRI). All these measures have in common that they do not rely on single neuron activity, but measure the collective effect of many neurons with potential contributions from glia, vasculature and other physiological parameters [15, 51]. These measures sacrifice spatial resolution, but have the advantage to sample activity from large portions or even the entire brain.

In the following, we first go over the well-known challenges that arise from subsampling or coarse sampling when testing for criticality in more details, and then show two recent advances to overcome the subsampling problem.

To this end, we first recapitulate the basic properties of a dynamical system at and around the critical point in a pedagogic manner, using the framework of a branching process. We then show that estimation of the control parameter of the branching process, the "branching parameter" m, is inherently and systematically biased under subsampling when using classical approaches. By uncovering the origin of the systematic bias, we devised a novel estimator that is capable to infer m in a unbiased manner—in principle even from sampling a single unit. The majority of this chapter is closely reproduced from [92].

We then turn to the problem of subsampling when assessing avalanche distributions, cluster size distributions, or node degree distributions of a graph. We first show for all these properties, that subsampling affects them in the same manner. Hence deriving the avalanche distribution or the degree distribution of the full system from a subsample is mathematically equivalent. We recapitulate how the distributions change under subsampling using probability generating functions, and then derive the method of "subsampling scaling", which allows us to infer the parent distribution from the subsampled one, and demonstrate how to distinguish critical from non-critical systems. This chapter is reproduced from [50].

2 Investigating Criticality in Experiments

An intriguing property of systems that operate at a critical point is the emergence of scale-free characteristics. Most importantly for neuroscience, scale-free distributions are expected for avalanche sizes and durations if the network is poised at a critical point. In the specific case of critical branching processes, which we introduce in all details later, the avalanche size distributions follows a power-law with exponent -1.5, $p(s) \sim s^{-1.5}$, and the duration distribution with exponent -2, $p(d) \sim d^{-2}$. Indeed, most of the experimental evidence for criticality in neuroscience is based on observing such power-laws, in particular power-law distributions of the sizes.

Subsampled avalanches. Owing to the scale-free nature of the avalanche distributions, it would be intuitive that a small subset of a critical system also shows scale-free avalanche distributions—but with a cutoff that depends on the number of neurons or electrodes sampled. This expectation is fueled by the theory of "finite-size scaling". Because of the finite size of any simulated system, the avalanche distributions cannot extend to infinity, but show a system-size dependent cutoff. Likewise, the susceptibility, dynamic range and correlation length do not diverge to infinity but are bounded by the system size. If a system, however, is at criticality, then the cutoff and finite-size effects show a clear scaling with the system size, and the scale-free behavior of an infinite-size system can be demonstrated by extrapolation ("finite-size scaling").

With this background of scale-freeness and finite-size theory, it came as a large surprise when it was discovered that a small subset of a large, critical system does not show the expected power-law distributions, but can exhibit strong deviations [67]. Depending on the specific critical model, and depending on the sampling configuration, a critical system can appear as if it was either subcritical or supercritical [67]. These results demonstrated that *not* finding a power law under subsampling does *not* prove that the system is *not* critical. In more detail, for the numerical exploration of subsampling effects, it was assumed that an array of n by n electrodes with a distance δ is used in a critical model to sample its activity. This approach was inspired by the typical multi-electrode configurations used for spike and LFP recordings. It was found that for close electrode spacing, the avalanche size distribution $p(s)$ of the Bak–Tang–Wiesenfeld Sandpile model [4] does not show the expected power law. Instead, $p(s)$ shows clear peaks at the multiples of units sampled ($c \times n^2$, $c = 1, 2 \ldots$). These $p(s)$ are reminiscent of supercritical systems. With increasing the electrode spacing, the peaks disappear and the $p(s)$ resemble distributions from subcritical systems [67, 69]. By now we know that a *random placement of electrodes* all over the system gives rise to an approximate power law with a cutoff that depends linearly on the number of electrodes or neurons sampled [50] (see Sect. 4). The resulting distribution of avalanche sizes depends both on the topology of the full network and placement of electrodes. Although true network structure of the neuronal tissue remains unknown, we argue that placing electrodes at random diminishes the local clustering effects and allows us to capture global dynamics. The same results hold for any sampling configuration on a system with random connections (no spatial structure in the topology). The resulting "subsampling scaling" under random sampling is reminiscent but clearly different from classical finite-size scaling. For a combined finite-size-subsampling-scaling ansatz, we refer to [50] (see Sect. 4). The difference between subsampling and finite-size scaling is in part routed in the fact that in a finite system, the boundaries are typically absorbing. Under subsampling, however, the hidden units contribute to the dynamics by enabling propagation of activity on a system much larger than the observed part. Hence the differences between a small system, and a small subset of a large system.

To investigate the effect of subsampling, it is essential to make a clear statement about the meaning of "whole system". For each question and each system, there could be a different definition of "whole system" and "system size". For the cultures examined with the help of MEAs and multi-unit activity, "system size" is the number of neurons in the culture. However, the "whole system" might also be all neurons in the layer 2/3 of rats cortex [9], in a different approach, it could be the whole brain as seen in the MRI studies [83]. Careful and explicit definition of what the entire system represents is essential both for the correct interpretation of the results of any investigation and for accurate accounting for subsampling.

Quantitative subsampling studies. After the first reports of subsampling effects [67, 75], many experimental studies discussed their potential influence. A number of studies also performed systematic analyses of subsampling-induced changes in $p(s)$ and $p(d)$ by applying further subsampling, i.e. analyzing subsets of the recorded units, to name just the most detailed ones [45, 70, 76, 96]. All these approaches, however, were purely heuristic. They compared numerically obtained subsampling results from models with those of experiments. The first analytical treatments, to the best of our knowledge, were only achieved recently [50, 92]. As the approaches are readily applicable to recorded data, we discuss them in all detail in the following sections (Sects. 3 and 4).

Preceding investigations of subsampling in neuronal systems, a similar problem has been treated in the past in the context of graphs. When assessing very large networks, such as world wide web, social networks on Facebook, or gene regulatory networks, one either does not have access to all data, or the data set is so large that for practical reasons only a subnetwork can be treated. Thus in the latter case, a central question is how to select a subnetwork, but preserve all features of the full system [47], and how to visualize it truthfully [73]. Assuming that sampling is constrained by the experimental accessibility, a central question is to test whether graphs are scale-free, i.e. whether their degree distribution follows a power-law. Interestingly, the scale-free property is lost under subsampling [81]. This has been derived under the assumption of random sampling, i.e. every node is sampled with a given probability p. In contrast, when subsampling Erds–Renyi, exponential and truncated-normal graphs, they all can be mistaken as scale-free under strong subsampling [35, 43, 60]. Hence wrong inferences about the full system are easily drawn when naively extrapolating from the subsampled graph to the full one. In the context of graph theory, the subsampling theory has already provided a number of remarkable results, and is still a field of very active research (see for example [1, 30, 47, 57, 74, 82]). Hence translating the principles discovered for graph theory to neuroscience and specifically to the analysis of neural avalanches promises very interesting future insights.

In addition to subsampling, experimental assessment of criticality is hindered further by a number of challenges that originate from the complex nature and spatial reach of dynamics at criticality. In the following we discuss the challenge to unambiguously separate one avalanche from the next one, the effects of coarse-sampling, and the effects of potential non-stationarities in the input.

Separation of timescales. Are neural avalanches clearly separated in time and in space, or do they overlap? If they overlap, how many avalanches start in a given time window, how many are running in parallel, and what happens under collision of two avalanches? The aim here is not to discuss potential collision-based computing, but to draw the attention to a difference between neural avalanches as observed in vivo, and those of classical "self-organized critical" (SOC) systems. In SOC systems, the pauses between any two avalanches are much longer than the avalanches proper. This condition is termed "separation of time scales" (STS) and is considered necessary for SOC [70, 72]. Hence in SOC systems, even under subsampling the STS would enable us to clearly separate the events that belong to one avalanche from those that belong to the subsequent one. In neural recordings in vivo, however, the separation of two subsequent avalanches is not straightforward, because no clear pauses are observed. As a practical solution, avalanches are obtained by applying temporal binning to the time series of events [7]. Any empty bin by definition separates one avalanche from the next one. If binning is applied to a system that shows a STS, then the specific choice of the bin size does not change the avalanche distribution for a wide range of bin sizes [50, 68]. For neural recordings, however, an invariance of $p(s)$ or $p(d)$ under changes in bin size is an exception [50, 70]. Most studies, including the seminal study by John Beggs and Dietmar Plenz, observed clear changes with changing the bin size [7, 8, 32, 67, 69, 70, 95]. The bin-size dependence does not only complicate an unambiguous separation of avalanches, it also in principle contradicts the SOC hypothesis. Therefore, a parsimonious solution is offered by assuming a working-point of neural systems that is close to criticality, but in the subcritical regime, which does not necessitate a STS. In more detail, the further a system is away from the critical point, the less large avalanches it produce [36]. With the absence of system-spanning avalanches, several avalanches can evolve in parallel, splitting up, merging, and vanishing again. From this melange of avalanches, individual ones cannot be distinguished anymore, but overall they can generate sustained activity with large fluctuations, long correlations in time and space, and with a spike rate as observed in vivo. In fact, in vivo spiking activity is very well approximated by a model in a subcritical state, namely a branching model that is poised about 2% away from criticality [70, 92, 93]. With keeping a 2% safety-margin to criticality, the brain not only avoids the necessary pauses in processing associated with a STS and criticality, it also maintains a safety-margin from the supercritical state, which has been associated with epilepsy [56, 58, 94]. Studies of visual processing also demonstrated preference for the slightly subcritical state [85]. Thus overall, a distance of about 2% to criticality may allow the brain to still profit from the computational advantages associated with criticality, while avoiding instabilities that can occur when tipping over to supercriticality.

Coarse sampled measures. Subsampling is most prominent when recordings spikes. When recording instead LFP, ECoG, BOLD or EEG signals, the coverage is much better—but at the cost of spatial and temporal resolution. Rather than sampling spikes, a summed signal of synaptic and membrane currents as well as glia, metabolic and other contributions is recorded. Thus instead or in addition to subsampling, one obtains a "coarse sampling" of the system. Again, because of the scale-free nature of critical phenomena, it could be expected that the coarse sampled system should have similar properties to the original one. For example when applying a specific approach of coarse sampling to the Ising model at critical temperature, namely coarse graining or block-spin renormalization, the estimated magnetization is unaffected. Importantly, the invariance only holds when the system is critical, and it only holds for a particular coarse graining scheme, i.e. one tiles the entire 2D Ising model into adjacent subsets of k by k spins. k is typically a small number for numerical reasons, and importantly the "tiles" are all non-overlapping and directly adjacent [65]. This condition is violated when considering e.g. a LFP signal: The LFP also samples the signals in its vicinity, but first of all not all neurons in the vicinity contribute equally, and second two adjacent electrodes have typically overlapping sampling regions ("field of view"). Thus even if all neural activity was independent, the overlap of the field of view would introduce correlations between the LFP signals of two electrodes [65]. As a consequence, correlations between LFP signals are in general expected to be larger than correlations between the respective "blocked" signals. The coarse-sampling-induced increase in correlation compared to the underlying system obviously affects the observed avalanche distributions. Hence the intuition that a "coarse sampled" critical system should have the same properties as the original underlying system, does not hold. As a consequence, it is difficult to draw unambiguous conclusions from a coarse-sampled system about the underlying dynamics. In general, the measurement-correlations induced by coarse sampling (e.g. LFP, EEG, BOLD) tends to make a system appear closer to critical then it truly is [65].

Nonstationary input. A common challenge one meets whenever assessing dynamical properties of neural systems, is the challenge to deal with non-stationarities or unknown common input. Both can lead to false-positive observations. For example, false-positive transfer of information can be found between units that are not directly coupled, but receive input from the same (unknown) source, potentially with different delays or with different noise levels [89]. Two brain areas may seem to be coupled by "cross-frequency coupling" simply because of common non-stationarities in their activity [2], and when imposing the same long-range correlated input to uncoupled neurons, then the spiking activity obviously is long-range correlated owing to the

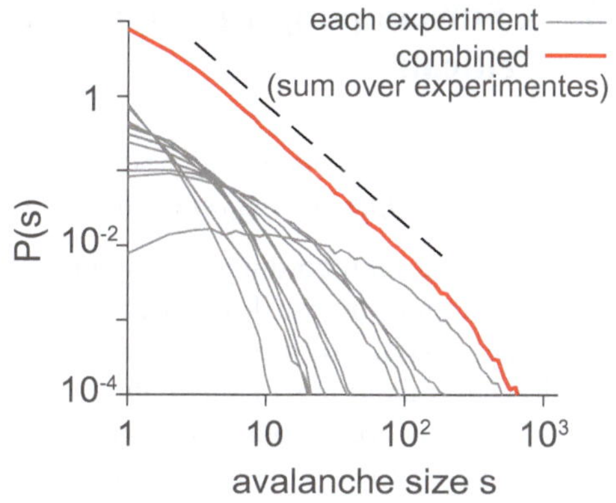

Fig. 1 Spuriously scale-free avalanches. Apparent scale-free avalanches can appear by pooling multiple experiments. While avalanche size distributions of each experiment individually are indicative of the underlying subcritical systems, the sum over all experiments exhibits a power-law. Figure reproduced from [68]

drive [86]. Last but not least, if a brain areas would receive its main input from a driving brain area that is in a critical state, one can expect that the receiving brain area also shows signatures of criticality. Hence the nature of the input clearly matters. However, while it is clear that imposing long-range correlated drive to a system can impose critical-like properties, the question is whether non-stationary input in general can give rise to apparent criticality [68]. With Oren Shiki, we showed very recently that already very mild conditions, e.g. a slow change in input strength by a factor 10, can lead to approximate power-law distributions with slopes characteristic for a critical branching process, i.e. -1.5 and -2 for the size and duration distribution respectively [68]. In more detail, we modeled the neurons as independent Poisson units with non-stationary rate. Assuming that the spike rate of all neurons is changed by some stimulation or some neuromodulation between e.g. 1 and 10 Hz over the course of the recording, then the obtained avalanche distributions are not approximate exponential, as expected for stationary Poisson neurons. Instead, they show power-laws covering almost two orders of magnitude, and resemble the $p(s)$ and $p(d)$ observed in many experiments. We also derived analytically how power laws without cutoff could be obtained. However, such conditions cannot be realized physiologically [68]. An important technical consequence is that the same power-laws can arise when combining avalanche distributions from various experiments or sessions. If the different sessions have different spike or event rate, and each session individually would show "subcritical" distributions, then averaging over all these sessions can generate power-law distributions (Fig. 1). The principle behind this is that the sum of exponential distributions $p(s) \sim e^{-\alpha s}$ with different exponents α can give rise to power-laws. In conclusion, non-stationary rates (or different rates across recording sessions) can give rise to approximate power-laws with the expected exponents. However, when taking into account further indications of criticality, such as the relationship between mean avalanche size and duration, estimated spike count ratio and so forth, apparent criticality can often be distinguished from true criticality [68].

3 A Subsampling-Invariant Estimator for the Distance to Criticality

In this section, we recapitulate the derivation a novel, subsampling invariant estimator [92]. This *multiple regression* estimator is useful for the following reasons:

1. It is based on recordings of ongoing dynamics and therefore does not rely on a separation of timescales, which seems unlikely in many applications.
2. It has analytically been shown to be invariant even under strong subsampling, down to recordings of single units.
3. It is readily applicable to subsampled data, because it only requires a sufficiently long time series, without even knowing the system size or the number of sampled units n.

We will first introduce branching processes as a useful description of spike propagation in neuronal networks. We will then review established estimators for these branching processes and derive the novel, subsampling invariant estimator. Finally, we will briefly discuss applications to two real-world systems: disease spread and cortical dynamics.

3.1 Branching Process: A Minimal Model of Spike Propagation

In this section, we first introduce an intuitive motivation of activity propagation, and then we introduce the mathematical framework of the branching process formally. We state the main results in the beginning of each section. We then show the detailed derivation, marked by (*), which can be skipped by the reader not interested in these details.

3.1.1 Approximating Activity Propagation on Neural Networks

To gain an intuitive understanding of our mathematical approach, make a thought experiment in your favorite spiking network: apply one additional spike to an excitatory neuron, in analogy to the approach by London and colleagues [53]. How does the network respond to that perturbation? As a first order approximation, one quantifies the number of spikes that are triggered by this perturbation *additionally* in all postsynaptic neurons. This number may vary from trial to trial, depending on the membrane potential of each postsynaptic neuron; however, what interests us most is m, the *mean number of spikes triggered by the one extra spike*. Taking a mean-field approximation and assuming that the perturbation indeed is small, any of these triggered spikes in turn trigger spikes in their postsynaptic neurons in a similar manner, and thereby the perturbation may cascade through the system. Mathematically, such cascades can be described by a branching process [36, 38, 61]. In the next step, assume that perturbations are started continuously at rate h, for example through afferent input from other brain areas or sensory modalities. As neurons presumably do not distinguish whether a postsynaptic potential was elicited from a neuron from within the network, or from afferent input, all spikes are assumed to have on average the same impact on the network dynamics. Together, this leads to the mathematical framework of a branching process [7, 34, 48, 69, 70, 75], which can generate dynamics spanning asynchronous irregular and critical states, and hence is well suited to probe network dynamics in vivo. Most importantly, this framework allows to infer m and other properties from the ongoing activity proper, because one treats any single spike as a minimal perturbation on the background activity of the network. Mathematical approaches to infer m are long known if the full network is sampled [40, 88, 92]. Under subsampling, however, the estimate of m can be strongly biased [92]. Below, we show how to infer m in an unbiased manner, even if only a tiny fraction of neurons is sampled.

The novel estimator for m is readily applicable to subsampled data, because it only requires a sufficiently long time series of sampled activity a_t, and the assumption that *on expectation* a_t is proportional to a total activity A_t. Hence, in general it suffices to sample the system randomly, without even knowing the system size N, the number of sampled units n, or any moments of the underlying process. Importantly, one can obtain a consistent estimate of m, even when sampling only a very small fraction of the system, under homogeneity even when sampling only one single unit. This robustness makes the estimator readily applicable to any system that is well-approximated by a branching process.

The framework of branching processes can be interpreted as a stochastic description of spike propagation on networks, as outlined above. It can alternatively be taken as a *strictly phenomenological* approximation to network dynamics that enables us to infer details of network *statistics* despite subsampling. Independent of the perspective, the dynamics of the network is mainly governed by m. If an action potential only rarely brings any postsynaptic neuron above threshold ($m \approx 0$), external perturbations quickly die out, and neurons spike independently and irregularly, driven by external fluctuations h. In general, if one action potential causes less than one subsequent action potential on average ($m < 1$), perturbations die out and the network converges to a stable distribution, with increasing fluctuations and variance the closer m is to unity. If $m > 1$, perturbations may grow infinitely, potentially leading to instability. The critical state ($m = m_C = 1$) separates the stable (subcritical) from the unstable (supercritical) phase. When approaching this critical state from below, the expected size $\langle s \rangle$ and duration $\langle d \rangle$ of individual cascades or avalanches diverge: $\langle s \rangle \sim \frac{1}{m_C - m}$. Therefore, especially close to criticality, a correct estimate of m is vital to assess the risk that the network develops large, potentially devastating cascades, which have been linked to epileptic seizures [56], either generically or via a minor increase in m.

3.1.2 Branching Processes

The previously motivated approximation of spike propagation will now be formalized into the mathematical model of branching processes. After defining the branching process, we then derive its three distinct stability regimes, its mean activity, and its autocorrelogram.

Definition. In a branching process, we denote the number of neurons activated at each time step t (in terms of bins of length Δt) by A_t. Every single active neuron i activates a random number $y_{\mathrm{i,t}}$ of neurons in the following time step. In addition, the neurons become spontaneously active or are activated by external input h_t with mean h:

$$A_{t+1} = \sum_{i=1}^{A_t} y_{t,i} + h_t, \tag{1}$$

The random variables $y_{\mathrm{i,t}}$ can take any value from zero to the number of postsynaptic neurons. The precise distribution of $y_{\mathrm{i,t}}$ does not impact the overall network dynamics and stability, because the network stability is determined by m, the mean number of neurons activated by a single extra spike: $m = \langle y_{\mathrm{i,t}} \rangle$, which is the control parameter of the process.

Three stability regimes. We analyze the stability of the branching process by investigating its long-term behavior. To this end, we evaluate the *expected* evolution. Given a current state A_t, the expected activity in the next time step is given by the conditional expectation for A_{t+1} in Eq. (1), which yields

$$\langle A_{t+1} | A_t \rangle = m A_t + h. \tag{2}$$

To obtain the expected long-term behavior, we iterate this equation for k time steps into the future:

$$\langle A_{t+k} | A_t \rangle = m^k A_t + h \frac{m^k - 1}{m - 1}. \tag{3}$$

From these equations, two central observations follow. First, the activity at the following time step is linear in the activity in the preceding time step, and second, the same holds for k time steps into the future. The general linear factor for any k is m^k.

Furthermore, Eq. (3) reveals three distinct regimes:

Subcritical, $m < 1$. In the subcritical regime, $m^k \to 0$ as k increases. Hence, $\langle A_{t+k} | A_t \rangle \to h/(1-m)$, i.e. regardless of the current network state, activity is expected to relax to an equilibrium $h/(1-m)$ when waiting long enough (we will later show that this is in fact the mean activity of the process).

Critical, $m = 1$. When $m = 1$, the activity is expected to grow linearly at rate h, $\langle A_{t+k} | A_t \rangle = A_t + k\,h$ by L'Hopital's rule.

Supercritical, $m > 1$. When $m > 1$, the activity is expected to grow exponentially, $\langle A_{t+k} | A_t \rangle \propto m^k$. because this term dominates in Eq. (3).

Therefore, the subcritical regime is stable and activity converges to a stationary distribution. In contrast, the critical and supercritical regimes are unstable and non-stationary, the activity increases infinitely, unless $h = 0$.

Mean activity. We find the mean activity of a branching process by searching a fixed point of Eq. (2), i.e. a solution to $\langle A_{t+1} | A_t \rangle = A_t$. For the subcritical regime, this fixed point is given by

$$\langle A_t \rangle = \frac{h}{1-m} \tag{4}$$

and determines the mean activity, which is proportional to the external input, and increases when approaching the critical point at $m = 1$. In contrast, if $m \geq 1$, Eq. (2) has no fixed point any more. Instead, the process grows from one step to the next and is non-stationary. A mean firing rate does not exist any more.

Autocorrelogram of branching processes. We will now show that a branching process in the subcritical regime $m < 1$, like any process with an autoregressive representation in Eq. (2), has an exponential autocorrelogram with one single characteristic timescale τ:

$$C(t\,\Delta t) = m^k = \exp\left(-\frac{k\Delta t}{\tau}\right), \qquad \tau = -\frac{\Delta t}{\log m}. \tag{5}$$

Here, we identified the characteristic timescale τ in terms of the branching ratio m. It clearly diverges, i.e. $\tau \to \infty$, when approaching the critical point $m \to 1^-$.

(*) In order to calculate the autocorrelogram $C(k\,\Delta t)$, we evaluate the Pearson correlation coefficient r_k for pairs A_t and A_{t+k}, $r_k = \mathrm{Cov}[A_t, A_{t+1}]/s_{A_t}\, s_{A_{t+1}}$. In the subcritical regime, stationarity dictates that the standard deviations s_{A_t} and $s_{A_{t+k}}$ are equal, such that the correlation becomes

$$C(k\,\Delta t) = r_k = \frac{\mathrm{Cov}[A_t, A_{t+k}]}{\mathrm{Var}[A_t]}. \tag{6}$$

In order to calculate the covariance, we have to evaluate $\langle A_t\, A_{t+k}\rangle$. We do so using the law of total expectation:

$$\langle A_t\, A_{t+k}\rangle = \langle\langle A_t\, A_{t+k} \mid A_t\rangle\rangle_{A_t},$$

where the inner expectation is taken with respect to the joint distribution of A_t and A_{t+k} and the outer one with respect to A_t. We can now make use of the known result for the evolution of the branching process from Eq. (3). Substituting $\langle A_{t+k} \mid A_t\rangle$ and taking the expectation with respect to A_t gives

$$\begin{aligned}
\langle A_t\, A_{t+k}\rangle &= \langle A_t \left(m^k\, A_t + h\frac{1-m^k}{1-m} \right) \rangle_{A_t} \\
&= m^k\, \langle A_t^2 \rangle + h\frac{1-m^k}{1-m} \langle A_t \rangle \\
&= m^k\, \langle A_t^2 \rangle + \frac{h}{1-m}\, (1-m^k)\, \langle A_t \rangle \\
&= m^k\, \langle A_t^2 \rangle + (1-m^k)\, \langle A_t \rangle^2,
\end{aligned} \tag{7}$$

identifying $\langle A_t \rangle$ according to Eq. (4) in the last step. Finally, we obtain an expression for the covariance [31]:

$$\begin{aligned}
\mathrm{Cov}[A_{t+k}, A_t] &= \langle A_{t+k}\, A_t \rangle - \langle A_{t+k} \rangle \langle A_t \rangle \\
&= m^k\, \langle A_t^2 \rangle + (1-m^k)\, \langle A_t \rangle^2 - \langle A_t \rangle^2 \\
&= m^k\, \left(\langle A_t^2 \rangle - \langle A_t \rangle^2 \right) \\
&= m^k \mathrm{Var}[A_t].
\end{aligned} \tag{8}$$

Inserting this evaluation of the covariance into the correlation coefficient from Eq. (6) yields the exponential autocorrelogram in Eq. (5).

3.2 Estimating the Branching Parameter Under Full Sampling

Under full sampling, estimation theory for the branching parameter has been well established. In particular, a universal estimator which does not depend on the special distribution of the triggered spikes $y_{t,i}$ is given by linear regression [40, 88].

Linear regression estimator. Equation 2 directly indicates how to estimate the branching parameter m from the activity time series A_t, namely by linear regression between pairs A_t and A_{t+1}. The slope of linear regression directly returns a consistent estimate of m, for example by means of the least square estimator for the regression slope:

$$\hat{m} = \frac{\mathrm{Cov}[A_t, A_{t+1}]}{\mathrm{Var}[A_t]}. \tag{9}$$

For the stationary process ($m < 1$), this equals the correlation $\mathrm{Cov}[A_t, A_{t+1}]/s_{A_t}\, s_{A_{t+1}}$, because the standard deviations of A_t and A_{t+1} are equal. Hence, linear regression of A_t and A_{t+1} returns an estimate for the branching parameter m under full sampling.

Mean ratio estimator. This estimation of the branching parameter relates to the one introduced by Beggs and Plenz [7]. They denoted the branching parameter m by σ and estimated it as $\sigma_{BP} = \langle \frac{A_{t+1}}{A_t} \rangle_{A_t > 0}$, where the expected value is taken over all

$A_t > 0$ to avoid division by zero. Using the same steps as before, we can evaluate

$$\begin{aligned}\sigma_{BP} &= \langle \frac{A_{t+1}}{A_t} \rangle_{A_t>0} = \langle \langle \frac{A_{t+1}}{A_t} \mid A_t \rangle \rangle_{A_t>0} \\ &= \langle \frac{mA_t + h}{A_t} \rangle_{A_t>0} \\ &= m + \langle \frac{h}{A_t} \rangle_{A_t>0}. \end{aligned} \tag{10}$$

Hence, if $h \to 0$, i.e. under the condition of infinitesimal input, and if the system shows non-zero activity, then the σ_{BP} and m are identical. Else, σ_{BP} returns larger results than m. In particular, the estimate σ_{BP} will be close to the real value of m if $A_t \gg h$. According to Eq. (4), this implies that the estimate tends to be valid if the system is close to critical, whereas the overestimation is larger the greater the distance from criticality.

3.3 *Estimating the Branching Parameter Under Subsampling*

In general, one cannot record the full activity A_t but only a subsampled time series a_t (Fig. 2a). We first define an implementation of subsampling in our stochastic framework. We then show that the established estimators fail to infer the correct branching ratio under subsampling. Finally, we introduce a novel estimator, which returns the correct $\hat{m}$ even under subsampling.

3.3.1 Approximating Subsampling in Neural Networks

We introduced a plausible implementation of subsampling in our stochastic framework. When n out of N neurons are sampled, one can approximate subsampling by assuming that each spike is observed with a probability $p = n / N$. Then at each time t the subsampled activity is a sample of a binomial distribution, $a_t \sim \text{Bin}(A_t, p)$. From this, three important observations about our stochastic implementation of subsampling follow: (i) If the activity at two different times is equal, $A_t = A_{t'}$, the subsampled activities a_t and $a_{t'}$ are samples of the same distribution, but do not necessarily take the same value. (ii) *On average*, a_t is proportional to A_t, $\langle a_t \mid A_t \rangle = pA_t$. (iii) Assuming a separation of timescales, the subsampled avalanche sizes follow the same principles, $s_{\text{sub}} \sim \text{Bin}(s, p)$ by the additivity of binomial distributions. This observation is used in Sect. 5.

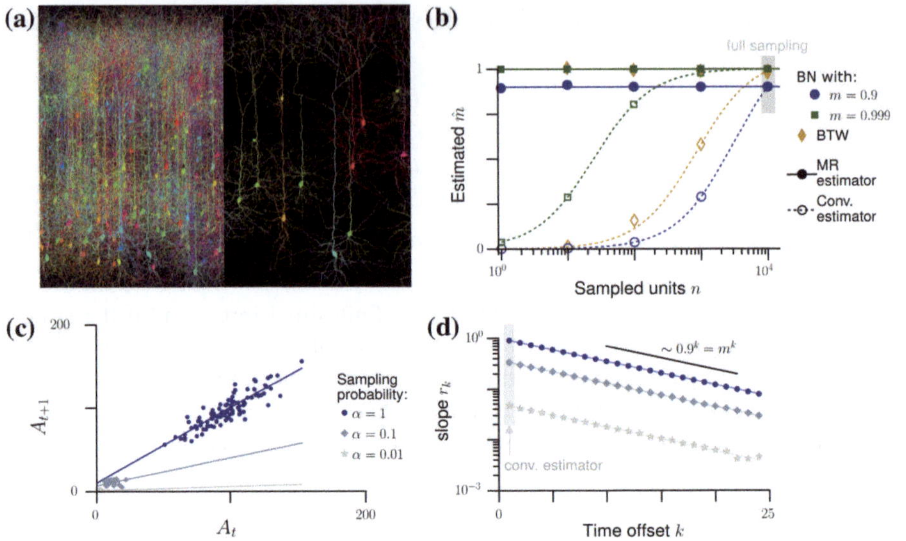

Fig. 2 Subsampling and multiple regression estimator. **a** In complex networks, such as the brain, often only a small subset of all units can be sampled (spatial subsampling); figure reproduced from [92]. **b** In recurrent networks (BN, Bak–Tang–Wiesenfeld model (BTW)), the conventional estimator (empty symbols) substantially underestimates the branching ratio m when less units n are sampled, as theoretically predicted (dashed lines). The novel multistep regression (MR) estimator (full symbols) always returns the correct estimate, even when sampling only 10 or 1 out of all $N = 10^4$ units. **c** MR estimation is exemplified for a subcritical branching process ($m = 0.9$, $h = 10$), where active units are observed with probability α. Under subsampling (gray), the regression slopes r_1 are smaller than under full sampling (blue). **d** While conventional estimation of m relies on the linear regression r_1 and is biased under subsampling, MR estimation infers $\hat{m}$ from the exponential relation $r_k \propto m^k$, which remains invariant under subsampling

3.3.2 The Established Estimators Are Biased Under Subsampling

We will now show that the linear regression estimator, which is consistent under full sampling, strongly underestimates m under subsampling. Under subsampling the linear regression estimator yields

$$\hat{m} = \frac{\mathrm{Cov}[a_t, a_{t+1}]}{\mathrm{Var}[a_t]} = p^2 \frac{\mathrm{Var}[A_t]}{\mathrm{Var}[a_t]} m = b\, m, \quad \text{with } b = p^2 \frac{\mathrm{Var}[A_t]}{\mathrm{Var}[a_t]}. \tag{11}$$

The bias factor b depends on the details of the investigated system and subsampling and is, in general, not known. Therefore, the subsampling bias is unknown and cannot be directly corrected for (Fig. 2b, c). We then show that the estimator proposed by Beggs and Plenz is also biased and overestimates m in the presence of input:

$$\sigma_{BP} = m + \langle \frac{h}{A_t} \rangle_{a_t > 0}. \tag{12}$$

Interestingly, the bias of this estimator is largely independent of subsampling. The subsampling here only enters through the probability to observe $a_t > 0$, because the average is only taken over such values. As before, the estimator is only reliable for $h = 0$.

(*) In order to gauge the performance of linear regression under subsampling, we need to evaluate the least square estimator $\text{Cov}[a_t, a_{t+k}] \,/\, \text{Var}[a_t]$ when applied to the subsampled time series a_t. We do so using the law of total expectation:

$$\langle a_t\, a_{t+1}\rangle = \langle\langle a_t\, a_{t+1} \mid A_t,\, A_{t+1}\rangle\rangle_{A_{t+1}, A_t},$$

with the inner expectation value being taken with respect to the joint distribution of a_{t+1} and a_t, and the outer with respect to the joint distribution of A_{t+1} and A_t. Conditioning on both A_t and A_{t+1} makes $(a_t \mid A_t)$ and $(a_{t+1} \mid A_{t+1})$ independent because of our stochastic implementation of subsampling. Hence, the expectation factorizes. By definition, $\langle a_t \mid A_t = j\rangle = p\, j$ and

$$\begin{aligned}\langle a_t\, a_{t+1}\rangle &= \langle (pA_{t+1})\,(pA_t)\rangle_{A_{t+1}, A_t}\\ &= p^2 \langle A_t\, A_{t+1}\rangle. \end{aligned} \tag{13}$$

Recalling $\langle A_t\, A_{t+1}\rangle = m\, \text{Var}[A_t]$ from Eq. (8), we find the linear regression estimator in Eq. (11):

$$\hat{m} = \frac{\text{Cov}[a_t, a_{t+1}]}{\text{Var}[a_t]} = p^2\, \frac{\text{Var}[A_t]}{\text{Var}[a_t]}\, m = b\, m, \quad \text{with } b = p^2\, \frac{\text{Var}[A_t]}{\text{Var}[a_t]}.$$

We evaluate the estimator proposed by Beggs and Plenz by similar steps to those before:

$$\begin{aligned}\sigma_{BP} = \langle \frac{a_{t+1}}{a_t} \rangle_{a_t > 0} &= \langle\langle \frac{a_{t+1}}{a_t} \mid A_t,\, A_{t+1}\rangle\rangle_{a_t>0}\\ &= \langle \frac{pA_{t+1}}{pA_t} \mid A_t,\, A_{t+1}\rangle_{a_t>0}\\ &= \langle\langle \frac{A_{t+1}}{A_t} \mid A_t\rangle\rangle_{a_t>0}\\ &= \langle \frac{mA_t + h}{A_t}\rangle_{a_t>0}\\ &= m + \langle \frac{h}{A_t}\rangle_{a_t>0}. \end{aligned} \tag{14}$$

3.3.3 Multistep Regression Estimator

We now introduce a novel estimator, which is unbiased even under subsampling (Fig. 2b). Instead of a single linear regression, we treat the general case of linear regression between pairs a_t and a_{t+k}. The least square estimator for the slope of this regression is denoted by $\hat{r}_k$. We show that each $\hat{r}_k$ returns a biased slope $b\, m^k$ under subsampling and not m^k as expected from Eq. (3). However, even though b is unknown, we showed that it is identical for all k. Therefore, the bias can be partialed out as described below.

Definition of the estimator. Derived from these observations, we proposed the following estimator:

1. Estimate multiple regression slopes $\hat{r}_k$ of linear regression between pairs a_t and a_{t+k} for different lags $k = 1, \ldots, k_{\max}$.
2. Fit these slopes to an exponential model $\hat{r}_k = \hat{b} \cdot \hat{m}^k$ in the parameters $\hat{b}$ and $\hat{m}$. The resulting estimate $\hat{m}$ is independent of the bias $\hat{b}$ and therefore consistent even under subsampling (Fig. 2b, d).

This estimator is called *multiple regressions* (MR) estimator.

Bias of individual regressions. We now explicitly derive that the biased slopes of regression between a_t and a_{t+k}, which are at the core of MR estimation, are given by

$$\hat{r}_k = \frac{\mathrm{Cov}[a_t, a_{t+k}]}{\mathrm{Var}[a_t]} = p^2\, \frac{\mathrm{Var}[A_t]}{\mathrm{Var}[a_t]}\, m^k, \tag{15}$$

i.e. the bias factor is precisely $\hat{b} = p^2\, \mathrm{Var}[A_t] \,/\, \mathrm{Var}[a_t]$. Note that the special case $k = 1$ corresponds to the conventional, biased linear regression estimator shown before.

(*) In order to calculate the covariance, we evaluate $\langle a_t\, a_{t+k}\rangle$ and $\langle a_{t+k}\rangle$ separately. Following the steps of Eqs. (7) and (13), we find

$$\langle a_t\, a_{t+k}\rangle = p^2 \langle\langle A_t\, A_{t+k} \mid A_t\rangle\rangle_{A_t} = p^2\, \left(m^k\, \langle A_t^2\rangle + (1 - m^k)\, \langle A_t\rangle^2\right) \tag{16}$$

and similarly

$$\langle a_{t+k}\rangle = \langle a_t\rangle = \langle\langle a_t \mid A_t\rangle\rangle_{A_t} = p\langle A_t\rangle = p\frac{h}{1-m}. \tag{17}$$

Finally, we combine Eqs. (16) and (17) to obtain an expression for the covariance:

$$\begin{aligned} \mathrm{Cov}[a_{t+k}, a_t] &= \langle a_{t+k}\, a_t\rangle \;-\; \langle a_{t+k}\rangle\, \langle a_t\rangle \\ &= p^2\, \left(m^k\, \langle A_t^2\rangle + (1 - m^k)\, \langle A_t\rangle^2\right) - p^2 \langle A_t\rangle^2 \\ &= p^2 m^k \mathrm{Var}[A_t]. \end{aligned} \tag{18}$$

Inserting this evaluation of the covariance into the definition of the least square estimator for the regression slope yields the result from Eq. (15):

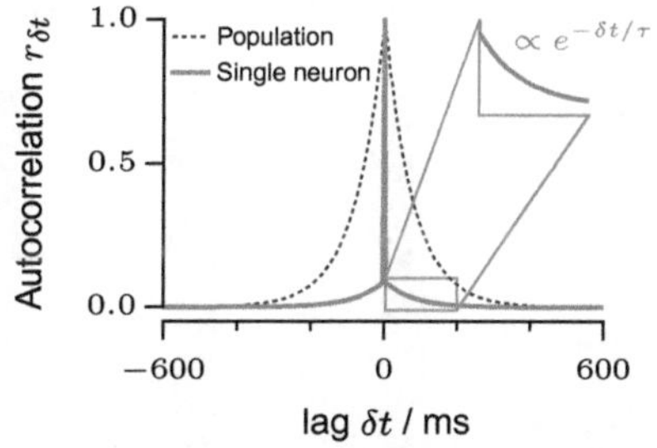

Fig. 3 A branching process has an exponential autocorrelogram of the population activity with decay time τ (blue dotted line). In contrast, the autocorrelogram of single neurons shows a sharp drop from $C(0) = 1$ at lag $\delta t = 0\,\text{s}$ to the next lag $C(\pm\Delta t)$ (gray solid line). We showed that this drop is simply a subsampling-induced bias. When ignoring the zero-lag value, the autocorrelation strength is decreased, but the autocorrelation time of the network activity is preserved in the activity of single neurons (inset). Panel reproduced from [93]

$$\hat{r}_k = \frac{\text{Cov}[a_t, a_{t+k}]}{\text{Var}[a_t]} = p^2 \frac{\text{Var}[A_t]}{\text{Var}[a_t]} m^k. \tag{19}$$

Autocorrelogram under subsampling. As seen before, the linear regression slopes are equal to the autocorrelogram in the subcritical case, $C(k\,\Delta t) = \hat{r}_k$. From the derivation of the MR estimator, it becomes obvious that (Fig. 3)

1. In contrast to the exponential autocorrelation function $C(k\,\Delta t) = \exp\left(-\frac{k\Delta t}{\tau}\right)$ of the fully sampled process, the autocorrelation of the subsampled activity typically exhibits a sharp drop from zero lag ($C(0) = 1$) to the first bin ($C(\Delta t) = b\,m$, $b \ll 1$).
2. Even though the autocorrelation *strength* is reduced by a factor b, the autocorrelation *time* is conserved in the floor of the autocorrelogram, $C(k\,\Delta t) = b\,\exp\left(-\frac{k\Delta t}{\tau}\right)$ for $k \neq 0$.

Overall, branching processes offer a well-understood approach to describing systems on a continuous spectrum, including a critical phase transition, by a single control parameter m. While estimating m is straight-forward when the activity of all neurons is known, established approaches fail under subsampling. The novel MR estimator infers m even under strong subsampling. It relies on the fact that, even though the autocorrelation *strength* is reduced, the autocorrelation *time* is preserved in the floor of the autocorrelogram.

3.4 Application to Real-World Time Series

We demonstrated the bias of conventional estimation and the robustness of MR estimation at the example of two real-world applications: epidemiological case counts and cortical population activity [92].

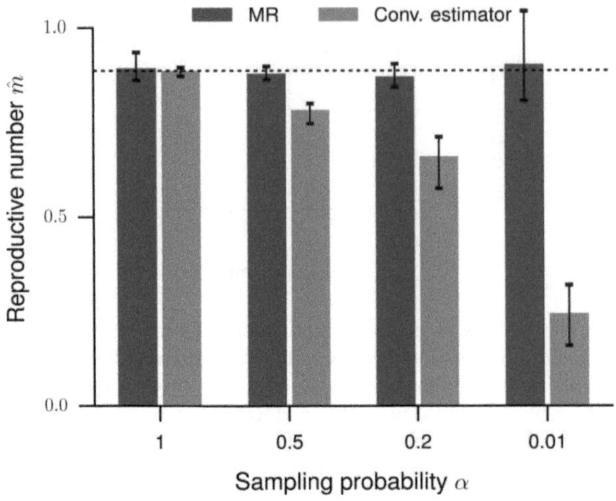

Fig. 4 In epidemic models, the reproductive number m can serve as an indicator for the infectiousness of a disease within a population, and predict the risk of large incidence bursts. We have estimated $\hat{m}$ from incidence time series of measles in Germany, which can be assumed to be fully sampled because of the clarity of symptoms and the strict reporting policies. When artificially subsampling the measles recording (under-ascertainment), conventional estimation underestimates $\hat{m}_C$, while MR estimation still returns the correct value. Both estimators return the same $\hat{m}$ under full sampling. Panel reproduced from [92]

3.4.1 Spread of Measles

We used the MR estimator to infer the "reproductive number" m from incidence time series of different diseases [21]. Disease propagation represents a nonlinear, complex, real-world system often approximated by branching processes [14, 25]. Here, m determines the disease spreading behavior and has been deployed to predict the risk of epidemic outbreaks [27]. However, the problem of subsampling or *under-ascertainment* has always posed a challenge [37, 63].

As a first step, we cross-validated the novel against the conventional estimator using the spread of measles in Germany, surveyed by the Robert-Koch-Institute (RKI). We chose this reference case, because we expected case reports to be almost fully sampled owing to the strict reporting policy supported by child care facilities and schools [39, 90], and to the clarity of symptoms. Indeed, the values for $\hat{m}$ inferred with the conventional and with the novel estimator, coincided (Fig. 4). In contrast, after applying artificial subsampling to the case reports, thereby mimicking that each infection was only diagnosed and reported with probability $p < 1$, the conventional estimator severely underestimated the spreading behavior, while MR estimation always returned consistent values. This shows that the MR estimator correctly infers the reproductive number m directly from subsampled time series, without the need to know the degree of under-ascertainment α.

3.4.2 Cortical Spiking Activity

We applied the MR estimator to cortical spiking activity in vivo (Fig. 5). Analyzing in vivo spiking activity from Macaque monkey prefrontal cortex during a memory task, anesthetized cat visual cortex with no stimulus (Figs. 5a, b), and rat hippocampus during a foraging task returned $\hat{m}$ to be between 0.963 and 0.998 (median $\hat{m} = 0.984$, Fig. 5d), corresponding to autocorrelation times between 100 and 2000 ms. This clearly suggests that spiking activity in vivo is neither AI-like ($m = 0$), nor consistent with a critical state ($m = 1$), but in a reverberating state that shows autocorrelation times of a few hundred milliseconds. We call this range of the dynamical states *reverberating*, because input reverberates for a few hundred millisecond in the network, and therefore enables integration of information [16, 41, 59, 91]. A branching network set up in this reverberating regime reproduced experimental avalanche size distributions better than AI-like or near-critical ($m = 0.9999$) networks (Fig. 5f).

When choosing random subsets of n neurons from the total of 50 recorded single units, even for single neurons, MR estimation returned about the same median $\hat{m}$ (Fig. 5c). In contrast, the conventional estimator misclassified neuronal activity by strongly underestimating $\hat{m}$: instead of $\hat{m} = 0.984$, it returned $\hat{m}_C = 0.271$ for the activity of all 50 neurons.

4 Assessing Network Dynamics Using Neuronal Avalanches

The classical approach to test for criticality is to assess the distributions of neuronal avalanches. Avalanche size and the duration distributions following a power law are taken as strong evidence that the underlying system is indeed critical. This approach to assess criticality has been inspired by the study of Peer Bak, Chao Tang and Kurt Wiesenfeld, who demonstrated power-law distributed avalanches or cascades in a simple cellular automaton model [4]. That cellular automaton approximated the dynamics on the slope of a generic sand-pile: Adding a single grain could trigger other grains to "topple", and thereby a cascade or avalanche of activity spread over the system. The sizes and the durations of the cascades are power-law distributed, meaning that most avalanches only show one or maybe two topplings, and only very few spread the entire system. However, there is no "typical" avalanche size, meaning that mean and the variance are not sufficient to describe the distribution. Because the emergence of these power laws is independent of initial conditions, they termed the resulting dynamical state "self-organized critical", stressing that the critical state is an attractor. Finding that a critical state can be a stable attractor had major impact, because it promised to explain the emergence of power laws in many very diverse biological and physical systems, from earthquakes, to forest fires and solar flares [5, 22, 54]. Potentially the first researches to point out that networks of (non-leaky) integrate-and-fire neurons readily display criticality, were Swen Dunkelmann and Günter Radons in 1994 [24]. They also revealed that the emerging long-range correlations can be exploited to enhance computational properties. In the following

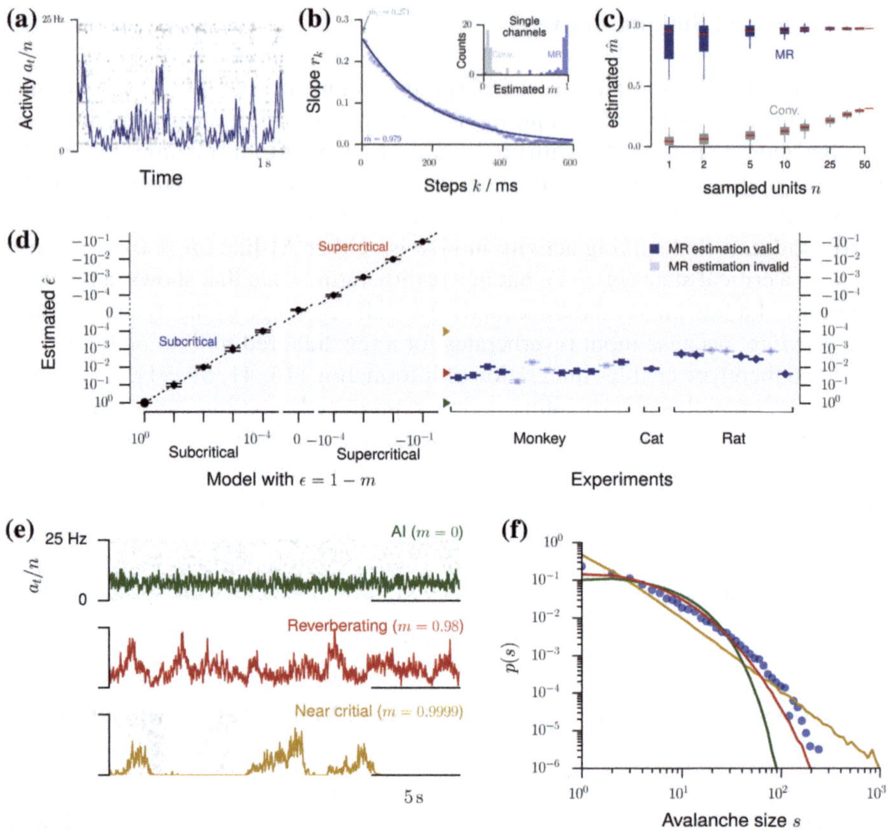

Fig. 5 **Animal spiking activity in vivo**. In neuroscience, m denotes the mean number of spikes triggered by one spike. We estimated $\hat{m}$ from spiking activity recorded in vivo in monkey prefrontal cortex, cat visual cortex, and rat hippocampus. **a** Raster spike plot and population rate a_t of 50 single units illustrated for cat visual cortex. **b** MR estimation based on the exponential decay of the autocorrelation of r_k of a_t. Inset: Comparison of conventional and MR estimation results for single units (medians $\hat{m}_C = 0.057$ and $\hat{m} = 0.954$ respectively). **c** $\hat{m}$ estimated from from further subsampled cat recordings, estimated with the conventional and MR estimator. Error bars indicate variability over 50 randomly subsampled n out of the recorded 50 channels. **d** Avalanche size distributions for cat visual cortex (blue) and the networks with AI, reverberating and near-critical dynamics in panel **f**. **e** For all simulations, MR estimation returned the correct distance to instability (criticality) $\varepsilon = 1 - m$. In vivo spike recordings from rat, cat, and monkey, clearly differed from critical ($\varepsilon = 0$) and AI ($\varepsilon = 1$) states (median $\hat{m} = 0.98$, error bars: 16–84% confidence intervals, note that some confidence intervals are too small to be resolved). Opaque symbols indicate that MR estimation was rejected. Green, red, and yellow arrows indicate ε for the dynamic states shown in panel **f**. **f** Population activity and raster plots for AI activity, reverberating, in vivo-like, and near critical networks. All three networks match the recording from cat visual cortex with respect to number of recorded neurons and mean firing rate. Figure reproduced from [92]

decade, a number of neural network models with increasing degrees of physiological detail demonstrated criticality [12, 13, 17–19, 26, 42], soon followed by first experimental evidence [52, 94]. The most prominent experimental demonstration of criticality in neural networks is arguably the study by John Beggs and Dietmar Plenz in 2003, who demonstrated that neural avalanches extracted from population spikes in acute slices display clear power-laws for their sizes and durations [7]. Since then, the analysis of neuronal avalanches has become a standard approach to test for criticality in the brain.

Extracting neuronal avalanches from neuronal recordings is in principle straightforward. The binary data from all observed units is collapsed to one point process and binned with the fixed bin width Δt. Units used for binning could be spikes [64], thresholded LFPs [7], thresholded BOLD signal [83], or MEG activity [62]. Then each empty bin signifies the pause between the avalanches. The sequence of the non-empty bin surrounded by the empty ones is called an avalanche. The size of the avalanche can be measured, for example, as the sum of all bin-values within the avalanche. The duration is the number of bins that avalanche is occupying. As we see, there is one main parameter, Δt that specifies how the avalanches will be defined. In the ideal case, sizes and durations are invariant to the particular choice of Δt within a certain (preferably, large) intervals. This situation will correspond to the time-scale separation between process starting the avalanche and the propagation of activity within the avalanche. However, often bin-size plays important role in defining the avalanche distributions [7].

5 Subsampling Scaling: A Theory About Inferring the Avalanche Distribution of the Full System

Spatial subsampling can alter the observed avalanche distribution, and inferring the distribution of the full system can be very difficult (see Sect. 2). To overcome the bias, the approach of "subsampling scaling" has been derived [50]. Subsampling scaling is applicable to different observables, including distributions of neuronal avalanches, of number of people infected during an epidemic outbreak, and of node degrees. It allows one to infer the "parent distribution", i.e. the distributions of the full system, from the subsampled one. It is applied to distinguish critical from subcritical systems, and to disentangle subsampling and finite size effects. Last but not least, by analyzing spike recordings from developing neural networks, we showed that only mature, but not young networks follow power-law scaling, indicating self-organization to criticality during development

Examples of subsampling biases, some of them dramatic, have already been demonstrated in numerical studies. Subsampling has been shown to affect avalanche distributions in various ways, which can make a critical system appear sub- or super-critical [67, 69, 70, 75, 76, 92, 96]. When inferring not avalanches, but the topology of networks, it has been derived that, contrary to common intuition, a subsample from a scale-free network is not itself scale-free [81]. Similarly, sampling from a locally connected network can make the network appear "small-world" [29]. Importantly, these biases are not due to limited statistics (which could be overcome by collecting more data, e.g. acquiring longer recordings, or more independent subsamples of a system), but genuinely originates from observing a small fraction of the system, and then making inferences including unobserved parts. Although the existence of subsampling effects are known, in the literature there is so far no general analytical understanding of how to overcome them.

A derivation of subsampling effects for degree distributions of a network has been conducted by Stumpf and colleagues, who provided a first analytical framework, stating the problem of subsampling bias [81]. Interestingly, we could show that these results in general can be translated to subsampling effects on avalanches, and this allowed us to then derive an analytical framework that enables us to infer the 'parent distribution', i.e. the avalanche distribution of the full system, from the subsampled one [50]. The analytical approach is outline below, together with numerical validations as well as applications to spike recordings of developing neural networks. Together, these results introduced a novel, rigorous approach to study under-observed systems.

5.1 Theory of Subsampling Scaling

To derive how spatial subsampling affects the avalanche size distribution p, we define a minimal model of "mathematical subsampling". The avalanche size s is the total number of events or spikes within a cascade. In general, the cluster size is described by a discrete, non-negative random variable X. Let X be distributed according to a probability distribution $P(X = s) = P(s)$. For subsampling, assume for each avalanche that each of its events is independently observed with probability p (or missed with probability $1 - p$). Then X_{sub} is a random variable denoting the number of *observed* events of an avalanche, and $X - X_{\mathrm{sub}}$ the number of missed events. This subsampling scheme assumes that a random set of the neurons in the network is sampled, and X_{sub} represents the number of all events generated by the *observed* units within one avalanche on the full system. Note, that this definition translates one cluster in the full system to exactly one cluster under subsampling (potentially of size zero), and that this definition does not require explicit binning [50]. The probability distribution of X_{sub} is called in the following the "subsampled distribution" $P_{\mathrm{sub}}(s)$. Importantly, the framework does not only apply to (neural) avalanches, but an analogous treatment can be applied to e.g. graphs. There a "cluster" represents the set of (directed) connections of a specific node, and thus X is the degree of that node. Under subsampling, i.e. considering a random subnetwork, only connections between *observed* nodes are taken into account, resulting in the subsampled degree X_{sub}.

As each event is observed independently, the probability of $X_{\mathrm{sub}} = s$ is the sum over probabilities of observing clusters of $X = s + k$ events, where k denotes the missed events and s the sampled ones (binomial sampling):

$$P_{\text{sub}}(s) = P(X_{\text{sub}} = s) = \sum_{k=0}^{\infty} P(s+k) \binom{s+k}{s} p^s (1-p)^k. \tag{20}$$

This equation holds for any discrete $P(s)$ defined on $\mathbb{N}_0$, the set of non-negative integers. To infer $P(s)$ from $P_{\text{sub}}(s)$, we develop in the following a novel "subsampling scaling" that allows to parcel out the changes in $P(s)$ originating from spatial subsampling. A correct scaling ansatz collapses the $P_{\text{sub}}(s)$ for any sampling probability p.

Two specific families of distributions are of particular importance in the context of neuroscience, namely exponential distributions $P(s) = C_\lambda e^{-\lambda s}$ with $\lambda > 0$, and power laws $P(s) = C_\gamma s^{-\gamma}$ with $\gamma > 1$. These two families are known to show different behaviors under subsampling [81] (see also [50] for more details):

1. For exponential distributions, $P(s)$ and $P_{\text{sub}}(s)$ belong to the same class of distributions, only their parameters change under subsampling. Notably, this result generalizes to positive and negative binomial distributions, which include Poisson distributions.
2. Power-laws or scale-free distributions, despite their name, are not invariant under subsampling. Namely, if $P(s)$ follows a power-law distribution, then $P_{\text{sub}}(s)$ is not a power law but only approaching it in the limit of large cluster size ($s \to \infty$).

In more detail, for exponential distributions, $P(s) = C_\lambda e^{-\lambda s}$, $s \in \mathbb{N}_0$, subsampling with probability p results in an exponential distribution with decay parameter λ_{sub} that can be expressed as a function of λ and p (full analytical derivation uses probability generating function discussed in details in supplementary information of [50]):

$$\lambda_{\text{sub}} = \ln\left(\frac{e^\lambda + p - 1}{p}\right) \Leftrightarrow \lambda = \ln((e^{\lambda_{\text{sub}}} - 1)p + 1). \tag{21}$$

Likewise, changes in the normalizing constant $C_\lambda = 1 - e^{-\lambda}$ of $P(s)$ are given by:

$$C_\lambda / C_{\lambda_{\text{sub}}} = 1 - e^{-\lambda} + pe^{-\lambda} = \frac{e^{-\lambda_{\text{sub}}} + p - pe^{-\lambda_{\text{sub}}}}{p}. \tag{22}$$

These two relations allow to derive explicitly a subsampling scaling for exponentials, i.e. the relation between $P(s)$ and $P_{\text{sub}}(s)$:

$$\begin{aligned} P(s) &= \frac{C_\lambda}{C_{\lambda_{\text{sub}}}} P_{\text{sub}}\left(\frac{\lambda}{\lambda_{\text{sub}}} s\right) = \frac{e^{-\lambda_{\text{sub}}} + p - pe^{-\lambda_{\text{sub}}}}{p} P_{\text{sub}}\left(\frac{\ln\left(e^{\lambda_{\text{sub}}} p - p + 1\right)}{\lambda_{\text{sub}}} s\right) \\ &= \left(1 - e^{-\lambda} + pe^{-\lambda}\right) P_{\text{sub}}\left(\frac{\lambda}{\ln\left(\frac{e^\lambda + p - 1}{p}\right)} s\right) = G(p,\lambda) P_{\text{sub}}(sF(p,\lambda)). \end{aligned} \tag{23}$$

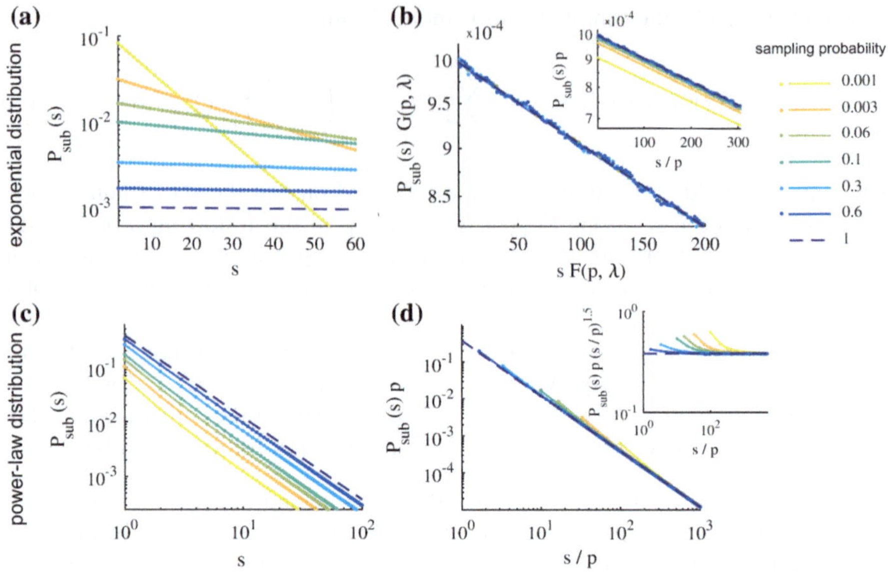

Fig. 6 Mathematical subsampling of exponential and power-law distributions. **a** Subsamplings of an exponential distribution with exponent $\lambda = 0.001$. **b** Collapse of subsampled exponential distributions by subsampling scaling derived in Eq. 23. Inset: same with p-scaling (Eq. 25). **c** Subsampled power-law distributions with exponent $\gamma = 1.5$. **d** Collapse of the same distributions by p-scaling (Eq. 25); inset: flattened version. Note the log-linear axes in **a**, **b**, and the double-logarithmic axes in **c**, **d**. Solid lines are analytical results (Eq. 20), dots are numerical results from subsampling 10^7 avalanches (realizations of the random variable X) of the corresponding original distribution. Colors indicate the sampling probability p

Thus given an exponential distribution $P(s)$ of the full system, all distributions under subsampling can be derived. Vice versa, given the observed subsampled distribution $P_{\text{sub}}(s)$, the full distribution can be analytically derived if the sampling probability p is known. Therefore, for exponentials, the scaling ansatz above allows to collapse all distributions obtained under subsampling with any p (Fig. 6a, b).

The presented formalism is analogous to the one proposed by Stumpf et al. [81]. They studied which distributions changed and which preserved their classes under subsampling. In the following we extend that study, and then develop a formalism that allows to extrapolate the original distribution from the subsampling, also in the case where an exact solution is not possible.

For power-law distributions of X, X_{sub} is not power-law distributed, but only approaches a power law in the tail ($s \to \infty$). An approximate scaling relation, however, collapses the tails of distributions as follows (mathematical derivation see in supplementary information of [50]). For $s \to \infty$, a power law $P(s) = C_\gamma s^{-\gamma}$ and the distributions obtained under subsampling can be collapsed by:

$$P(s) = p^a P_{\text{sub}}(p^b s), \quad \text{for any } a, b \in \mathbb{R} \text{ with } a - b\gamma = 1 - \gamma. \tag{24}$$

For any a, b satisfying the relation above, this scaling collapses the tails of power-law distributions. The "heads", however, deviate from the power law and hence cannot be collapsed (see deviations at small s, Fig. 6d). These deviations decrease with increasing p, and with $\gamma \rightarrow 1^+$ [81]. We call these deviations "*hairs*" because they "grow" on the "heads" of the distribution as opposed to the tails of the distribution. In fact, the hairs allow to infer the system size from knowing the number of sampled units if the full systems exhibits a power-law distribution (for discussion see supplementary information of [50]).

The different behavior of exponential and power-law distributions under subsampling poses a considerable challenge if one aims at inferring the parent distribution for neural avalanches. This is because avalanche distributions from neural data (but also many other systems) are often characterized by a power law with an an exponential tail, which represents finite-size effects or subcriticality [4, 23, 70, 80, 85]. The different subsampling behavior of the tail of the distribution (which is typically exponential), and the head (which shows a power law) in principle hinders a joint treatment. However, for the the case that is typical for finite size critical systems, namely a power law that transits smoothly to an exponential around $s = s^{\text{cutoff}}$, we could identify an approximate scaling ansatz, as follows.

Under subsampling, $s_{\text{sub}}^{\text{cutoff}}$ depends linearly on the sampling probability: $s_{\text{sub}}^{\text{cutoff}} = p \cdot s^{\text{cutoff}}$. Hence, the only solution to the power-law scaling relation (Eq. 24) that collapses (to the best possible degree), both, the power-law part of distributions *and* the onsets of the cutoff is the one with $a = b = 1$:

$$P(s) \approx pP_{\text{sub}}(p \cdot s). \tag{25}$$

As this scaling is linear in p, we call it *p-scaling*. A priori, p-scaling is different from the scaling for exponentials (Eq. 23). However, p-scaling is a limit case of the scaling for exponentials under the assumption that $\lambda \ll p$ [50]. Thus p-scaling collapses power laws with exponential tail if λ is small, and also much smaller than the sampling probability. This condition is typically met in critical, but not in subcritical systems. Indeed, when applying subsampling to critical models, p-scaling collapsed the avalanche size distributions $P_{\text{sub}}(s)$ obtained under subsampling (Fig. 7a, b). This is not the case for the $P_{\text{sub}}(s)$ obtained from sub-critical models (Fig. 7c). Thereby, p-scaling promises to distinguish critical from non-critical systems in experiments as well.

5.2 *Assessing Criticality from Spike Recordings: Learning More by Sampling Less*

We applied p-scaling to neural recordings of developing networks in vitro to investigate whether their avalanches indicated a critical state. To this end, we evaluated recordings from $N = 58$ multi-units (see Methods, [87]). This is only a small

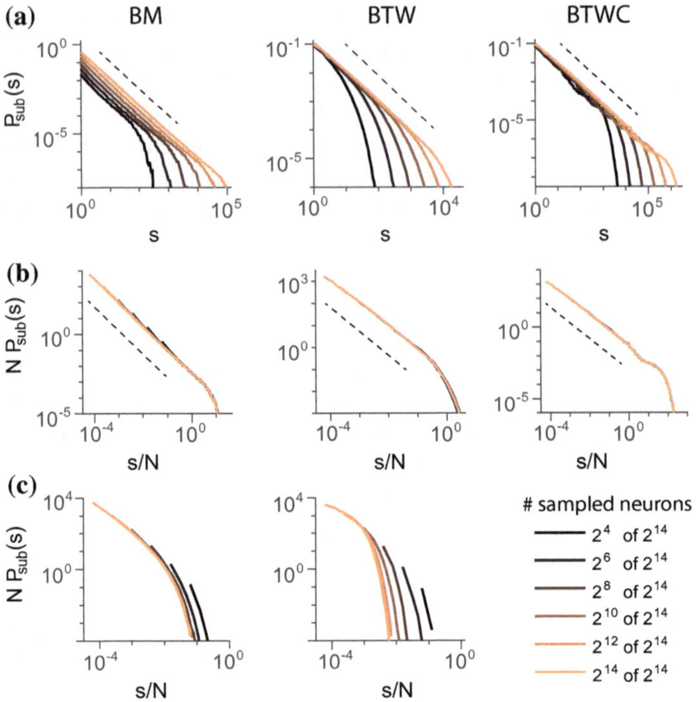

Fig. 7 Subsampling scaling in critical and subcritical models. The three columns show results for the branching model (BM), the Bak–Tang–Wiesenfeld model (BTW) [4], and the BTW with periodic boundary conditions (BTWC). **a** Avalanche size distribution $P_{\text{sub}}(s)$ for different degrees of subsampling, as denoted in the legend. **b** Same distributions as in A, but with p-scaling. (Note that scaling by N leads to a collapse equivalent to scaling by $p = N/M$ at fixed system size M). **c** Scaled distributions from subcritical versions of the models. Here, results for the BTWC are extremely similar to those of the BTW and are thus omitted. Dashed lines indicate power-law slopes of -1.5 and -1.1 for the BM and BTW/BTWC, respectively, for visual guidance

fraction of the entire neural network, which comprised $M \approx 50{,}000$ neurons. Thus the avalanche size distribution obtained from the whole analyzed data is already a subsampled distribution $P_{\text{sub}}(s)$. To apply p-scaling, we generated a family of distributions by *further subsampling*, i.e. evaluating a subset $N' < N$ of the recorded units. The intuition behind this approach is the following: When randomly sampling the recording electrodes, we create different subsamples from the original system, similar to what would happen if the recording system was smaller. If these artificially subsampled systems behave very different one from another and allow for no scaling, we cannot argue about the distribution in the entire system even based on the recording from the whole MEA. However, if any scaling collapses the family of these subsampled arrays, we can extrapolate the distribution from what we recorded. In particular, in critical systems, p-scaling is expected to collapse this family of distributions if avalanches are defined unambiguously.

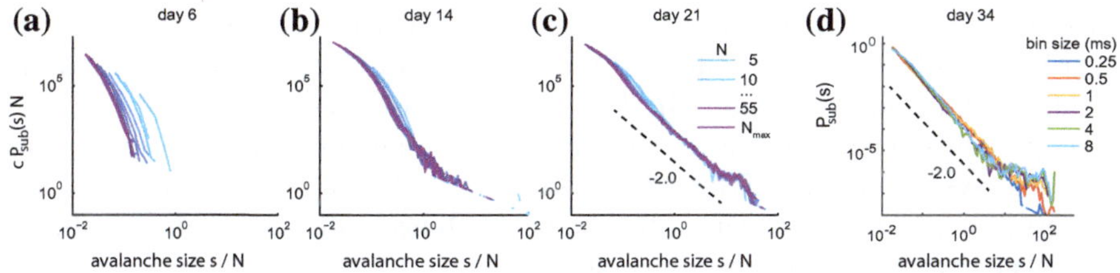

Fig. 8 Avalanche size distributions $c \cdot P_{\text{sub}}(s)$ (in absolute counts) of spiking activity of developing neural networks in vitro. **a**: For young cultures, $P_{\text{sub}}(s)$ did not collapse under p-scaling, indicating that the full network does not show a power-law distribution for $P_{\text{sub}}(s)$. **b, c** More mature networks show a good collapse, allowing to extrapolate the distribution of the full network. In panels **a, b**, and **c** the bin size is 1 ms, and c is a total number of recorded avalanches in the full system, in **a** $c = 53{,}803$, in **b** $c = 307{,}908$, in **c** $c = 251{,}156$. The estimated number of neurons in the cultures is $M \approx 50{,}000$. **d** $P_{\text{sub}}(s)$ from sampling spikes from all electrodes but evaluated with different bin sizes (see legend); the approximate invariance of $P_{\text{sub}}(s)$ against changes in the bin size indicates a separation of time scales in the experimental preparation. Figure is reproduced from [50], for distribution of all recording days of all experiments see SI of [50]

Interestingly, for early stages of neural development, p-scaling does not collapse $P_{\text{sub}}(s)$, but for the more mature networks we found a clear collapse (Fig. 8). Thus developing neural networks start off with collective dynamics that is not in a critical state, but with maturation approach criticality [64, 84]. Some of the mature networks show small bumps in $P_{\text{sub}}(s)$ at very large avalanche sizes ($s \approx 5000 \Leftrightarrow s/N \approx 60$). These very large avalanches comprise only a tiny fraction of all avalanches (about 2 in 10,000). At first glance, the bumps are reminiscent of supercritical systems. However, supercritical neural models typically show bumps at system or sampling size ($s = N$), not at those very large sizes. We suggest that the bumps are more likely to originate from neurophysiological finite size effects [50, 97].

For the full, mature network, our results predict that $P(s)$ would extend not only over three orders of magnitude as here, but over six, because $p \approx 10^{-3}$. Our analysis of neural recordings illustrates how further spatial subsampling allows to infer properties of the full system, even if only a tiny fraction of its collective dynamics has been observed, simply by sampling even less ($N' < N$) of the full system.

5.3 *Subsampling Versus Finite Size Scaling*

In the real world we are often confronted with data affected by both subsampling and finite system size effects, i.e. observations originated from a small part of a large, but not infinite system. Thus we need to deal with a combination of both: subsampling effects as a result of incomplete data acquisition and finite-size effects inherited from the full system. To disentangle influences from system size and system dynamics, finite size scaling (FSS) has been introduced [49, 71]. It allows to infer the behavior of an infinite system from a set of finite systems. At a first glance, finite size and subsampling effects may appear to be very similar. However, if they were, then distributions obtained from sampling N units from any system with $N \leq M$ would be identical, i.e. independent of M. This is not the case, as shown in [50].

Importantly, it is possible to combine FSS and subsampling scaling [50]: Consider a critical system, where FSS is given by: $M^{\beta} P(sM^{\nu}; M) = g(s)$, here $g(s)$ is a universal scaling function. Then FSS can be combined with subsampling scaling to obtain a universal subsampling-finite-size scaling:

$$NM^{\beta-1} P_{\text{sub}}(sNM^{\nu-1}; M, N) = g(s). \tag{26}$$

Using Eq. 26 allows to infer the distribution for arbitrary subsampling (N) of any system size (M).

5.4 *Open Questions*

Subsampling scaling provided an understanding of how distributions of avalanches change under spatial subsampling, and how to infer the distribution of the fully sampled system from the subsampled one. To this end, it is essential that the avalanches are extracted unambiguously, i.e. one avalanche in the full system translates to exactly one avalanche (potentially of size zero) under subsampling. This condition is fulfilled easily if the system shows a separation of time scales (STS), i.e. the pauses between subsequent avalanches are much longer than the avalanches themselves. Given a STS, temporal binning [7] can be used to unambiguously extract avalanches under subsampling. However, the chosen bin size must neither be too small nor too large: If too small, a single avalanche on the full system can be "cut" into multiple ones when entering, leaving, and re-entering the recording set. This leads to steeper $P_{\text{sub}}(s)$ with smaller bin size (Fig. 8 d). In contrast, if the bin size is too large, subsequent avalanches can be "merged" together. For a range of intermediate bin sizes, however, $P_{\text{sub}}(s)$ is invariant to changes in the bin size. If a system, however, lacks a STS, then $P_{\text{sub}}(s)$ is expected to change for any bin size. This may underlie the frequently observed changes in $P_{\text{sub}}(s)$ in neural recordings [3, 7, 32, 67, 69, 75, 80], as discussed in [70]. The first step toward relaxing a time-scale separation constrain is to investigate systems where external input can be delivered during the avalanche, but separation of avalanches is preserved. In this case power-law scaling remains, but the exponent is changed [20]. Hence an important open question is how to cutting and merging of avalanches affect subsampling scaling.

The present study analytically treats subsampling scaling for power laws (with cutoff), exponential distributions, and negative and positive binomial distributions. For all other distributions, utmost care has to be taken when aiming at inferences about the full system from its subsampling. One potential approach is to identify a scaling ansatz numerically, i.e. minimizing the distance between the different $P_{\text{sub}}(s)$ numerically, in analogy to the approach for avalanche shape collapse [3, 28, 55, 69, 78]. We found that for our network simulations such a numerical approach identified the same scaling parameters as our analytic derivations. However, given the typical noisiness of experimental observations, a purely numerical approach should be taken with a grain of salt, as long as it is not backed up by a circular form analytical solution.

Our analytical derivations assumed annealed sampling, which in simulations was well approximated by pre-choosing a random subset of neurons or nodes for sampling. Any sampling from randomly connected networks is expected to lead to the same approximation. However, in networks with e.g. local connectivity, numerical results depend strongly on the choice of sampled units [67]. For example, for windowed subsampling (i.e. sampling a local set of units) a number of studies reported strong deviations from the expected power laws in critical systems or scale free networks [29, 67, 75]. In contrast, random subsampling, as assumed here for our analytical derivations, only leads to minor deviations from power laws (hairs). Thus to diminish corruption of results by subsampling, future experimental studies on criticality should aim at implementing random instead of the traditional windowed sampling, e.g. by designing novel electrode arrays with pseudo-random placement of electrodes on the entire area of the network. In this case, we predict deviations from power laws to be minor, i.e. limited to the "hairs" and the cutoff.

We present here first steps towards a full understanding of subsampling. With our analytical, mean-field-like approach to subsampling we treat two classes of distributions and explore corresponding simulations. In future, extending the presented approach to a window-like sampling, more general forms of correlated sampling, and to further classes of distributions will certainly be of additional importance to achieve unbiased inferences from experiments and real-world observations.

References

1. Ali, W., Wegner, A.E., Gaunt, R.E., Deane, C.M., Reinert, G.: Comparison of large networks with sub-sampling strategies. Sci. Rep. **6**, 28955 (2016)
2. Aru, J., Aru, J., Priesemann, V., Wibral, M., Lana, L., Pipa, G., Singer W., Vicente R.: Untangling cross-frequency coupling in neuroscience. Curr. Opin. Neurobiol. **31**(September 2014), 51–61, 2015
3. Arviv, O., Goldstein, A., Shriki, O.: Near-critical dynamics in stimulus-evoked activity of the human brain and its relation to spontaneous resting-state activity. J. Neurosci. **35**(41), 13927–13942 (2015)
4. Bak, P., Tang, C., Wiesenfeld, K.: Self-organized criticality: an explanation of $1/f$ noise. Phys. Rev. Lett. **59**, 381–384 (1987)
5. Bak, P., Chen, K., Creutz, M.: Self-organized criticality in the "Game of Life". Nature **342**(6251), 780–782 (1989)
6. Bédard, C., Kröger, H., Destexhe, A.: Does the 1/f frequency scaling of brain signals reflect self-organized critical states? Phys. Rev. Lett. **97**(11), 1–4 (2006)
7. Beggs, J., Plenz, D.: Neuronal avalanches in neocortical circuits. J. Neurosci **23**, 11167–11177 (2003)
8. Beggs, J., Plenz, D.: Neuronal avalanches are diverse and precise activity patterns that are stable for many hours in cortical slice cultures. J. Neurosci. **24**(22), 5216–5229 (2004)
9. Bellay, T., Klaus, A., Seshadri, S., Plenz, D.: Irregular spiking of pyramidal neurons organizes as scale-invariant neuronal avalanches in the awake state. eLife **4**, 1–25 (2015)

10. Bertschinger, N., Natschläger, T.: Real-time computation at the edge of chaos in recurrent neural networks. Neural Comput. **16**(7), 1413–1436 (2004)
11. Boedecker, J., Obst, O., Lizier, J.T., Mayer, N.M., Asada, M.: Information processing in echo state networks at the edge of chaos. Theory Biosci. **131**(3), 205–213 (2012)
12. Bornholdt, S., Rohlf, T.: Topological evolution of dynamical networks. Phys. Rev. Lett. **84**(26), 6114–6117 (2000)
13. Bornholdt, S., Röhl, T.: Self-organized critical neural networks. Phys. Rev. E **67**(6), 066118 (2003)
14. Brockmann, D., Hufnagel, L., Geisel, T.: The scaling laws of human travel. Nature **439**(7075), 462–5 (2006)
15. Buzsáki, G., Anastassiou, C.A., Koch, C.: The origin of extracellular fields and currents-EEG, ECoG, LFP and spikes. Nat. Rev. Neurosci. **13**(6), 407–420 (2012)
16. Chaudhuri, R., Knoblauch, K., Gariel, M.A., Kennedy, H., Wang, X.J.: A large-scale circuit mechanism for hierarchical dynamical processing in the primate cortex. Neuron **88**(2), 419–431 (2015)
17. Chen, D.-M., Wu, S., Guo, A., Yang, Z.: Self-organized criticality in a cellular automaton model of pulse-coupled integrate-and-fire neurons. J. Phys. A Math. Gen. **28**, 5177 (1995)
18. Corral, Á., Pérez, C.J., Díaz-Guilera, A.: Self-organized criticality induced by diversity. Phys. Rev. Lett. **78**(8), 1492–1495 (1997)
19. Da Silva, L., Papa, A.R.R., De Souza, A.M.C.: Criticality in a simple model for brain functioning. Phys. Lett. Sect. A Gen. At. Solid State Phys. **242**(6), 343–348 (1998)
20. Das, A., Levina, A.: Critical neuronal models with relaxed timescales separation (2018). arXiv:1808.04196
21. Diekmann, O., Heesterbeek, J.A.P., Metz, J.A.J.: On the definition and the computation of the basic reproduction ratio R0 in models for infectious diseases in heterogeneous populations. J. Math. Biol. **28**(4), 365–382 (1990)
22. Drossel, B., Schwabl, F.: Self-organized critical forest-fire model. Phys. Rev. Lett. **69**, 1629–1632 (1992)
23. Drossel, B.: Self-organized criticality and synchronization in a forest-fire model. Phys. Rev. Lett. **76**(6), 936–939 (1996)
24. Dunkelmann, S., Radons, G.: Neural networks and abelian sandpile models of self-organized criticality. In: Marinaro, M., Morasso, P.G. (eds.) ICANN '94 Proceedings of the International Conference on Artificial Neural Networks, p. 867. Springer, Berlin (1994)
25. Earn, D.J.: A simple model for complex dynamical transitions in epidemics. Science **287**(5453), 667–670 (2000)
26. Eurich, C.W., Herrmann, M., Ernst, U.: Finite-size effects of avalanche dynamics. Phys. Rev. E **66**, 066137–1–15 (2002)
27. Farrington, C.P., Kanaan, M.N., Gay, N.J.: Branching process models for surveillance of infectious diseases controlled by mass vaccination. Biostat. (Oxford, England) **4**(2), 279–95 (2003)
28. Friedman, N., Ito, S., Brinkman, B.A.W., Shimono, M., Lee DeVille, R.E., Dahmen, K.A., Beggs, J.M., Butler, T.C.: Universal critical dynamics in high resolution neuronal avalanche data. Phys. Rev. Lett. **108**(20), 208102 (2012)
29. Gerhard, F., Pipa, G., Lima, B., Neuenschwander, S., Gerstner, W.: Extraction of network topology from multi-electrode recordings: is there a small-world effect? Front. Comput. Neurosci. **5**, 4 (2011)
30. González-Bailón, S., Wang, N., Rivero, A., Borge-Holthoefer, J., Moreno, Y.: Assessing the bias in samples of large online networks. Soc. Netw. **38**, 16–27 (2014)
31. Haccou, P., Haccou, P., Jagers, P., Vatutin, V.A.: Branching Processes: Variation, Growth, and Extinction of Populations, vol. 5. Cambridge university press, Cambridge (2005)
32. Hahn, G., Petermann, T., Havenith, M.N., Yu, S., Singer, W., Plenz, D., Nikolic, D.: Neuronal avalanches in spontaneous activity in vivo. J. Neurophysiol. **104**(6), 3312–2222 (2010)
33. Hahn, G., Ponce-Alvarez, A., Monier, C., Benvenuti, G., Kumar, A., Chavane, F., Deco, G., Frégnac, Y.: Spontaneous cortical activity is transiently poised close to criticality. PLoS Comput. Biol. **13**(5), 1–29 (2017)

34. Haldeman, C., Beggs, J.: Critical branching captures activity in living neural networks and maximizes the number of metastable states. Phys. Rev. Lett. **94**, 058101 (2005)
35. Han, J.-D.J., Dupuy, D., Bertin, N., Cusick, M.E., Vidal, M.: Effect of sampling on topology predictions of protein-protein interaction networks. Nat. Biotechnol. **23**(7), 839 (2005)
36. Harris, T.E.: The Theory of Branching Processes. Springer, Berlin (1963)
37. Hauri, A.M., Westbrock, H.-J., Claus, H., Geis, S., Giernat, S., Forβbohm, M., Uphoff, H.: Electronic outbreak surveillance in Germany: a first evaluation for nosocomial norovirus outbreaks. PLoS One, **6**(3), e17341 (2011)
38. Heathcote, C.R.: A branching process allowing immigration. J. R. Stat. Soc. Ser. B (Methodol.) **27**(1), 138–143 (1965)
39. Hellenbrand, W., Siedler, A., Tischer, A., Meyer, C., Reiter, S., Rasch, G., Teichmann, D., Santibanez, S., Altmann, D., Claus, H., Kramer, M.: Progress toward measles elimination in Germany. J. Infect. Dis. **187**(Suppl 1), S208–S216 (2003)
40. Heyde, C.C., Seneta, E.: Estimation theory for growth and immigration rates in a multiplicative process. J. Appl. Probab. **9**(2), 235 (1972)
41. Jaeger, H., Haas, H.: Harnessing nonlinearity: predicting chaotic systems and saving energy in wireless communication. Science **304**(5667), 78–80 (2004)
42. Jing, H., Takigawa, M.: Topographic analysis of dimension estimates of EEG and filtered rhythms in epileptic patients with complex partial seizures. Biol. Cybern. **83**, 391 (2000)
43. Khanin, R., Wit, E.: How scale-free are biological networks. J. Comput. Biol. **13**(3), 810–818 (2006)
44. Kinouchi, O., Copelli, M.: Optimal dynamical range of excitable networks at criticality. Nat. Phys. **2**, 348–352 (2006)
45. Klaus, A., Shan, Y., Plenz, D.: Statistical analyses support power law distributions found in neuronal avalanches. PloS One **6**(5), e19779 (2011)
46. Langton, C.G.: Computation at the edge of chaos: phase transitions and emergent computation. Phys. D Nonlinear Phenom. **42**(1–3), 12–37 (1990)
47. Leskovec, J., Faloutsos, C.: Sampling from large graphs. In: Proceedings of the 12th ACM SIGKDD International Conference on Knowledge Discovery and Data Mining, pp. 631–636. ACM (2006)
48. Levina, A., Herrmann, J.M., Denker, M.: Critical branching processes in neural networks. PAMM, 7(1), 1030701–1030702 (2007)
49. Levina, A., Herrmann, J.M., Geisel, T.: Theoretical neuroscience of self-organized criticality: from formal approaches to realistic models. In: Criticality in Neural Systems, pp. 417–436. Wiley-VCH Verlag GmbH & Co. KGaA (2014)
50. Levina, A., Priesemann, V.: Subsampling scaling. Nat. Commun. **8**, 15140 (2017)
51. Lindén, H., Tetzlaff, T., Potjans, T.C., Pettersen, K.H., Grün, S., Diesmann, M., Einevoll, G.T.: Modeling the spatial reach of the LFP. Neuron **72**(5), 859–872 (2011)
52. Linkenkaer-Hansen, K., Nikouline, V.V., Palva, J.M., Ilmoniemi, R.J.: Long-range temporal correlations and scaling behavior in human brain oscillations. J. Neurosci. Off. J. Soc. Neurosci. **21**(4), 1370–1377 (2001)
53. London, M., Roth, A., Beeren, L., Häusser, M., Latham, P.E.: Sensitivity to perturbations in vivo implies high noise and suggests rate coding in cortex. Nature **466**(7302), 123–127 (2010)
54. Lu, E.T., Hamilton, R.J.: Avalanches and the distribution of solar flares. Astrophys. J. **380**, L89–L92 (1991)
55. Marshall, N., Timme, N.M., Bennett, N., Ripp, M., Lautzenhiser, E., Beggs, J.M.: Analysis of power laws, shape collapses, and neural complexity: new techniques and matlab support via the ncc toolbox. Front. Physiol. **7**, 250 (2016)
56. Meisel, C., Storch, A., Hallmeyer-Elgner, S., Bullmore, E., Gross, T.: Failure of adaptive self-organized criticality during epileptic seizure attacks. PLoS Comput. Biol. **8**(1), e1002312 (2012)
57. Middendorf, M., Ziv, E., Wiggins, C.H.: Inferring network mechanisms: the drosophila melanogaster protein interaction network. Proc. Natl. Acad. Sci. USA **102**(9), 3192–3197 (2005)

58. Milton, J.G.: Neuronal avalanches, epileptic quakes and other transient forms of neurodynamics. Eur. J. Neurosci. **36**(2), 2156–63 (2012)
59. Murray, J.D., Bernacchia, A., Freedman, D.J., Romo, R., Wallis, J.D., Cai, X., Padoa-Schioppa, C., Pasternak, T., Seo, H., Lee, D. and Wang, X.-J.: A hierarchy of intrinsic timescales across primate cortex. Nat. Neurosci. **17**(12), 1661–1663 (2014)
60. Orbanz, P.: Subsampling large graphs and invariance in networks (2017). arXiv:1710.04217
61. Pakes, A.G.: Branching processes with immigration. J. Appl. Probab. **8**(1), 32 (1971)
62. Palva, J.M., Zhigalov, A., Hirvonen, J., Korhonen, O., Linkenkaer-Hansen, K., Palva, S.: Neuronal long-range temporal correlations and avalanche dynamics are correlated with behavioral scaling laws. Proc. Natl. Acad. Sci. **110**(9), 3585–3590 (2013)
63. Papoz, L., Balkau, B., Lellouch, J.: Case counting in epidemiology: limitations of methods based on multiple data sources. Int. J. Epidemiol. **25**(3), 474–478 (1996)
64. Pasquale, V., Massobrio, P., Bologna, L.L., Chiappalone, M., Martinoia, S.: Self-organization and neuronal avalanches in networks of dissociated cortical neurons. Neuroscience **153**(4), 1354–1369 (2008)
65. Neto, J.P., Priesemann, V.: Coarse sampling bias inference of criticality in neural system. In preparation
66. Plenz, D., Niebur, E., Schuster, H.G.: Criticality in Neural Systems. Wiley, New York (2014)
67. Priesemann, V., Munk, M., Wibral, M.: Subsampling effects in neuronal avalanche distributions recorded in vivo. BMC Neurosci. **10**(1), 40 (2009)
68. Priesemann, V., Shriki, O.: Can a time varying external drive give rise to apparent criticality in neural systems? PLOS Comput. Biol. **14**(5), e1006081 (2018)
69. Priesemann, V., Valderrama, M., Wibral, M., Le Van Quyen, M.: Neuronal avalanches differ from wakefulness to deep sleep-evidence from intracranial depth recordings in humans. PLOS Comput. Biol. **9**(3), e1002985 (2013)
70. Priesemann, V., Wibral, M., Valderrama, M., Pröpper, R., Le Van, M., Quyen, T.G., Triesch, J., Nikolić, D., Munk, M.H.J.: Spike avalanches in vivo suggest a driven, slightly subcritical brain state. Front. Syst. Neurosci. **8**, 108 (2014)
71. Privman, V.: Finite-size scaling theory. In: Finite Size Scaling and Numerical Simulation of Statistical Systems, vol. 1 (1990)
72. Pruessner, G.: Self-organised criticality: theory, models and characterisation. Cambridge University Press, Cambridge (2012)
73. Rafiei, D.: Effectively visualizing large networks through sampling. In: IEEE Visualization. VIS 05, pp. 375–382. IEEE (2005)
74. Rezvanian, A., Meybodi, M.R.: A new learning automata-based sampling algorithm for social networks. Int. J. Commun. Syst. **30**(5) (2017)
75. Ribeiro, T.L., Copelli, M., Caixeta, F., Belchior, H., Chialvo, D.R., Nicolelis, M.A.L., Ribeiro, S.: Spike avalanches exhibit universal dynamics across the sleep-wake cycle. PLoS One **5**(11), e14129 (2010)
76. Ribeiro, T.L., Ribeiro, S., Belchior, H., Caixeta, F., Copelli, M.: Undersampled critical branching processes on small-world and random networks fail to reproduce the statistics of spike avalanches. PloS One **9**(4), e94992 (2014)
77. Sethna, J.: Statistical Mechanics: Entropy, Order Parameters, and Complexity, vol. 14. Oxford University Press, Oxford (2006)
78. Shaukat, A., Thivierge, J.-P.: Statistical evaluation of waveform collapse reveals scale-free properties of neuronal avalanches. Front. Comput. Neurosci. **10**, 29 (2016)
79. Shew, W.L., Yang, H., Petermann, T., Roy, R., Plenz, D.: Neuronal avalanches imply maximum dynamic range in cortical networks at criticality. J. Neurosci. **29**(49), 15595–15600 (2009)
80. Shriki, O., Alstott, J., Carver, F., Holroyd, T., Henson, R.N.A., Smith, M.L., Coppola, R., Bullmore, E., Plenz, D.: Neuronal avalanches in the resting meg of the human brain. J. Neurosci. **33**(16), 7079–7090 (2013)
81. Stumpf, M.P.H., Wiuf, C., May, R.M.: Subnets of scale-free networks are not scale-free: sampling properties of networks. PNAS **102**(12), 4221–4224 (2005)

82. Stumpf, M.P.H., Wiuf, C.: Sampling properties of random graphs: the degree distribution. Phys. Rev. E **72**, 36118 (2005)
83. Tagliazucchi, E., Balenzuela, P., Fraiman, D., Chialvo, D.R.: Criticality in large-scale brain fMRI dynamics unveiled by a novel point process analysis. Front. Physiol. **3**, 15 (2012)
84. Tetzlaff, C., Okujeni, S., Egert, U., Wörgötter, F., Butz, M.: Self-organized criticality in developing neuronal networks. PLoS Comput. Biol. **6**(12), e1001013 (2010)
85. Tomen, N., Rotermund, D., Ernst, U.: Marginally subcritical dynamics explain enhanced stimulus discriminability under attention. Front. Syst. Neurosci. **8**(August), 1–15 (2014)
86. Touboul, J., Destexhe, A.: Power-law statistics and universal scaling in the absence of criticality. Phys. Rev. E **95**(1), 1–15 (2017)
87. Wagenaar, D.A., Pine, J., Potter, S.M.: An extremely rich repertoire of bursting patterns during the development of cortical cultures. BMC Neurosci. **7**(1), 1 (2006)
88. Wei, C.Z.: Convergence rates for the critical branching process with immigration. Stat. Sin. **1**, 175–184 (1991)
89. Wibral, M., Lizier, J.T., Priesemann, V.: Bits from brains for biologically inspired computing. Front. Robot. AI **2**(March), 1–25 (2015)
90. Wichmann, O., Siedler, A., Sagebiel, D., Hellenbrand, W., Santibanez, S., Mankertz, A., Vogt, G., Van Treeck, U., Krause, G.: Further efforts needed to achieve measles elimination in Germany: results of an outbreak investigation. Bull. World Health Organ. **87**(2), 108–115 (2009)
91. Wilting, J., Dehning, J., Neto, J.P., Rudelt, L., Wibral, M., Zierenberg, J., Priesemann, V.: Operating in a reverberating regime enables rapid tuning of network states to task requirements. Front. Syst. Neurosci. **12**, 55 (2018)
92. Wilting, J., Priesemann, V.: Inferring collective dynamical states from widely unobserved systems. Nat. Commun. **9**(1), 2325 (2018)
93. Wilting, J., Priesemann, V.: On the ground state of spiking network activity in mammalian cortex (2018). arXiv:1804.07864
94. Worrell, G.A., Cranstoun, S.D., Echauz, J., Litt, B.: Evidence for self-organized criticality in human epileptic hippocampus. NeuroReport **13**(16), 2017–2021 (2002)
95. Yaghoubi, M., De Graaf, T., Orlandi, J.G., Girotto, F., Colicos, M.A., Davidsen, J.: Neuronal avalanche dynamics indicates different universality classes in neuronal cultures. Sci. Rep. **8**(1), 1–11 (2018)
96. Yu, S., Klaus, A., Yang, H., Plenz, D.: Scale-invariant neuronal avalanche dynamics and the cut-off in size distributions. PloS One **9**(6), e99761 (2014)
97. Zierenberg, J., Wilting, J., Priesemann, V.: Homeostatic plasticity and external input shape neural network dynamics. Phys. Rev. X **8**(3), 031018 (2018)

Viola Priesemann studied physics and neuroscience at TU Darmstadt, the MPI for Brain Research, Caltech and Ecole Normale Superieure. Briefly after obtaining her Ph.D. in 2013, she became independent group leader at the MPI for Dynamics and Self-Organization, investigating the interplay between funtion and dynamics. She employs tools from statistical physics and information theory, and in particular develops approaches that enable inference about the full system when only a tiny fraction is sampled.

Anna Levina studied mathematics at Saint-Petersburg State University, Russia and received her Ph.D. in mathematics from Göttingen University, Germany in 2008. She has worked on modelling cortical activity as a Postdoctoral researcher at the Max Planck Institute for Mathematics in Sciences and at the Institute of Science and Technology Austria, before starting her group as a Junior Professor at the University of Tübingen. Her research interests include self-organization of neuronal systems, mathematical modeling, and laws of cortical optimization.

Jens Wilting studied Physics and Mathematics at the Universities of Bonn and Göttingen and Imperial College London. He is currently a doctoral student at the Max-Planck-Institute for Dynamics and Self-Organization (Göttingen). His research interests are focused on collective dynamics of cortical networks, their assessment under subsampling, and their relation to machine learning.

The Role of Criticality in Flexible Visual Information Processing

Nergis Tomen and Udo Ernst

Abstract Dynamical systems close to a critical state have the ability to spontaneously engage large numbers of units in collective events called avalanches–but how can this property be actively employed by the brain in order to perform meaningful computations under realistic circumstances? In our study we investigate this question by focusing on the visual system which has to meet a major challenge: to rapidly integrate information from a large number of single channels, and in a flexible manner depending on behavioral and external context. In this framework we are going to discuss two distinct examples, the first a bottom-up figure-ground segregation scenario and the second a top-down enhancement of object discriminability under selective attention. Both scenarios make explicit use of critical states for information processing, while formally extending the concept of criticality to inhomogeneous systems subject to a strong external drive.

1 Introduction

Complex systems with numerous nonlinear, coupled elements may perpetually self-organize to a dynamical state close to a phase transition [2]. Such a system can be described as a branching process and, at the critical state, is characterized by power-law distributions of a variety of observables [76]. Considered a signature of neuronal networks poised at criticality, scale-free organization of neuronal activity has been observed in a wide range of data including in acute slices of rat cortex and organotypic cultures [5, 17], dissociated cultures [48], awake monkeys [49], human MEG and EEG [47, 63] and human intracranial depth recordings [52].

Neuronal networks operating at a critical state have been of interest to neuroscientists due to the proposed computational benefits [61]. It has been suggested

N. Tomen (✉) · U. Ernst
University of Bremen, Cognium, Hochschulring 18, 28359 Bremen, Germany
e-mail: nergis@neuro.uni-bremen.de

U. Ernst
e-mail: udo@neuro.uni-bremen.de

N. Tomen et al. (eds.), *The Functional Role of Critical Dynamics in Neural Systems*, Springer Series on Bio- and Neurosystems 11,
https://doi.org/10.1007/978-3-030-20965-0_12

that critical dynamics may play a functional role in supporting complex computations [6, 35], maximizing information diversity [46] and the number of metastable states [25], and improving object representation [71] with favourable dynamic range characteristics [20, 32, 62].

The abundance of power-law scaling in the brain, taken together with the theoretical functional benefits of networks operating close to a critical state forms the basis of the criticality hypothesis [4, 9, 29, 50], which suggests that critical dynamics is an emergent property of cortical networks, supporting optimal information processing.

While the enhanced information processing capabilities close to a phase transition are an important argument for this hypothesis, demonstrations of them often rely on rather 'abstract' theoretical measures, without realistic encoding or read-out scenarios in mind. Furthermore, such demonstrations employ relatively homogeneous models without structured functional connectivity or strong external drive. This creates an exciting opportunity to explore more concrete links between the abstract computational scenarios and the biologically realistic cortical dynamics under cognitive load.

In addition, the majority of the experimental studies reporting avalanche dynamics in neuronal networks analyze recordings of spontaneous activity *in vitro* or of the resting state *in vivo*, although recent studies attempt to understand the emergence of criticality in the cortex under sensory stimulation [1, 60]. Nevertheless, perfect scale-free organization does not seem to emerge under all conditions, especially in the strongly driven regime. Previous findings indicate that signatures of criticality are progressively disturbed during sustained wakefulness [44] and may diminish with increasing cognitive load [70].

Similarly, avalanche dynamics *in vivo* exhibit differences between wakefulness and deep sleep [52] and seem to be poised at a slightly subcritical state [53].

In an attempt to unify these points, in this chapter we explore the idea that flexible and dynamic information processing in the cortex, which requires fast responses to both incoming sensory input as well as rapid top-down modulations due to changing task requirements, may be boosted by critical dynamics. We investigate the hypothesis that in the active brain engaged in a cognitive task not all subpopulations are tuned to a perfect critical state at all times. Instead, the local activity regime can be actively modulated by properties of the external input or changing task demands.

In order to make a concrete link to cortical function, we will focus specifically on the visual system and investigate how critical dynamics may offer functional benefits based on established experimental scenarios. In general, we will present a paradigm in which avalanche dynamics of local subnetworks, heavily involved in processing of a sensory stimulus, may be actively tuned towards the critical point, ultimately contributing to the detection or discrimination of the stimulus by higher visuocortical areas.

2 Collective Dynamics and Information Integration in the Visual System

In this section, we will present a brief introduction to the visual system, its dynamics, and the typical computational problems it has to solve, as well as establish the general concepts and terminology which will be relevant for the detailed examples demonstrating how critical dynamics might support cortical function.

Visual scenes. A visual scene is usually composed of a multitude of 'objects' such as four animals in front of a house with a window (Fig. 1). However, their representations on the retina or the primary visual cortex is distributed and fragmented: Every neuron has access to only a tiny part of the visual field (small blue circles), while objects typically occupy a much larger region (large blue frame). Hence for making sense of a visual scene, the brain has to process these elementary pieces

Fig. 1 A natural scene from northern Germany. For segmenting this image into meaningful parts, our brain has to integrate local feature information into global figure representations: For example, orientation-selective cells in V1 might detect the presence of horizontally oriented bars inside their classical receptive fields (blue circles), which are then linked into a representation of the whole window (blue rectangle) by subsequent stages in visual cortex. Thereby representations of different "figures" (such as the dog and donkey) might share subsets of features (contours, outlined in dark and bright red). More complex feature combinations such as whole shapes can only be successfully integrated if attention is directed towards the corresponding location in the visual field (dashed blue circles, head of dog and cat) [40]

of information in relation to each other, thus enabling perception and, ultimately, a successful interaction with the environment.

Nevertheless, information integration in vision is computationally demanding. Objects in a scene might be partially occluded or possess fragmented outlines, thus complicating the integration process. Scenes can be cluttered, requiring to search for the behaviorally relevant information in a background of noise. Low level representation of an object (e.g. in pixels) can vary dramatically, for example trees seem different when viewed from varying angles or under different lighting conditions, however, they still belong to the same class of objects. Moreover, scenes are dynamic: their content can change rapidly, making a continuous and fast reassessment of the current visual input necessary. Realizing these different computational functions requires many different neuronal processes to contribute, and despite the amazing progress on deep convolutional neural networks in machine vision it is still largely unclear how the human visual system solves all of these challenges. But above all details, there emerges a fundamental need for two dynamic capabilities in cortical networks: the ability to engage a very large number of neurons in dependence on external evidence and behavioral context, and to do this very quickly and flexibly—a promising 'application' for critical networks.

Feature integration. Physiological studies have established that, apart from the relatively local processing in the retina and LGN, information integration is mainly performed in the cortical domain. Single neurons in the primary visual cortex (the first cortical area processing an incoming visual stimulus) preferentially respond to certain basic features of a local image patch, such as its orientation or spatial frequency [10, 30]. Since these neurons thus signal the presence or absence of specific features in a visual scene, information integration in vision is also termed feature integration. In general, perceptual grouping of visual features take place based on classical Gestalt principles, where factors such as proximity, similarity or good continuation of features determine whether they will collectively be processed as belonging together [75], thus leading to a salient percept. Salient figures in a cluttered scene are robustly and easily segregated from the background, in what is termed a "pop-out" effect [33]. For example in contour integration, fragmented outlines of potential object boundaries are bound into coherent percepts [15]. Such enhanced processing of features due to stimulus properties is often described as a bottom-up effect [77].

So what might be the neuronal correlate of feature integration? In many different settings, feature integration is accompanied by firing rate modulations (e.g., see [8, 51, 55]). Neurons which are stimulated with their preferred feature centered inside their classical receptive field (cRF) have been shown to exhibit lower or higher firing rates in dependence on a surrounding, contextual stimulus outside their cRF [37, 64]—in particular when the center-surround configuration matches Gestalt principles such as colinearity of oriented line segments [38]. The caveat of these rate modulations serving as an explanation for feature integration is both their magnitude and their temporal characteristics: for example, in contour integration the rate enhancement is often small [3], and occurs with a large latency (150 ms) after stimulus onset [21]. In contrast, other experiments show that human image classification can be extremely rapid, leading to consistent differences in EEG traces in an animal/no-animal discrimination task (where contour integration would only be one of the first steps in a chain of multiple processing stages) after only 150 ms [69].

As an alternative, information integration might rely on the relative timing of action potentials: Spatially and temporally correlated neuronal activity has long been hypothesized as a putative mechanism for representing parts or features of an object which belong together [27, 41]. In general, mutual synchronization between two neurons tends to become stronger if the stimulus components within their receptive fields are more likely to belong to one object [34]. Models using oscillatory synchrony for binding features have been proposed [16, 54], but establishing oscillations takes time and the emerging neural activity is too regular in comparison to synchronization observed in physiological studies of the visual system [66].

Selective processing. To address the problem that a visual scene typically contains a substantial amount of information at any given time, the visual system dynamically and flexibly allocates its limited resources, in a context and task-dependent manner. For example, attentional mechanisms may selectively enhance the processing of information about specific features or locations in the visual field. Electrophysiological studies have shown that rate modulations can gradually boost the representation of behaviourally relevant stimuli at the expense of irrelevant stimuli (e.g., see [39, 45, 73]). Furthermore, it was observed that such enhancements are accompanied by an increase in oscillatory cortical activity in the γ-band [19, 67]. In this context, γ-oscillations have been proposed to be the essential mechanism for information routing regulated by attention [18, 22]. Such oscillation based modulations provide the cortex with an effective mechanism capable of efficiently gating information flows [26] which would be difficult to achieve with the often much weaker rate modulations [23]. In contrast to feature integration, mechanisms of selective attention such as the enhancement of stimulus representations and the preferential routing of information are top-down effects [68].

In this chapter, we are going to focus on the general idea that near-critical dynamics may promote rapid and robust processing of dynamically modulated information streams. In order not to be constrained by the specifics of a certain mechanism, we are going to discuss two distinct examples, the first a bottom-up figure-ground segregation scenario and the second a top-down enhancement of object discriminability under selective attention (Fig. 2). We are going to argue that regardless of the specific mechanism, local emergence of critical dynamics in the active brain, either in a stimulus driven manner or through attentional modulations, may assist in selectively improved processing of local information. This is broadly achieved by the activity of local subnetworks, representing different objects in the visual field, being pushed towards a critical point and away from a subcritical regime. This framework

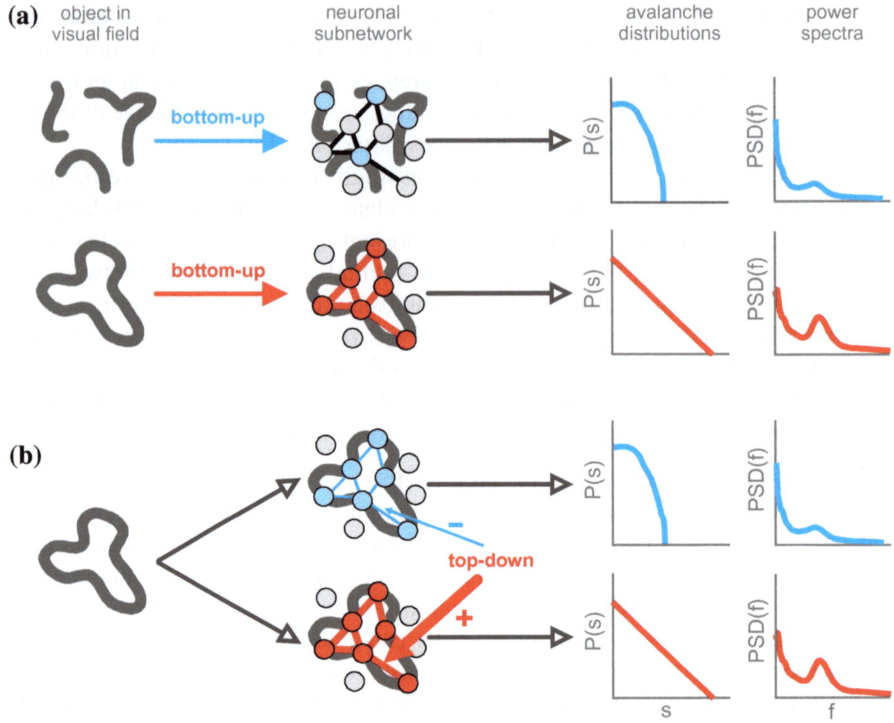

Fig. 2 Critical dynamics exhibited by local cortical subnetworks may be instrumental in flexible modulations of object representation. **a** Bottom-up modulation of the neuronal response: A complete set of features which constitute an object in the visual field is a more salient stimulus than the random co-occurrence of a few of its features. A subnetwork poised at the critical point will display scale-free dynamics over the full system size, only if all of its units are externally driven (red circles and links). Similarly, power spectral distribution of local population activity will display stronger oscillatory dynamics when the stimulus is more salient. **b** Top-down modulation of the neuronal response. According to changing behavioural needs, mechanisms such as selective visual attention may dynamically shift the subnetwork towards or away from the critical point. A shift towards the ordered regime is accompanied by enhanced oscillatory power, potentially assisting in enhancing the communication between different areas

brings together the concepts of oscillations in population activity, synchronization at the spiking level and phase transitions, and suggests a universal functional role for correlated activity: flexible processing of information in a dynamically changing world.

3 Critical Subnetworks and Figure-Ground Segregation

In this section we will investigate the idea that the correlations in the activity of local, functionally connected subpopulations of cortical neurons may be modulated in a bottom-up manner by the presence of objects in the visual field. In particular we will focus on how critical dynamics may emerge in a subnetwork with the presentation of a salient stimulus, contributing to enhanced figure-ground segregation performance when using a coincidence detector as a read-out mechanism.

3.1 A Critical Model of Feature Integration

Motivation. As described in Sect. 2, particular feature combinations ('figures') are processed by our visual system selectively, or in a privileged manner, even in the absence of top-down instructions to do so. Such stimuli are commonly accompanied by a "pop-out" effect, making these feature combinations highly salient in perception even if they are embedded into a "background" of distractor elements (for an example, see Fig. 3a). "Pop-out" is generally described as a stimulus-directed bottom-up effect. For example, a single local feature, such as a line segment whose orientation deviates from all other segments present in a scene will "pop-out" to an observer [72]. Interestingly, evidence suggests that both context [31] as well as the global stimulus structure [33] play a role in modulating the processing of an object by low-level, local feature detectors, thus requiring rapid co-processing and integration of multiple streams of information on different spatial scales.

Considering both the general notion that a dynamical regime close to a phase transition may boost the capacity of a network to sustain rapid and efficient modulations, as well as the dependence of perceptual grouping tasks on integration of spatially extended information, we hypothesize that spatially and temporally correlated neuronal activity may be conducive to bottom-up "pop-out" effects, which are helpful in figure-ground segregation.

How local features are integrated to form global percepts, in order to segregate relevant figures from a background, is extensively studied, especially in psychophysical investigations on Gestalt perception [75]. For example, research on contour integration focuses on how the visual system identifies the outlines of shapes or figures by grouping colinearly or cocircularly aligned edge elements into coherent percepts (Fig. 1, blue circles and red outlines). For illustrating our modeling paradigm, we will use contour integration as a proxy for all processes where binding of local features ultimately contributes to visual perception.

A typical visual stimulus employed in psychophysical contour integration experiments consists of oriented Gabor patches. Other than a large number of randomly oriented distractor elements in the background, a stimulus may contain a target contour made up of appropriately placed and mutually aligned edge elements [15] (Fig. 3a). Due to the adherence of the Gabor patches to cRF shapes reflecting the

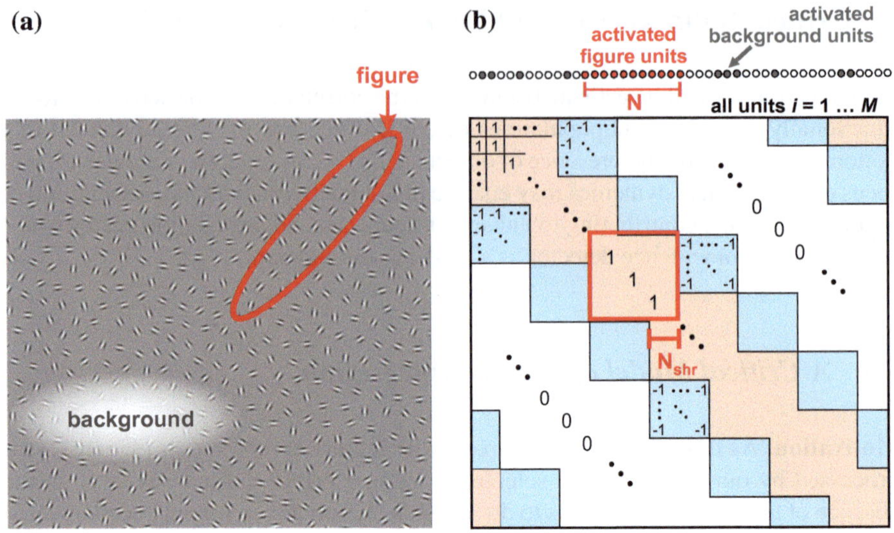

Fig. 3 **a** A typical stimulus from a contour integration experiment. It contains one figure consisting of colinearly aligned, oriented Gabor patches, which are hidden among randomly oriented background elements. **b** The coupling matrix for the network model. This example consists of $N_{sub} = 7$ subnetworks of N mutually excitatorily coupled EHE units each (regions shaded in red). Every subnetwork shares N_{shr} units with its two nearest neighbors. The stimulus externally drives all N units in one subnetwork (filled red circles) and $N - N_{shr}$ background units randomly selected from the remaining subnetworks (filled gray circles). Optionally, one can add inhibitory synaptic connections between subnetworks sharing units (regions shaded in blue). In our investigations, we consider both, purely excitatory and mixed excitatory-inhibitory networks. In the dynamics of the model (Eq. 2), the value of w_{ji} displayed in this example is scaled with the coupling strength α, meaning that the absolute efficacies of all excitatory and inhibitory connections are equal

orientation selectivity of neurons in the early visual system, we think of each edge element as driving a population of local feature detectors. Contour integration has been shown to be an efficient and fast process, which is robust within a large range of stimulus parameters both in humans [12, 28] and in animal experiments [42]. Overall, the speed and accuracy with which contours can be detected in psychophysical experiments lead us to believe that a quick spread of activity and a subsequent swift employment of a large number of feature detectors which are involved in encoding the figure (or alternatively a swift suppression of background units) is necessary for effective contour integration. Therefore, we next investigate whether and how large scale avalanches spanning subnetworks processing a figure input may be beneficial in figure-ground segregation tasks.

Model dynamics. As model system, we adapt the Ernst–Herrmann–Eurich (EHE) model, which is an analytically well-understood network model displaying critical dynamics [13]. The units in the original network are globally connected ($w_{ji} = 1$ $\forall j, i$) non-leaky integrators and the coupling strength α between the units determines the distance of the dynamics from the critical state (see [13] for details).

In particular, each unit $i = 1, \ldots, N$ is described by a state variable $u_i \in [0, 1)$. At each discrete time step k, a randomly selected unit u_i is perturbed by external input Δu:

$$u_i(k+1) = u_i(k) + \Delta u. \tag{1}$$

If its state variable u_i exceeds a threshold, in this case 1, then unit i is reset, and all units j in the network including i receive recurrent input, scaled by the weight matrix w_{ji} and the coupling strength α:

$$u_i(k) \geq 1 \iff \begin{array}{l} u_i \to u_i - 1 \\ u_j \to u_j + \frac{\alpha w_{ji}}{N}. \end{array} \tag{2}$$

If any unit j then exceeds the threshold succeeding the recurrent input, the procedure in Eq. 2 is iterated until the avalanche terminates. Here we assume a separation of time scales, such that an avalanche of spikes is infinitely fast compared to the time scale on which the external, driving input arrives. This means that each avalanche is initiated by a single unit firing due to external input, and is completed within one discrete simulation time step. Such an implementation allows us to precisely quantify the avalanche size and duration distributions.

Couplings and subnetworks. For studying feature integration, we now assume that each unit represents a feature detector. If the corresponding feature, for example an oriented line segment, is present in a visual stimulus, the unit is activated by some external input. Detectors for features not present in the current visual stimulus receive no external drive. A figure is defined by an ensemble of features, which is a subset of all features represented in the network. We further require our network to be able to represent multiple figures, and that different figures might share subsets of features (Fig. 3b).

Unlike in the original model [13], we use a systematically structured coupling matrix for representing figures in network topology. We begin with a zero matrix of size $M \times M$ and successively embed N_{sub} subnetworks, each containing N mutually and excitatorily coupled units. We posit that each subnetwork overlaps with exactly two other subnetworks by sharing N_{shr} units (feature detectors), which yields a coupling structure resembling a one-dimensional figure 'chain' with periodic boundary conditions (Fig. 3b, regions shaded in red color), with a total network size of $M = N_{sub} \cdot (N - N_{shr})$ units. In this scheme, the units in each subnetwork collectively encode the presence of a figure in the visual field (such as a well-defined contour line as in Fig. 3a), and neighboring subnetworks may be activated by figures which share local features in their neuronal representation (Fig. 1, red contours).

Network activity. In order for the network to perform a figure detection task, we present it with a stimulus containing one whole figure and random background elements. Such a stimulus drives the N units of one subnetwork, and $N - N_{shr}$ randomly selected units from the remaining $M - N$ units with external input (filled circles at top of Fig. 3b). All other units receive no driving input. The single unit i which receives the external drive according to Eq. 1 is selected with uniform probability from the $N_{act} = 2N - N_{shr}$ externally activated units which make up the figure and background subpopulations.

For evaluating the dynamics of different subpopulations, we separately record the activity of a 'target' and a 'distractor' ensemble with N units each. The target ensemble is composed of the N activated figure units. The distractor ensemble is composed of the $N - N_{shr}$ externally activated background units plus N_{shr} units of the currently activated figure which are shared with one of the other, not activated figures. This is in consideration of the idea that units which belong to more than one figure may also be sampled as part of the background by higher areas.

Since in a setting with shared figure representations we expect some degree of activity to spread from externally driven units to the rest of the network, we would ideally like to assess the extent to which it would affect network dynamics, and ultimately task performance. To address this point, we extend our paradigm by introducing inhibitory connections between neighboring subnetworks (Fig. 3b, regions shaded in blue color), with the intention to alleviate cross-talk between them through the shared units, and compare the results to our original setup.

3.2 Results

For the results presented here, we use $N_{sub} = 7$ subnetworks, each consisting of $N = 225$ EHE units, and we drive the network with an external input of $\Delta u = 0.022$ for a total of $K = 10^7$ time steps. The spiking activity of the network is recorded, and the dynamics of different populations characterized as we vary the overlap between the subnetworks N_{shr} in addition to the coupling strength α. This gives us an idea about the plausibility of such a network topology under increasingly heavy load (i.e. where many units need to be shared between internal representations of different figures) as well as about the robustness of dynamics as we move closer to and beyond the critical state.

Synchronization and avalanches. When we investigate the avalanche dynamics in the purely excitatory model, we find, trivially, that the activity of the target ensemble becomes critical and subsequently supercritical with increasing α. Eventually, firing rates grow without bound and the network gets stuck in a regime of infinitely long avalanches. We identify the α for which subnetwork activity is critical by finding the avalanche size distributions for which the Kolmogorov–Smirnov (KS) statistic is minimized (Fig. 4a, white circles). For calculating the KS statistic [65], we used a reference power-law distribution with exponent -1.43. This was the exponent which yielded the minimum KS distance for a control network: a globally connected EHE network of size N, simulated for a total of K time steps.

For very small N_{shr}, the presentation of a stimulus with one complete figure embedded into random background elements splits the network into two distinct groups: the dynamics of the subnetwork representing the target figure quickly become

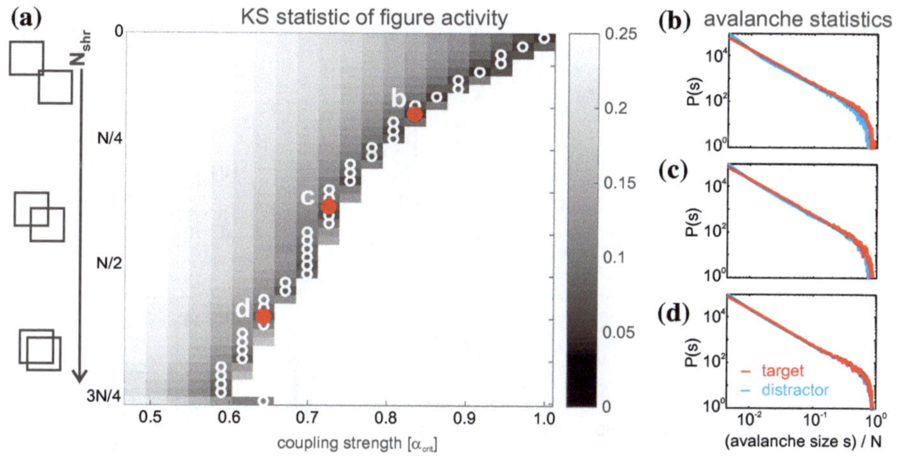

Fig. 4 **a** The KS statistic of the avalanche size distributions in the target population as a function of the subnetwork overlap N_{shr} and the coupling strength α, in the purely excitatory network. In the bottom-right white region, infinite avalanches emerge. The numerically calculated critical point for each N_{shr} is marked with a white circle. Coupling strength α is displayed in units of the theoretical critical value $\alpha_{crit} = 1 - 1/\sqrt{N}$ for a globally connected network of N units [13]. **b, c** and **d** show the avalanche size distributions in the target and distractor populations for the corresponding parameter combinations marked in red in **a**

synchronous as we increase the coupling strength, whereas the dynamics of the rest of the network as well as the distractor population lag slightly behind the target population in the sense that they comparatively lack large avalanches and highly correlated activity. Background activity instead exhibits a slightly larger amount of small sized avalanches (Fig. 4b).

As mentioned before, due to our choice of sampling of target and distractor activity, the dynamics of the two populations converge naturally as N_{shr} approaches N. Therefore, any existing discriminability of target avalanche activity from distractor activity is quickly diminished with increasing N_{shr} (Fig. 4b–d). However, this convergence is bolstered by runaway activity, or activity spreading out from the strongly synchronous target units. This cross-talk between the subnetworks is also a factor in why the critical value of α decreases as we increase N_{shr}.

Addition of inhibition. One way to circumvent runaway activity and alleviate the co-activation of background populations is to employ inhibitory interactions. Indeed, in the network with structured inhibition, we find that the coupling strength for which the target activity is critical is now robust against changes in N_{shr} (Fig. 5a), as well as against the number of subnetworks embedded in the network N_{sub} (not shown) up to an overlap of $N_{shr} = N/2$. After this point, non-neighboring subnetworks begin to share units, which is an effect our choice of inhibitory topology cannot offset. We can verify that the cross-talk between subnetworks is negligible by plotting the

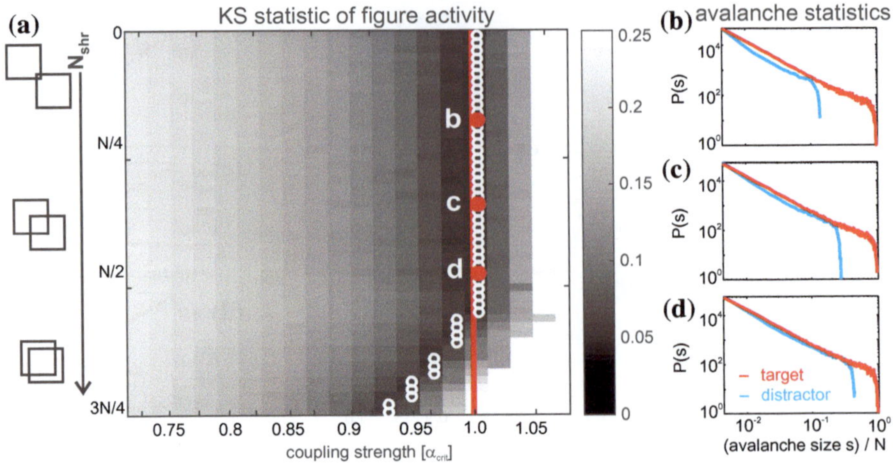

Fig. 5 **a** The KS statistic of the avalanche size distributions in the target population as a function of the subnetwork overlap N_{shr} and the coupling strength α, in the network with mixed excitation and inhibition. The red line marks the critical point α_{crit} of a globally coupled network with N units. Other annotations are the same as in Fig. 4. **b, c** and **d** show the avalanche size distributions in the target and distractor populations for the corresponding parameter combinations marked in **a**

analytical value of the critical α for a globally connected, stand-alone network of N units in comparison to the KS statistic minima (Fig. 5a, red line).

The reduction in activity spill-over from the target population as well as the suppression of the externally activated distractor units is also apparent in the corresponding avalanche size distributions (Fig. 5b–d). In comparison to the dynamics in the absence of inhibition, we find that synchrony in the distractor population is further quenched while overall background activity levels are lowered. This across-the-board decrease in background activity also contributes to the formation of a larger buffer region of supercriticality between the critical point, and the transition to the infinite avalanche regime. While an average α increase of about 4% is sufficient to bring the purely excitatory network from critical dynamics into the infinite avalanche regime, an average increase of about 8% in α is required for the network with inhibition.

Task performance. The discrepancy between the synchronization levels of target and distractor populations seems promising for putative read-out mechanisms to successfully perform figure-ground segregation. To capitalize on this finding, we use two leaky integrate-and-fire (LIF) neurons, acting as coincidence detectors, for distinguishing between figure and background. We feed the population activity of the target population to one LIF neuron, and activity of the distractor population to a second LIF neuron. In order to employ the LIF neurons as coincidence detectors and exclude information contained in the firing rates, we normalize the total input to each LIF neuron over the duration of a trial by its sum (Eq. 4). Specifically, the membrane potentials $V_{tar}(t)$ and $V_{dis}(t)$ of the respective read-out neurons evolve according to

$$\tau \frac{\mathrm{d}V_p(t)}{\mathrm{d}t} = -V_p(t) + f \sum_{k=1}^{K} \bar{s}_p(k)\,\delta(t - k\Delta t) \tag{3}$$

where $p \in \{tar, dis\}$, τ is a time constant with $\tau = 10$ ms, and f denotes an arbitrary gain factor. The ODE was numerically integrated using the Euler method, with the spiking threshold at $V_\theta = 1$.

Inputs $\bar{s}_p(k)$ are derived from the discrete population spike time series $s_p(k)$ which contain the size of the avalanche which took place in population p at simulation time step $k = 1 \ldots K$ of the discretized dynamics (Eq. 1). The avalanche size is defined as the number of units which elicited a spike in the same time step k due to recurrent input. Since we separately record the population activity of the target and distractor populations, $s_{tar}(k)$ and $s_{dis}(k)$, respectively, and each population contains N units in total, s assumes integer values in the interval $[0, N]$. As mentioned before, in order to eliminate a performance increase due to rate differences between the populations, $\bar{s}_p(k)$ is obtained by normalizing the population activity by its sum over time.

$$\bar{s}_p(k) = \frac{s_p(k)}{\sum_{k'=1}^{K} s_p(k')} \tag{4}$$

where once again $p \in \{tar, dis\}$. For feeding the network outputs $\bar{s}_p(k)$ into the LIF neurons, we choose the width of our simulation time step $\Delta t = 0.02$ ms, such that the average firing rate of a target figure unit, when the whole figure subnetwork was activated at the critical point, was approximately 40–50 Hz (depending on N_{shr}). This choice of Δt additionally revealed a periodic activity of large sized avalanches, or oscillatory behaviour, in the γ-frequency range—a realistic setup for a local subnetwork in the visual cortex [11].

Figure detection performance P_{fig} is then defined as the difference between the output rates of the two read-out neurons, which in addition to α and N_{shr} depend on the gain factor f. In other words, how the task performance behaves, and where in the phase space it exhibits a maximum, changes depending on the value of f.

In order to isolate the effects of N_{shr} and α on performance, we first assess whether a global maximum can be located for the performance as a function of f. This is carried out by first computing the task performance $P_{fig}(N_{shr}, \alpha, f)$ in the whole phase space spanned by N_{shr} and α for a given value of f, and then computing the maximum $\Sigma_{perf}(f) = \max\limits_{N_{shr},\alpha}[P_{fig}(N_{shr}, \alpha, f)]$ for each f. We then locate the factor f_{opt} at which $\Sigma_{perf}(f)$ reaches its maximum. To achieve the desired degree of precision with lower computation time, we implement a golden section search on the function $\Sigma_{perf}(f)$ in the interval $f \in [0, 0.02]$.

Within this range, we found that $\Sigma_{perf}(f)$ has a clear global maximum at f_{opt} (Fig. 6a). Typically the performance maxima are found at smallest values of N_{shr} and large, supercritical values of α, where target activity is high and distractor activity is almost completely suppressed (see e.g. Fig. 6b). The global maximum shown here is close to a rate difference of 150 Hz, but the value of the performance will vary

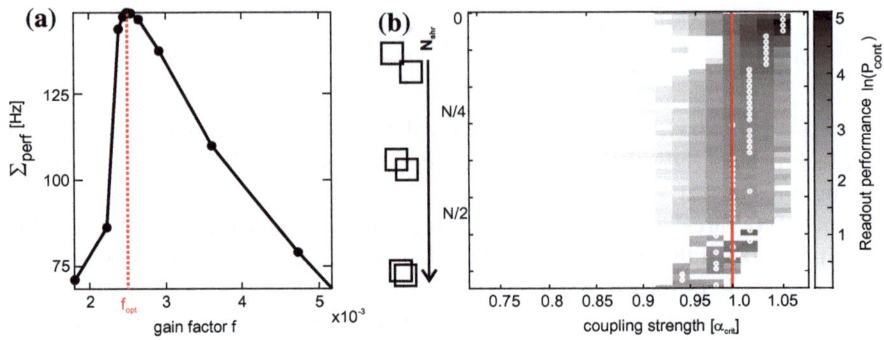

Fig. 6 **a** The maxima Σ_{perf} of the figure-ground segregation task's performance P_{fig} as a function of f displays a clear global maximum at f_{opt}. Since P_{fig} is defined as the output rate difference of the read-out LIF neurons, its units are in Hz. **b** Logarithm of P_{fig} as a function of subnetwork overlap N_{shr} and coupling strength α using the optimal gain factor f_{opt}. White circles mark the performance maxima for each value of N_{shr}. The red line shows the critical α as in Fig. 5. For the medium range of N_{shr} where the performance maxima converge to the critical point, the read-out rate difference P_{fig} is in the range of 12–20 Hz

depending on N_{shr} and read-out parameters such as the time constant τ of the LIF neurons (see Sect. 5.1 for a more in-depth discussion).

The value of f which maximizes $\Sigma_{perf}(f)$ appears to be relatively robust against changes in other parameters such as the number of externally driven background units and the read-out neuron time constant τ, when chosen within a biologically plausible range (see Sect. 5.1). Interestingly, when adopting the value $f = f_{opt}$, figure detection performance in the phase space peaks close to the critical α for most medium-sized values of N_{shr} (Fig. 6b). We find, as would be expected, that for a small number of shared units between subnetworks, where the risk of activity leak is low, it is most advantageous for the local figure population to display supercritical, strongly synchronous activity, in order to maximize coincidence detection. As N_{shr}, along with the overlap of the target and distractor populations, increases, figure detection performance P_{fig} drops and the location of the peak shifts towards the critical α. These findings suggest that the critical state provides the optimal regime in which to perform feature integration in our model for a range of medium to large subnetwork overlap N_{shr}.

3.3 Summary

In this section we have inspected the dynamics of an analytically well-understood network of EHE units, where mutually excitatorily coupled subnetworks are embedded into the coupling matrix. Each subnetwork can be interpreted as representing a visual object, or figure, defined by an ensemble of local features. As soon as these features are collectively present in a stimulus, the subnetwork will engage in synchronous activity, thus enabling coincidence detectors to optimally signal the presence of the figure when the dynamics are poised close to the critical point. Implementing structured inhibition between subnetworks which share units efficiently prevents cross-talk between neighboring populations, and makes the critical value of our control parameter α robust against changes in the number of shared units N_{shr} as well as the total number of subnetworks N_{sub}. Thus inhibition serves to make computation with critical dynamics structurally invariant, since little to no changes in the existing network topology would be necessary as the network expands and scales up.

Taken together, our results suggest that a task requiring fast and accurate integration of spatially extended information may benefit from figure representations in the form of critical subnetworks. When individual feature detectors in such subnetworks are simultaneously driven by a stimulus, large avalanches can cascade through them rapidly and in a synchronous manner. Capitalizing on this, our findings indicate that the critical state will provide the best figure-ground segregation results in a synchrony coding scenario, given realistic resource constraints such as a limited number of neurons. Overall, these results imply that bottom-up, stimulus-driven modulations of visual information processing, and well-known psychophysical phenomena such as the "pop-out" of more salient stimuli and suppression of distractors, may benefit from local subnetworks operating close to the critical state.

4 Attentional Modulations of Oscillatory States

In this section, we will talk about how operating close to a phase transition may benefit top-down modulations of information processing. In a specific example, we will focus on how the experimentally observed phenomena accompanying visual selective attention may be reproduced in a simple network model poised close to a transition boundary. We will also discuss how such an operational regime may support rapid and flexible enhancement of stimulus representation, as well as represent the information entropy maxima given certain read-out restrictions.

4.1 Selective Attention Improves Object Representation

Motivation. As described in Sect. 2, selective visual attentional mechanisms are associated with the emergence of γ-oscillations in visuocortical areas. In addition, attention modulated increases in γ-band power have also been observed accompanying increases in spike-field coherence [19], supporting our hypothesis that cascades of rapid spiking activity may play a role in mechanisms underlying selective visual attention. In particular, a top-down modulation of spike time correlation seems to be necessary to dynamically resolve which information will be gated in and which will be gated out by downstream visual areas [18, 23].

In this section we will probe in which ways improved synchrony in the γ-band under selective attention may contribute to an improvement in information processing in the visual cortex, and how such an improvement may coincide with a phase transition boundary in system dynamics.

Our model is motivated by an electrophysiology experiment [67] where a rhesus monkey (*Macaca mulatta*) had to perform a demanding delayed-match-to-sample task on one of two simultaneously presented, dynamic visual stimuli. These stimuli consisted of morphing shape sequences covering about $4^o \times 4^o$ of visual angle (for examples see Fig. 7). The monkey was instructed to attend to one of the stimuli, memorize the initial shape, and release a lever as soon as the initial shape reappeared in the morphing sequence. In parallel to this task, local field potential (LFP) recordings were performed with an epidural multielectrode array implanted over visual areas V1 and V4.

For three reasons, this experiment is interesting for our framework: First, recognizing the target shape required the monkey's visual system to integrate local information into a global percept. This had to be done rapidly and continuously, since the shapes were morphing and the target could appear at different times during a trial. Second, neural activity was recorded in an area (V4) which is believed to be central for shape perception, and with a method (epidural recordings) where a single electrode has access to a large population of neurons and is especially sensitive to large synchronous events. Third, the task was cognitively demanding and was not possible to perform without directing attention towards the behaviourally relevant stimulus.

Indeed the LFPs exhibited strong synchronization in the γ-band, which became even stronger when the stimulus inside the RF of the corresponding neural population was attended [67]. Although LFP recordings have a coarse spatial resolution, the γ-activity proved to be specific enough to successfully be employed in discriminating between different shapes in the RF [56]. Under attention, differences in synchronous activity between shapes were enhanced, hence improving classification performance of different visual stimuli [56].

This prompted us to ask: can these observations of neural representations of complex shapes and their modulation by selective attentional mechanisms be linked to correlated activity in the form of neuronal avalanches?

Model setup and dynamics. To answer this question we built a minimalistic model, where recurrently coupled local networks of leaky integrate-and-fire (LIF) neurons process the individual shapes (Fig. 7, for details of model setup and data analysis please refer to [71]). We assume that the epidural recording electrode averages over a large population of M neurons. Each of these neurons can be, as described and illustrated in the previous chapter (see Fig. 3), part of a subnetwork specialized in the processing of a particular shape. Presentation of a visual stimulus S will activate a subset of $N < M$ neurons (Fig. 7b) responding to the presence of the particular features contained in a specific shape, while the neurons representing features not contained in the current shape will remain silent. For simplicity, our simulations only consider the subpopulation of N neurons currently activated by the stimulus, and for

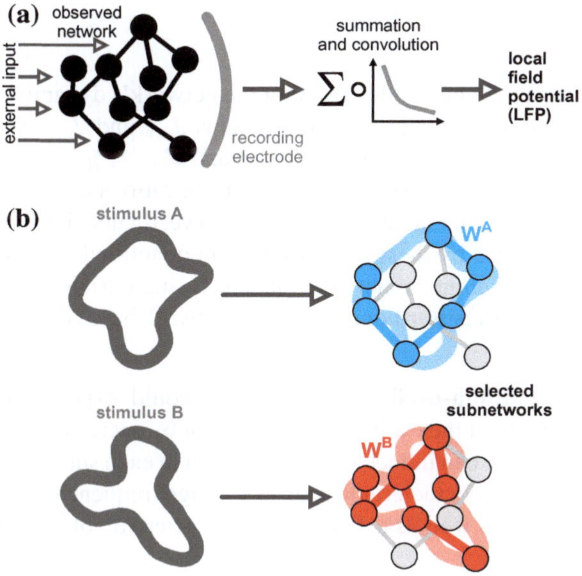

Fig. 7 **a** A randomly coupled recurrent network of mixed excitatory (80%) and inhibitory (20%) LIF neurons (left) generates action potentials which are observed by an (epidural) electrode, yielding local field potentials (LFPs) modeled as a sum over the pre- and postsynaptic network activity convolved with an exponentially decaying kernel. **b** Different stimuli in the visual field (shapes, on the left) activate different neuronal units according to their receptive field properties. The activated units select stimulus-specific subnetworks W^A or W^B (all of size N and with connection probability p, examples in blue and red), whose coupling structures differ slightly in graph theoretical measures

each shape we assume to have a *different* random coupling matrix W^S describing only the recurrent interactions among the *currently activated* subpopulation.

The shape-specific couplings W^S were generated with a sparse connection probability of $p = 0.05$ (Erdős–Rényi graph). With a finite N, such randomly generated couplings yield different graph theoretical parameters, such as mean in-degrees, mean out-degrees, mean betweenness centrality etc. The varying properties of the different subnetworks give rise to different dynamics, and are responsible for different stimuli producing different LFP spectra in our model (Fig. 7b). Neurons are driven externally by Poissonian input spike trains with constant rate $f_{max} = 10$ kHz. In order to analyze the activity of the subnetworks in a way similar to the analysis of the experimental data, we construct an LFP time series by smoothing the sum of the input and output spikes with an exponential filter. Spectral power of each time series is obtained, as in the experimental papers [56, 67], using a wavelet time-frequency decomposition. Attention is simulated by multiplicative modulation of the gain of all neurons.

As before (Sect. 3.1), we employ a separation of time scales, which allows us to record the size and duration of distinct avalanches exactly.

4.2 Results

The most important parameters determining the network's dynamics are the average excitatory and inhibitory synaptic input strengths J_{exc} and J_{inh}. We define J_{exc} as the absolute strength of an excitatory synapse and $J_{inh} = j_{inh} \cdot \varepsilon$, where j_{inh} is the absolute strength of an inhibitory synapse and ε is the ratio of the number of inhibitory neurons to the number of excitatory neurons. We investigated the model behaviour in the phase space spanned by these parameters, and quantified the discriminability of the LFP spectra of different subnetworks in short, single trials. Simultaneously, we determined the phase transition boundary of the network based on neuronal avalanche size distributions.

Avalanches and gamma-oscillations. As one would expect, by independently increasing the strength of the excitatory synapses, or by decreasing the strength of the inhibitory synapses, we can induce synchronization in each subnetwork (Fig. 8). The power spectra display a characteristic peak at γ-range frequencies, which grows with enhanced synchrony, as well as a $1/f$ offset, characteristic of neuronal background activity.

This transition to more synchronous activity is well reflected in the avalanche dynamics of the network. If we look at subnetworks with high inhibition and low excitation, we have exponential avalanche size distributions (blue curve in Fig. 9a), signifying, on average, small avalanche sizes and low overall correlated activity. As we increase excitation or decrease inhibition in the subnetwork, we observe supercritical size distributions (red curve in Fig. 9a), characterized by a large bump at large avalanche sizes, signifying highly correlated activity. A similarly abrupt

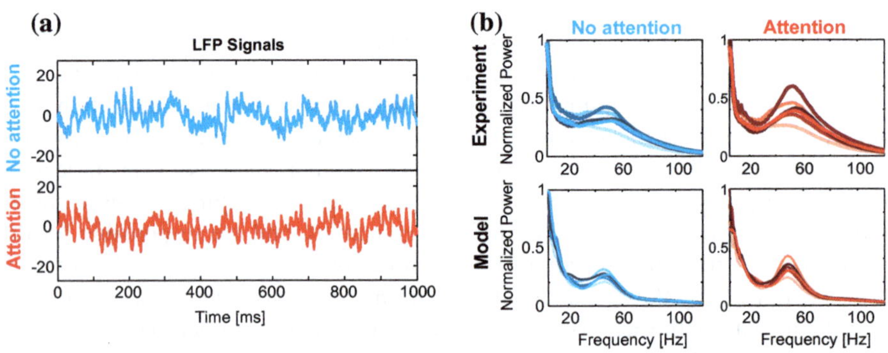

Fig. 8 **a** LFP time series generated by the model. We increase neural gain in the model in order to enhance synchrony under the attended condition. **b** Trial-averaged power spectra of the LFP signals during presentation of different shapes (indicated by different color shades) obtained from the model in comparison to the experimental data. For the model, different shapes are represented by the trial- and time-averaged power from different subnetworks. Experimental data shown is courtesy of Dr. Andreas Kreiter, Dr. Sunita Mandon and Katja Taylor [67]

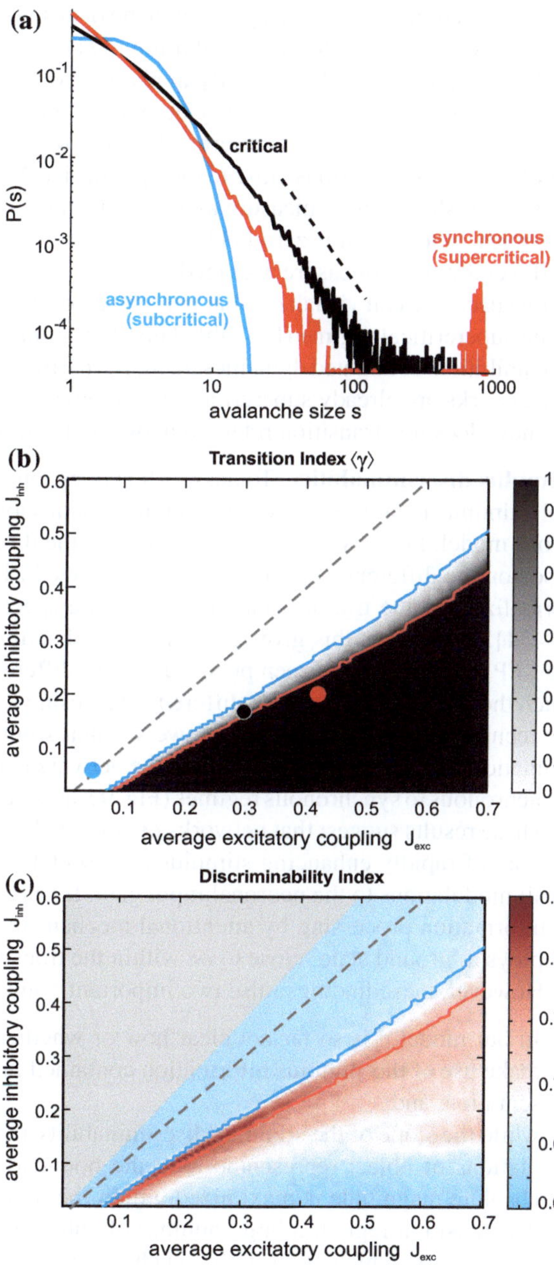

Fig. 9 **a** Avalanche size distributions for three different synaptic strength values producing asynchronous to synchronous spiking activity. Corresponding parameter values in phase space are marked in matching colors in panel **b**. **b** The γ-measure as a function of the excitatory and inhibitory coupling strengths. In the regions with $\langle\gamma\rangle = 1$ all subnetworks, representing all stimuli, are displaying supercritical dynamics, and in regions with $\langle\gamma\rangle = 0$ all subnetworks are displaying asynchronous, typically subcritical activity. The blue and red curves indicate the upper and lower boundaries of the transition region. **c** Discriminability index based on the LFP spectra of each stimulus, as a function of excitatory and inhibitory coupling strengths. Stimulus representation based on the oscillatory power of subnetworks peaks rapidly in a narrow region of the phase space. The transition region identified in **b** is indicated by the blue and red curves. Upper left triangular region in gray: parameter combinations not assessed in simulations (highly asynchronous regime). The dotted gray line in **b, c** marks parameter combinations with perfect balance between excitation and inhibition ($J_{exc} = J_{inh}$)

transition in dynamics is also apparent in the distributions of avalanche duration, and average avalanche size for a given duration [71].

We next identified where exactly in phase space model dynamics transition from asynchronous to regular oscillatory activity, using the avalanche distributions. We achieve this by using the γ-measure, which is a function of the relative weight of the tail of the size distributions close to the system size N. If this relative weight exceeds a given threshold, the γ-measure assumes the value of one, signifying supercritical dynamics, otherwise it is zero [71].

If we plot the γ-measure averaged over all subnetworks $\langle\gamma\rangle$ for each point in the phase space, we can visualize what percentage of the subnetworks have transitioned to the supercritical regime (Fig. 9b). This shows that between excitation-dominated and inhibition-dominated dynamics, there is a narrow transition region in which some subnetworks are already supercritical while others exhibit asynchronous dynamics. But how does this transition relate to network function?

Stimulus discriminability. To answer this question, we wanted to understand how well stimulus identity is represented in the synchronous activity of each subnetwork in our model. For this purpose, we quantified the discriminability of each stimulus based on the difference between the trial-averaged LFP spectra of different stimuli, normalized by the trial-to-trial variance of the spectra for each, individual stimulus [71]. In essence, this gives us a measure (discriminability index) of how much the LFP spectra vary between presentations of different stimuli as compared to how much the spectra vary between different trials using the same stimulus. Displayed in the recurrent coupling phase space, we see that stimulus discriminability increases dramatically in the small region where the network dynamics transition rapidly from asynchronous to synchronous regimes (Fig. 9c, area between the blue and red curves).

These results suggest that networks operating close to the transition boundary are capable of rapidly enhancing stimulus representation in their dynamics, with only small modulations to the neuronal input gain. If true, this means that flexible tuning of information processing by attentional mechanisms is optimal when the network employs a 'ground state' close to *or* within the transition region.

However, these findings raise two important points to consider:

1. In our model it is so far not clear how or whether downstream visual areas can make use of the stimulus information contained in the LFP spectral power representation, and,
2. while the slope of the stimulus discriminability index is favorable for rapid modulations of object representation at the boundaries of the transition region, its absolute magnitude is maximized within the transition region. However, within the transition region, a large number of subnetworks already display supercritical activity, which is strongly synchronous and exhibits a high degree of correlations. Ideally, we would like to have a better understanding of the trade-off between enhanced stimulus discriminability, and the loss of information at the single spike level due to increasing correlations, within and around the transition boundaries.

Spike pattern diversity. To address these points, we proceeded to compute the information entropy [59] within the coupling space, in order to assess the diversity of the spike patterns generated in response to stimuli. In doing so, we consider different scales of observation on which read-out of these patterns by neurons in downstream areas might take place. At the finest scale of observation, parametrized by $K = 1$, the read-out mechanism has access to complete information about the subnetwork spiking activity. In this case, it can discriminate between spikes originating from distinct neurons. At the coarsest observation scale, $K = N$, the read-out mechanism is not capable of observing every individual neuron, but rather integrates the spiking activity of all neurons at each discrete simulation time step (please see [71] for details).

This is implemented by calculating the information entropy $H(\mathbf{X})$ of the spike patterns $\mathbf{X}$, where each state X_i is based on a binary word of length N formed by the spiking activity at a given time step. For $K = 1$, the state variable $\mathbf{X}$ consists of N channels: each channel assumes a value of 1 if the corresponding neuron generated an action potential at time t, and 0 otherwise. The scale parameter K then reduces a spike pattern comprising spikes from N neurons to a representation of N/K channels with each channel containing the summed activity of K neurons (Fig. 10a).

Quantified this way, we find that information entropy declines drastically within the transition region for all K (Fig. 10b–c). As we increase K, however, the entropy maxima in the phase space are shifted towards the transition region and to higher coupling strengths (compare Fig. 10, panels b and c).

4.3 Summary

In this section we have shown that in a network of leaky IAF neurons, the transition region between asynchronous and highly correlated dynamics, identified on the basis of neuronal avalanches, correlates well with a peak in stimulus discriminability. However, such enhancement of object representation comes at the cost of dramatically reduced information entropy in spike patterns as a larger number of subnetworks begin to display supercritical dynamics with increasing relative excitation. This dynamical effect implies the existence of an optimum between linking features together and losing detailed stimulus information. How far this optimal working point for the brain is from criticality strongly depends on the nature of the function being computed, and the corresponding read-out mechanisms. In particular in a passive cognitive state, the cortex might be further from criticality, but ready for top-down mechanisms of selective attention to kick in and rapidly enhance object representation as necessary.

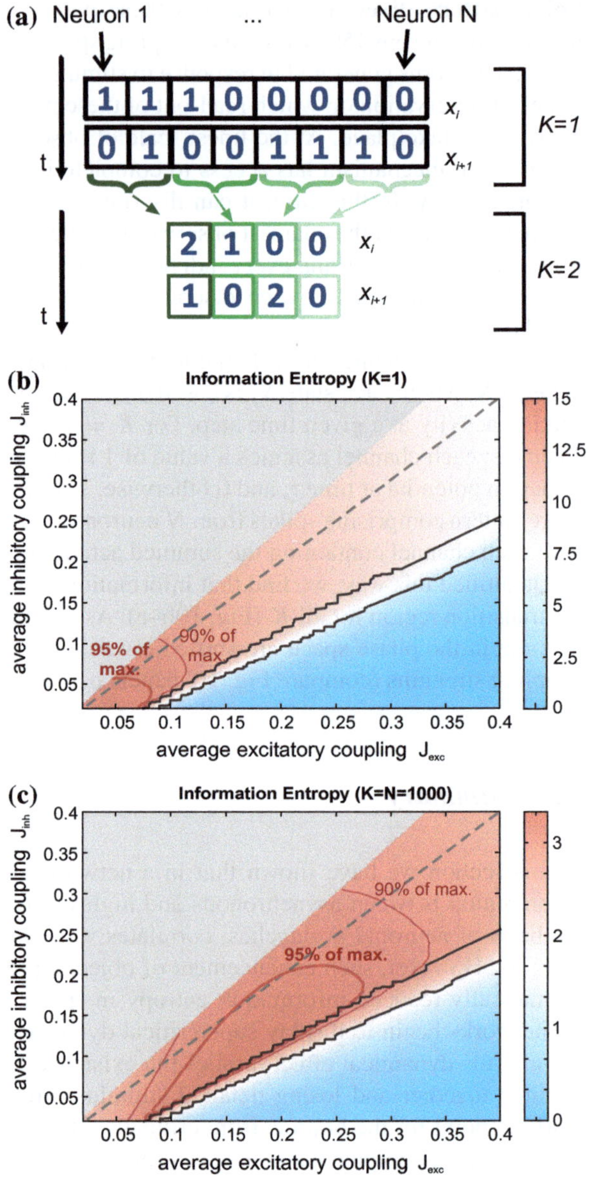

Fig. 10 **a** Information entropy of the spike patterns is calculated using state variables X_i. At the finest observation scale ($K = 1$), X_i consist of N-dimensional binary vectors, which represent whether each neuron fired a spike (1) or not (0) at a given point in time. For larger K, the activity of K adjacent cells is summed to construct X_i. **b** Information entropy in coupling space for the finest observation scale ($K = 1$). **c** Information entropy in coupling space for the coarsest observation scale ($K = N$). For both **b** and **c**, the results are averaged over all subnetworks. Regions where entropy is largest, namely 90 and 95% of the maximum entropy, are indicated by thin and thick red lines, respectively. Solid black lines mark the transition region and the dashed gray line the perfect weight balance as in Fig. 9c

5 Discussion

In this chapter we have investigated a specific role for critical dynamics in the cortex, namely, the rapid dynamic modulation of visual information processing. In order to demonstrate the versatility of our approach, we have focused on two distinct examples: one where changing stimulus salience modulates local subnetwork dynamics in a bottom-up manner, and one where top-down selective attentional mechanisms regulate oscillatory activity in local circuits. In both cases we have shown that system dynamics transition rapidly as networks get closer to a critical point, and thus being poised at the critical state provides us with an optimal regime in which to flexibly modulate network activity, with either small modifications of stimulus properties or changing task demands.

In particular, we have shown that both an increase in stimulus salience and attentional shifts may push local subnetworks processing a behaviourally relevant figure or object in the visual field towards a critical state. In the bottom-up scenario, such a push is accompanied by an improvement in figure detection task performance (Fig. 6) and in the top-down example by an enhancement of object representation in the oscillatory dynamics of these subnetworks (Fig. 9). Taken together, the two distinct scenarios, both benefiting from critical dynamics, hint at a generic correlation between fast and flexible information processing and an operating regime close to a phase transition, which is not restricted to a single mechanism. Importantly, these results provide us with concrete examples of how local emergence of criticality may play a functional role in thoroughly studied visuocortical phenomena.

In light of these findings we argue that the active cortex is not globally operating in a critical regime at all times but that the differential in correlations, both over time and space, is an effective tool for the cortex to selectively enhance the processing of relevant sensory information. In general, our findings imply that selective, preferential or otherwise dynamic processing of visual information in a continuously changing visual environment may be supported by local subnetworks operating close to a critical state, and our results show that in models under realistic constraints, the critical state may yield optimal performance for downstream read-out mechanisms.

5.1 *Robustness of Results*

We limited the models presented in this chapter to minimalistic systems capable of demonstrating the relevant effects. The resulting lack of a large number of free parameters facilitated confirming that our results don't critically depend on a very specific choice of parameter combinations among many.

In the feature integration model, we have run parameter scans over most of the free parameters. We have confirmed that the robustness of the critical coupling parameter against subnetwork overlap N_{shr} in the network with inhibition does not depend on the number of externally activated units N_{act}, subnetwork size N or the number of subnetworks N_{sub}. Concerning the read-out results, we have checked that varying the LIF neuron time constant τ within a biologically plausible range (5–20 ms) does not alter our conclusions qualitatively. However, we found the position of the task performance maxima in the phase space spanned by N_{shr} and α to strongly depend on our choice of Δt, when Δt is varied in isolation. Nevertheless, changes in Δt also heavily influence the value of f_{opt}. Therefore, repeating the golden section search and using the appropriate value of f_{opt} for a given Δt, we observed the performance maxima always close to the critical point (within the range $\Delta t = 0.01$–0.02 ms).

We have used the non-leaky integrator units due to the analytical understanding of how the coupling strength modulates the avalanche size distributions. More complex neuron models (e.g. the addition of leak currents or conductances) may lead to varying results, although network behaviour may be reproduced qualitatively given small leaks and adjusted parameters [13]. Nevertheless, with the appropriate choice of network topology, which ensures synchronous activity in subnetworks and properly structured inhibition to abate cross-talk between subnetworks, we would expect similar results for the task performance. Our specific choice of the read-out mechanism (LIF with normalized input) is arbitrary and we believe any coincidence detector scheme should produce a similar effect.

In the attention model, our conclusions depend mainly on the facts that in our model: (1) the emergence of synchronous spiking activity can be described by a phase transition as a function of an excitability parameter, and (2) synchronizability of the network depends implicitly on the topography of its connections. Therefore, we believe that as long as these requirements are met, discriminability enhancement will correlate with a narrow choice of parameters which generate near-critical dynamics.

5.2 *Physiological Plausibility*

In this chapter we restricted ourselves to relatively simple models in order to ensure the demonstrated effects do not require fine-tuning of a large number of free parameters. In general, many modelling studies which similarly put forward computational benefits of criticality rely on comparatively abstract measures or are restricted to homogeneous or slowly-driven networks [6, 32, 35, 61]. In order to better understand the dynamics of the awake, active cortex, where such computational benefits will presumably be most conspicuous, we explore the dynamics of highly structured (Sect. 3), strongly-driven (Sect. 4) networks, performing realistic tasks, with biologically plausible read-out mechanisms.

As a read-out mechanism in the feature integration model, we used coincidence detection. We normalized the inputs into the read-out neurons such that the input from the neuronal population processing a figure and the input from the neuronal population processing the distractors in the background would be indistinguishable using simply the rate information. This is equivalent to adjusting stimulus contrast, which contour detection performance is known to be resistant against [28]. In addition, although firing rate is a good discriminator in our specific, simple model, in other models which incorporate higher biological complexity, such as detailed balance [74] or inhibitory normalization [7], rates might carry little to no information useful for figure detection. Taken together with our results, this may imply that the cortex transmits local stimulus information using rates, and retains information about global object configurations in the time domain, thus implementing a multiplexed coding scheme.

In the brain, variability in connectivity of neurons in a local population is not random or homogeneous, but signifies a highly structured global network. As such, the biological cortex is not a random graph as was used in our attention model. However, the model captures the general idea that local feature detectors collectively process an object in the visual field, and that in an electrophysiology experiment the activity of a random selection of such local detectors is represented in the recordings. Furthermore, the fact that different subnetworks have different critical points in our phase space originates from modest topological differences between the randomly generated coupling matrices. This means that the demonstrated transition region shrinks and ultimately approaches a transition 'line' at the limit $N \rightarrow \infty$, while discriminability decreases to chance level. One may alternatively choose to extend the model to encompass a more structured topology established, for example, by a biologically inspired learning rule. Hence, enhancing topological differences between subnetworks in the attention model would increase the size of the transition region and make our results more robusts against increasing system size. In addition, the general effects observed in our attention model are compatible with other tentative mechanisms relating to how attention may intervene with feedforward processing such as rate increases [18, 67].

For the feature integration model, the 1-dimensional and symmetric coupling structure may be replaced by a more random assignment of units to multiple subnetworks or learned dynamical attractors with strong recurrent excitation within a subnetwork. With this altered topology, we would still expect a coincidence detector to display high performance around the critical point, as long as there's mutual inhibition between units belonging to different subnetworks. This is because the most important ingredient for the maximized figure detection performance with coincidence detection is synchronous activity within a target subnetwork which is absent between random background units.

Finally, our investigations raise one important question: Is our scheme only useful for a single stage of processing, or would it still be applicable in a multilayer model? This was not explored in this chapter, however, one important theoretical property of critical networks is their branching parameter being unity, often thought of as being beneficial for dynamics where activity should neither die out nor explode over time. This idea is reinforced by recent investigations which show that arbitrarily deep neural networks may only be trained close to criticality [58].

5.3 Critical Dynamics, Structure, and Function

Currently, there is a growing interest in bridging the gap between the abstract concepts surrounding critical phenomena in the thermodynamical sense and biological concepts such as structure and function which are of interest to neuroscientists. For example, new studies are aiming to establish links between more subtle effects of different connectivity structures on criticality [43] as well as investigating the capacity of networks to support replay of learned spatio-temporal patterns near a phase transition [57].

In our work, we intend to contribute to this growing body of literature, by providing examples which directly relate to experimental settings at the behavioural level. We suggest that critical dynamics may be favorable for supporting rapid changes in network activity, necessary for flexibly modulating information processing based on changing stimulus properties and task requirements. In order to achieve this flexibility, one factor plays a crucial role in our models: that the active cortex, strongly stimulated by sensory input, is not globally and constantly at a critical point. Similar observations were made previously, demonstrating progressive disturbance of critical dynamics with sustained wakefulness [44] and cognitive load [70]. Similarly, it has been suggested that avalanche dynamics *in vivo* display differences between wakefulness and deep sleep [52] and may be poised at a slightly subcritical state [53]. In an attempt to explain many of these observations, we suggest the tuning of local subnetworks towards and away from criticality as a generic mechanism to modulate sensory information processing.

Some recent literature suggests that a shift in neuronal dynamics away from the critical state may be linked to focused attention as well as active preparation for stimulus processing, seemingly at odds with our predictions. For example, it has been shown that a scale-free distribution of neuronal activity cascades in human EEG is associated with the resting state while focused attention shifts the dynamics towards a subcritical regime [14]. The conclusions we present in this chapter, however, are based on dynamics of local subnetworks, spanning much smaller spatial and temporal scales than a whole-brain EEG. Furthermore, cascade analysis of continuous data, such as EEG signals, typically depends on a thresholding of the signal in time domain, and will mainly capture the high amplitude oscillatory activity in the lower frequency bands, such as α-waves, which are known to decrease with more vigilant mental states. Similarly, power-law distributions in spiking data have been linked in awake monkeys with eye closure and asynchronous dynamic states have been observed when the subjects had their eyes open in a dark setting [24]. However, due to the lack of stimulus evoked activity in the experimental paradigm, the presented dataset lacked discernible oscillations in the γ-band, which is the dominant synchronization frequency in our models. Therefore, we do not believe these results are in contradiction with our predictions. In fact, inhibitory suppression of cortical subnetworks which are irrelevant for the processing of a visual stimulus is a fundamental part of our feature integration model, which may be related to the asynchronous dynamics observed in these studies.

From an experimental perspective, we hypothesize that signatures of criticality in stimulus-evoked activity and under attention may be readily and consistently found *in vivo* only when analyzing the activity of functionally connected subnetworks, during the presentation of a suitable stimulus. Currently, both subsampling of relevant populations as well as the separation of avalanche and external drive time scales present challenges for experimental tests of this hypothesis. However, we believe that both the efforts to correct for subsampling effects in data analysis [36] and emerging developments in multielectrode recording techniques and optogenetic methods will be instrumental in advancing our understanding of critical dynamics in the cortex.

Acknowledgements The authors would like to thank Dr. Andreas Kreiter and his group for providing the data shown in Fig. 8b. This work has been supported by the Bundesministerium für Bildung und Forschung (BMBF, Bernstein Award Udo Ernst, Grant No. 01GQ1106).

References

1. Arviv, O., Goldstein, A., Shriki, O.: Near-critical dynamics in stimulus-evoked activity of the human brain and its relation to spontaneous resting-state activity. J. Neurosci.: Off. J. Soc. Neurosci. **35**(41), 13,927–42 (2015). https://doi.org/10.1523/JNEUROSCI.0477-15.2015. http://www.ncbi.nlm.nih.gov/pubmed/26468194
2. Bak, P., Tang, C., Wiesenfeld, K.: Self-organized criticality: an explanation of the 1/f noise. Phys. Rev. Lett. **59**(4), 381–384 (1987). https://doi.org/10.1103/PhysRevLett.59.381
3. Bauer, R., Heinze, S.: Contour integration in striate cortex. Classic cell responses or cooperative selection? Exp. Brain Res. **147**(2), 145–152 (2002)
4. Beggs, J.M.: The criticality hypothesis: how local cortical networks might optimize information processing. Philos. Trans. Ser. A, Math., Phys., Eng. Sci. **366**(1864), 329–343 (2008). https://doi.org/10.1098/rsta.2007.2092. http://www.ncbi.nlm.nih.gov/pubmed/17673410
5. Beggs, J.M., Plenz, D.: Neuronal avalanches in neocortical circuits. J. Neurosci.: Off. J. Soc. Neurosci. **23**(35), 11,167–77 (2003). https://doi.org/10.1523/JNEUROSCI.23-35-11167.2003. http://www.ncbi.nlm.nih.gov/pubmed/14657176
6. Bertschinger, N., Natschläger, T.: Real-time computation at the edge of chaos in recurrent neural networks. Neural Comput. **16**(7), 1413–1436 (2004). https://doi.org/10.1162/089976604323057443
7. Carandini, M., Heeger, D.J., Movshon, J.A.: Linearity and normalization in simple cells of the macaque primary visual cortex. J. Neurosci.: Off. J. Soc. Neurosci. **17**(21), 8621–8644 (1997). https://doi.org/10.1523/JNEUROSCI.17-21-08621.1997
8. Chen, M.G., Yan, Y., Gong, X.J., Gilbert, C.D., Liang, H.L., Li, W.: Incremental integration of global contours through interplay between visual cortical areas. Neuron **82**(3), 682–694 (2014)
9. Chialvo, D.R.: Emergent complex neural dynamics. Nat. Phys. **6**(10), 744–750 (2010). https://doi.org/10.1038/nphys1803
10. Dayan, P., Abbott, L.F.: Theoretical Neuroscience: Computational and Mathematical Modeling of Neural Systems, 1st edn. Massachusetts Institute of Technology Press, Massachusetts (2001)
11. Eckhorn, R., Bauer, R., Jordan, W., Brosch, M., Kruse, W., Munk, M., Reitboeck, H.J.: Coherent oscillations: a mechanism of feature linking in the visual cortex? Biol. Cybern. **60**(2), 121–130 (1988). https://doi.org/10.1007/BF00202899
12. Ernst, U.A., Mandon, S., Schinkel Bielefeld, N., Neitzel, S.D., Kreiter, A.K., Pawelzik, K.R.: Optimality of human contour integration. PLoS Comput. Biol. **8**(5), e1002,520 (2012). https://doi.org/10.1371/journal.pcbi.1002520. http://dx.plos.org/10.1371/journal.pcbi.1002520

13. Eurich, C.W., Herrmann, J.M., Ernst, U.A.: Finite-size effects of avalanche dynamics. Phys. Rev. E **66**(6), 066,137 (2002). https://doi.org/10.1103/PhysRevE.66.066137. https://link.aps.org/doi/10.1103/PhysRevE.66.066137
14. Fagerholm, E.D., Lorenz, R., Scott, G., Dinov, M., Hellyer, P.J., Mirzaei, N., Leeson, C., Carmichael, D.W., Sharp, D.J., Shew, W.L., Leech, R.: Cascades and cognitive state: focused attention incurs subcritical dynamics. J. Neurosci.: Off. J. Soc. Neurosci. **35**(11), 4626–4634 (2015). https://doi.org/10.1523/JNEUROSCI.3694-14.2015
15. Field, D.J., Hayes, A., Hess, R.F.: Contour integration by the human visual system: evidence for a local association field. Vis. Res. **33**(2), 173–193 (1993). https://doi.org/10.1016/0042-6989(93)90156-Q
16. Finger, H., Knig, P.: Phase synchrony facilitates binding and segmentation of natural images in a coupled neural oscillator network. Front. Comput. Neurosci. **7**, 195 (2014). https://doi.org/10.3389/fncom.2013.00195
17. Friedman, N., Ito, S., Brinkman, B.A.W., Shimono, M., DeVille, R.E.L., Dahmen, K.A., Beggs, J.M., Butler, T.C.: Universal critical dynamics in high resolution neuronal avalanche data. Phys. Rev. Lett. **108**(20), 208,102 (2012). https://doi.org/10.1103/PhysRevLett.108.208102. https://link.aps.org/doi/10.1103/PhysRevLett.108.208102
18. Fries, P.: A mechanism for cognitive dynamics: neuronal communication through neuronal coherence. Trends Cogn. Sci. **9**(10), 474–480 (2005). https://doi.org/10.1016/J.TICS.2005.08.011
19. Fries, P., Reynolds, J.H., Rorie, A.E., Desimone, R.: Modulation of oscillatory neuronal synchronization by selective visual attention. Science (New York, N.Y.) **291**(5508), 1560–1563 (2001). https://doi.org/10.1126/science.291.5508.1560. http://www.ncbi.nlm.nih.gov/pubmed/11222864
20. Gautam, S.H., Hoang, T.T., McClanahan, K., Grady, S.K., Shew, W.L.: Maximizing sensory dynamic range by tuning the cortical state to criticality. PLOS Comput. Biol. **11**(12), e1004,576 (2015). https://doi.org/10.1371/journal.pcbi.1004576. http://dx.plos.org/10.1371/journal.pcbi.1004576
21. Gilad, A., Meirovithz, E., Slovin, H.: Population responses to contour integration: early encoding of discrete elements and late perceptual grouping. Neuron **78**(2), 389–402 (2013)
22. Grothe, I., Neitzel, S.D., Mandon, S., Kreiter, A.K.: Switching neuronal inputs by differential modulations of gamma-band phase-coherence. J. Neurosci.: Off. J. Soc. Neurosci. **32**(46), 16,172–80 (2012). https://doi.org/10.1523/JNEUROSCI.0890-12.2012. http://www.ncbi.nlm.nih.gov/pubmed/23152601
23. Grothe, I., Rotermund, D., Neitzel, S.D., Mandon, S., Ernst, U.A., Kreiter, A.K., Pawelzik, K.R.: Attention selectively gates afferent signal transmission to area V4. J. Neurosci.: Off. J. Soc. Neurosci. **38**(14), 3441–3452 (2018). https://doi.org/10.1523/JNEUROSCI.2221-17.2018
24. Hahn, G., Ponce-Alvarez, A., Monier, C., Benvenuti, G., Kumar, A., Chavane, F., Deco, G., Frégnac, Y.: Spontaneous cortical activity is transiently poised close to criticality. PLOS Comput. Biol. **13**(5), e1005,543 (2017). https://doi.org/10.1371/journal.pcbi.1005543. https://dx.plos.org/10.1371/journal.pcbi.1005543
25. Haldeman, C., Beggs, J.M.: Critical branching captures activity in living neural networks and maximizes the number of metastable states. Phys. Rev. Lett. **94**(5), 058,101 (2005). https://doi.org/10.1103/PhysRevLett.94.058101. https://link.aps.org/doi/10.1103/PhysRevLett.94.058101
26. Harnack, D., Ernst, U.A., Pawelzik, K.R.: A model for attentional information routing through coherence predicts biased competition and multistable perception. J. Neurophysiol. **114**(3), 1593–1605 (2015). https://doi.org/10.1152/jn.01038.2014
27. Hebb, D.O.: The Organization of Behavior: A Neuropsychological Theory. Wiley, New York (1949)
28. Hess, R., Hayes, A., Field, D.: Contour integration and cortical processing. J. Physiol.-Paris **97**(2–3), 105–119 (2003). https://doi.org/10.1016/J.JPHYSPARIS.2003.09.013

29. Hesse, J., Gross, T.: Self-organized criticality as a fundamental property of neural systems. Front. Syst. Neurosci. **8**, 166 (2014). https://doi.org/10.3389/fnsys.2014.00166
30. Hubel, D.H., Wiesel, T.N.: Receptive fields and functional architecture of monkey striate cortex. J. Physiol. **195**(1), 215–243 (1968). https://doi.org/10.1113/jphysiol.1968.sp008455
31. Jones, H.E., Grieve, K.L., Wang, W., Sillito, A.M.: Surround suppression in primate V1. J. Neurophysiol. **86**(4), 2011–2028 (2001). https://doi.org/10.1152/jn.2001.86.4.2011
32. Kinouchi, O., Copelli, M.: Optimal dynamical range of excitable networks at criticality. Nat. Phys. **2**(5), 348–351 (2006). https://doi.org/10.1038/nphys289
33. Kovács, I., Julesz, B.: A closed curve is much more than an incomplete one: effect of closure in figure-ground segmentation. Proc. Natl. Acad. Sci. U. S. A. **90**(16), 7495–7497 (1993). https://doi.org/10.1073/PNAS.90.16.7495
34. Kreiter, A.K., Singer, W.: Stimulus-dependent synchronization of neuronal responses in the visual cortex of the awake macaque monkey. J. Neurosci.: Off. J. Soc. Neurosci. **16**(7), 2381–2396 (1996). https://doi.org/10.1523/JNEUROSCI.16-07-02381.1996
35. Langton, C.G.: Computation at the edge of chaos: phase transitions and emergent computation. Phys. D: Nonlinear Phenom. **42**(1–3), 12–37 (1990). https://doi.org/10.1016/0167-2789(90)90064-V
36. Levina, A., Priesemann, V.: Subsampling scaling. Nat. Commun. **8**, 15,140 (2017). https://doi.org/10.1038/ncomms15140. http://www.nature.com/doifinder/10.1038/ncomms15140
37. Levitt, J.B., Lund, J.S.: Contrast dependence of contextual effects in primate visual cortex. Nature **387**, 73–76 (1997)
38. Li, W., Pich, V., Gilbert, C.D.: Contour saliency in primary visual cortex. Neuron **50**(6), 951–962 (2006)
39. Luck, S., Chelazzi, L., Hillyard, S., Desimone, R.: Neural mechanisms of spatial selective attention in areas v1, v2 and v4 of macaque visual cortex. J. Neurophysiol. **77**, 24–42 (1997)
40. Mack, A.: Inattentional blindness: looking without seeing. Curr. Dir. Psychol. Sci. (2003). https://doi.org/10.1111/1467-8721.01256
41. von der Malsburg, C.: The correlation theory of brain function. In: Models of Neural Networks, pp. 95–119. Springer, New York (1994). https://doi.org/10.1007/978-1-4612-4320-5. http://link.springer.com/10.1007/978-1-4612-4320-5_2
42. Mandon, S., Kreiter, A.K.: Rapid contour integration in macaque monkeys. Vis. Res. **45**(3), 291–300 (2005). https://doi.org/10.1016/J.VISRES.2004.08.010
43. Massobrio, P., Pasquale, V., Martinoia, S.: Self-organized criticality in cortical assemblies occurs in concurrent scale-free and small-world networks. Sci. Rep. **5**(1), 10,578 (2015). https://doi.org/10.1038/srep10578. http://www.nature.com/articles/srep10578
44. Meisel, C., Olbrich, E., Shriki, O., Achermann, P.: Fading signatures of critical brain dynamics during sustained wakefulness in humans. J. Neurosci.: Off. J. Soc. Neurosci. **33**(44), 17,363–72 (2013). https://doi.org/10.1523/JNEUROSCI.1516-13.2013. http://www.ncbi.nlm.nih.gov/pubmed/24174669. http://www.pubmedcentral.nih.gov/articlerender.fcgi?artid=PMC3858643
45. Moran, J., Desimone, R.: Selective attention gates visual processing in the extrastriate cortex. Science **229**(4715), 782–784 (1985). https://doi.org/10.1126/science.4023713
46. Nykter, M., Price, N.D., Larjo, A., Aho, T., Kauffman, S.A., Yli-Harja, O., Shmulevich, I.: Critical networks exhibit maximal information diversity in structure-dynamics relationships. Phys. Rev. Lett. **100**(5), 058,702 (2008). https://doi.org/10.1103/PhysRevLett.100.058702. https://link.aps.org/doi/10.1103/PhysRevLett.100.058702
47. Palva, J.M., Zhigalov, A., Hirvonen, J., Korhonen, O., Linkenkaer-Hansen, K., Palva, S.: Neuronal long-range temporal correlations and avalanche dynamics are correlated with behavioral scaling laws. Proc. Natl. Acad. Sci. U. S. A. **110**(9), 3585–3590 (2013). https://doi.org/10.1073/pnas.1216855110
48. Pasquale, V., Massobrio, P., Bologna, L.L., Chiappalone, M., Martinoia, S.: Self-organization and neuronal avalanches in networks of dissociated cortical neurons. Neuroscience **153**(4), 1354–1369 (2008). https://doi.org/10.1016/j.neuroscience.2008.03.050

49. Petermann, T., Thiagarajan, T.C., Lebedev, M.A., Nicolelis, M.A.L., Chialvo, D.R., Plenz, D.: Spontaneous cortical activity in awake monkeys composed of neuronal avalanches. Proc. Natl. Acad. Sci. U. S. A. **106**(37), 15,921–6 (2009). https://doi.org/10.1073/pnas.0904089106. http://www.ncbi.nlm.nih.gov/pubmed/19717463. http://www.pubmedcentral.nih.gov/articlerender.fcgi?artid=PMC2732708
50. Plenz, D.: Criticality in cortex: neuronal avalanches and coherence potentials. In: Criticality in Neural Systems, pp. 5–42. Wiley-VCH Verlag GmbH & Co. KGaA, Weinheim, Germany (2014). https://doi.org/10.1002/9783527651009.ch2. http://doi.wiley.com/10.1002/9783527651009.ch2
51. Polat, U., Sagi, D.: Lateral interactions between spatial channels: suppression and facilitation revealed by lateral masking experiments. Vis. Res. **33**(7), 993–999 (1993)
52. Priesemann, V., Valderrama, M., Wibral, M., Le Van Quyen, M.: Neuronal avalanches differ from wakefulness to deep sleep evidence from intracranial depth recordings in humans. PLoS Comput. Biol. **9**(3), e1002,985 (2013). https://doi.org/10.1371/journal.pcbi.1002985. http://dx.plos.org/10.1371/journal.pcbi.1002985
53. Priesemann, V., Wibral, M., Valderrama, M., Pröpper, R., Le Van Quyen, M., Geisel, T., Triesch, J., Nikolić, D., Munk, M.H.J.: Spike avalanches in vivo suggest a driven, slightly subcritical brain state. Front. Syst. Neurosci. **8**, 108 (2014). https://doi.org/10.3389/fnsys.2014.00108
54. Ritz, R., Gerstner, W., Fuentes, U., van Hemmen, J.: A biologically motivated and analytically soluble model of collective oscillations in the cortex. ii. Application to binding and pattern segmentation. Biol Cybern. **71(4)**, 349–358 (1994)
55. Roelfsema, P.R.: Cortical algorithms for perceptual grouping. Annu. Rev. Neurosci. **29**, 203–227 (2006)
56. Rotermund, D., Taylor, K., Ernst, U.A., Kreiter, A.K., Pawelzik, K.R.: Attention improves object representation in visual cortical field potentials. J. Neurosci. **29**(32), 10120–10130 (2009). https://doi.org/10.1523/JNEUROSCI.5508-08.2009
57. Scarpetta, S., de Candia, A.: Neural avalanches at the critical point between replay and non-replay of spatiotemporal patterns. PLoS ONE **8**(6), e64,162 (2013). https://doi.org/10.1371/journal.pone.0064162. http://dx.plos.org/10.1371/journal.pone.0064162
58. Schoenholz, S.S., Gilmer, J., Ganguli, S., Sohl-Dickstein, J.: Deep Information Propagation (2016). http://arxiv.org/abs/1611.01232
59. Shannon, C.E.: A mathematical theory of communication. Bell Syst. Tech. J. **27**(3), 379–423 (1948). https://doi.org/10.1002/j.1538-7305.1948.tb01338.x
60. Shew, W.L., Clawson, W.P., Pobst, J., Karimipanah, Y., Wright, N.C., Wessel, R.: Adaptation to sensory input tunes visual cortex to criticality. Nat. Phys. **11**(8), 659–663 (2015). https://doi.org/10.1038/nphys3370
61. Shew, W.L., Plenz, D.: The functional benefits of criticality in the cortex. The Neuroscientist **19**(1), 88–100 (2013). https://doi.org/10.1177/1073858412445487
62. Shew, W.L., Yang, H., Petermann, T., Roy, R., Plenz, D.: Neuronal avalanches imply maximum dynamic range in cortical networks at criticality. J. Neurosci.: Off. J. Soc. Neurosci. **29**(49), 15,595–600 (2009). https://doi.org/10.1523/JNEUROSCI.3864-09.2009. http://www.ncbi.nlm.nih.gov/pubmed/20007483. http://www.pubmedcentral.nih.gov/articlerender.fcgi?artid=PMC3862241
63. Shriki, O., Alstott, J., Carver, F., Holroyd, T., Henson, R.N.A., Smith, M.L., Coppola, R., Bullmore, E., Plenz, D.: Neuronal avalanches in the resting MEG of the human brain. J. Neurosci.: Off. J. Soc. Neurosci. **33**(16), 7079–7090 (2013). https://doi.org/10.1523/JNEUROSCI.4286-12.2013
64. Sillito, A.M., Grieve, K.L., Jones, H.E., Cudeiro, J., Davls, J.: Visual cortical mechanisms detecting focal orientation discontinuities. Nature **378**, 492–496 (1995)
65. Smirnov, N.: Table for Estimating the Goodness of Fit of Empirical Distributions. Ann. Math. Stat. **19**(2), 279–281 (1948). https://doi.org/10.1214/aoms/1177730256
66. Softky, W., Koch, C.: The highly irregular firing of cortical cells is inconsistent with temporal integration of random EPSPs. J. Neurosci. **13**(1), 334–350 (1993). https://doi.org/10.1523/JNEUROSCI.13-01-00334.1993

67. Taylor, K., Mandon, S., Freiwald, W., Kreiter, A.: Coherent oscillatory activity in monkey area V4 predicts successful allocation of attention. Cereb. Cortex **15**(9), 1424–1437 (2005). https://doi.org/10.1093/cercor/bhi023
68. Theeuwes, J.: Topdown and bottomup control of visual selection. Acta Psychol. **135**(2), 77–99 (2010). https://doi.org/10.1016/J.ACTPSY.2010.02.006
69. Thorpe, S., Fize, D., Marlot, C.: Speed of processing in the human visual system. Nature **381**, 520–522 (1996). https://doi.org/10.1038/381520a0
70. Tinker, J., Velazquez, J.L.P.: Power law scaling in synchronization of brain signals depends on cognitive load. Front. Syst. Neurosci. **8**, 73 (2014). https://doi.org/10.3389/fnsys.2014.00073
71. Tomen, N., Rotermund, D., Ernst, U.: Marginally subcritical dynamics explain enhanced stimulus discriminability under attention. Front. Syst. Neurosci. **8**, (2014). https://doi.org/10.3389/fnsys.2014.00151
72. Treisman, A.M., Gelade, G.: A feature-integration theory of attention. Cogn. Psychol. **12**(1), 97–136 (1980). https://doi.org/10.1016/0010-0285(80)90005-5
73. Treue, S., Maunsell, J.: Attentional modulation of visual motion processing in cortical areas mt and mst. Nature **382**, 539–541 (1996)
74. Vogels, T.P., Sprekeler, H., Zenke, F., Clopath, C., Gerstner, W.: Inhibitory plasticity balances excitation and inhibition in sensory pathways and memory networks. Science (New York, N.Y.) **334**(6062), 1569–1573 (2011). https://doi.org/10.1126/science.1211095. http://www.ncbi.nlm.nih.gov/pubmed/22075724
75. Wagemans, J., Elder, J.H., Kubovy, M., Palmer, S.E., Peterson, M.A., Singh, M., von der Heydt, R.: A century of Gestalt psychology in visual perception: I. Perceptual grouping and figure-ground organization. Psychol. Bull. **138**(6), 1172–1217 (2012). https://doi.org/10.1037/a0029333. http://www.ncbi.nlm.nih.gov/pubmed/22845751 http://www.pubmedcentral.nih.gov/articlerender.fcgi?artid=PMC3482144
76. Zapperi, S., Lauritsen, K.B., Stanley, H.E.: Self-organized branching processes: mean-field theory for avalanches. Phys. Rev. Lett. **75**(22), 4071–4074 (1995). https://doi.org/10.1103/PhysRevLett.75.4071
77. Zhang, X., Zhaoping, L., Zhou, T., Fang, F.: Neural activities in V1 create a bottom-up saliency map. Neuron **73**(1), 183–192 (2012). https://doi.org/10.1016/J.NEURON.2011.10.035

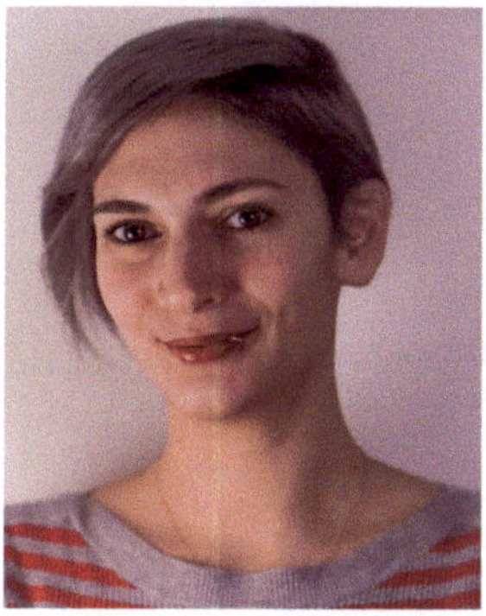

Nergis Tomen studied physics in Jacobs University, Germany, before getting her master's degree in Neuroscience in University of Bremen. She joined the group of Bernstein Prize winner Dr. Udo Ernst to pursue a Ph.D. in computational neurosience, in the Institute for Theoretical Physics, University of Bremen, Germany. Her research interests include critical phenomena in neuronal systems, attentional and context-dependent modulations of information processing in the visual system, nonlinear dynamics, oscillations and other forms of emergence in complex systems.

Udo Ernst studied physics at Frankfurt University, Germany and at the Max-Planck Institute for Dynamics and Self-Organization in Göttingen, Germany. He received his Ph.D. in physics from Frankfurt University in 1999 and continued being Postdoctoral researcher in the University of Bremen, Germany and at École Normale Supérieure in Paris, France. In 2010, Udo Ernst received the Bernstein Award in Computational Neuroscience which allowed him to establish an independent research group at Bremen University. His research interests include visual information processing and feature integration, selective attention and flexible neural computation, and understanding collective dynamics in neural networks such as criticality and synchronization.

Statistical Models of Neural Activity, Criticality, and Zipf's Law

Martino Sorbaro, J. Michael Herrmann and Matthias Hennig

Abstract We discuss the connections between the observations of critical dynamics in neuronal networks and the maximum entropy models that are often used as statistical models of neural activity, focusing in particular on the relation between *statistical* and *dynamical* criticality. We present examples of systems that are critical in one way, but not in the other, exemplifying thus the difference of the two concepts. We then discuss the emergence of Zipf laws in neural activity, verifying their presence in retinal activity under a number of different conditions. In the second part of the chapter we review connections between statistical criticality and the structure of the parameter space, as described by Fisher information. We note that the model-based signature of criticality, namely the divergence of specific heat, emerges independently of the dataset studied; we suggest this is compatible with previous theoretical findings.

1 Introduction

The debate about criticality in neural systems began with the observation of power laws in a number of experimentally measured variables related to neural activity. The first experimental observation of neuronal avalanches [6] found that their size distribution follows a power law with exponent of about $-3/2$, and their duration distribution follows one of exponent near -2 in cortical slices. These values are compatible with the exponents expected in critical branching processes—a well-studied topic in the field of complex systems physics [2]. Similar observations have been consistently reported in literature; moreover, the presence of power-law avalanche statistics was found to be theoretically justified by functional arguments on numerous occasions [5, 12, 42, 43, 49], and was shown to differ in different brain states [15, 39]. For an excellent high-level discussion of the topic, see [7].

M. Sorbaro (✉) · J. M. Herrmann · M. Hennig
School of Informatics, University of Edinburgh, 10 Crichton St., Edinburgh EH8 9AB, UK
e-mail: martopix@gmail.com

N. Tomen et al. (eds.), *The Functional Role of Critical Dynamics in Neural Systems*, Springer Series on Bio- and Neurosystems 11,
https://doi.org/10.1007/978-3-030-20965-0_13

An equally interesting instance of a power law is the finding that, in the population statistics of a neural network's activity, the rank of a state (the first being the most frequently observed, and so on) and its frequency are inversely proportional. This phenomenon, known as Zipf's law, was first observed by Auerbach in 1913: "If one sorts individuals by a given property in a descending fashion and stops doing so at rank n_1, or at n_2, or generally at rank n_x, where the property has gone down to values p_1, p_2, p_x, then a certain law exists between n_x and p_x. In our case, this law is especially simple, it is expressed by the formula: $n_x \cdot p_x = \text{constant}$" [3]. This author already alluded to the possibility of more complex forms of the same law (e.g. for the distribution of wealth), but did not speculate why it assumes its simplest form in the studied data. In the mid-1930s, the American linguist George Kingsley Zipf discovered that the frequency of occurrence of words in Joyce's Ulysses and American newspapers follows the same law [52], which is today called Zipf's law: the frequency of each word decays as a power law of its frequency rank. After Auerbach's original example, city sizes [10, 22], Zipf's law was confirmed in a variety of fields, including citation counts in scientific literature [40], earthquake magnitudes, wealth, solar flare size, number of emails and phone calls received, and many others [33].

Over the years several attempts have been made to understand Zipf's law. Zipf himself explains it by the *principle of least effort*: If words are stored in a linear array, then the low-frequency items are optimally located in a more distant place than more often used ones. The product of distance and frequency can be considered as a measure of the effort necessary to retrieve the word which he claims to be a constant. However, the assumption of an array, where the effort needed for retrieving an item is linear, which is necessary in order to obtain an inverse relationship rather than a general power law, seems unnatural when considering how items are stored in a neural network.

Another potential cause for the law can be seen in the idea of preferential attachment [4]. If, for example, the probability to move to or away from a city is assumed to be independent of its size, then Zipf's law for city sizes emerges. Other assumptions have been discussed and been shown to provide a better match for the distribution of city sizes [50], but again this may not easily carry over to states in a neural network.

Li [26] demonstrated that the words of an artificial language that simply consists of randomly chosen letters including a space sign tend to obey Zipf's law. However, the 'space' sign, which separates words in Li's approach, plays no such role in the analysis of neural data.

The authors of Ref. [1] aim at an explanation of Zipf's law by the existence of latent variables. Differently from the above attempts, this study is directly relevant for the analysis of neural activity. It also subsumes the scheme proposed by Li [26].

Although all of these attempts have their interest, there is some agreement that a deeper understanding is still lacking. In addition, there seems to be no clear justification on why criticality in the *statistical* sense and Zipf's law have been observed in neural data, or what brain function might benefit from it. It is interesting in this context that Zipf's law is a system property, i.e. it depends on the number of elements in the system and does not automatically apply to subset or unions of Zipfian sets. It can not be reduced to the mere presence of a particular probability distribution (such as $P(x) \sim x^{-2}$), but requires a conditional sampling procedure to be reproduced in a simulation [8]. The observations in neural data as well as a number of unsolved problems with this subject make it a very interesting subject of further investigation.

In what follows, we will discuss the connections between the observations of critical dynamics and maximum entropy models that are often used as statistical models of neural activity, reviewing the recent literature on the matter, and debate the possible relationship between this *statistical* concept of criticality and the *dynamical* criticality related to avalanche statistics. First, we will illustrate the concept of Zipf's distribution, its origin, and its applicability to neural data. In Sect. 2, we will introduce maximum entropy models, and show their connection to criticality and Zipf's law. In Sect. 3, we will make three observations that emphasise the difference between statistical and dynamical criticality: (Sect. 3.1) a system that shows dynamical, but not statistical criticality, (Sect. 3.2) the process of fitting an energy-based model, and (Sect. 3.3) the application of the theory of a large-scale corpus of biological data, where Zipf's law appears to hold, although the system is not dynamically critical. In Sect. 4, we will show connections between statistical criticality and the structure of the parameter space, as described by Fisher information. Finally, in Sect. 5, we will return to the question debate whether there is a relationship between statistical and dynamical criticality and conclude with an outlook on the problem.

2 Statistical Description of Spike Trains

2.1 Zipf's Law in Neural Data

For the specific case of neural activity, Zipf's law refers to the rank-probability law for the occurrence of each possible *pattern* of activity, which has been observed to follow a power law in the same sense as for words in the English language [48]. To understand what we mean by *pattern* or *state*, we need to adopt a simplified way of representing spike trains that we can call *digital*: discretising time in bins of equal size δt, we can define a Boolean variable

$$\sigma_n(t) = \begin{cases} 1 & \text{if neuron } n \text{ spikes between } t \text{ and } t + \delta t \\ 0 & \text{otherwise.} \end{cases} \tag{1}$$

At any given time, then, the population activity is described by a *codeword*

$$\boldsymbol{\sigma}(t) = (\sigma_1(t), \ldots, \sigma_N(t))$$

which describes, up to a precision of δt, the spiking state of the N neurons considered (Fig. 1). Modelling the system statistically, in this framework, means giving a full account of the probability of each possible codeword to appear. Note that, typically, we are not concerned with the dynamics of the system, and we disregard temporal

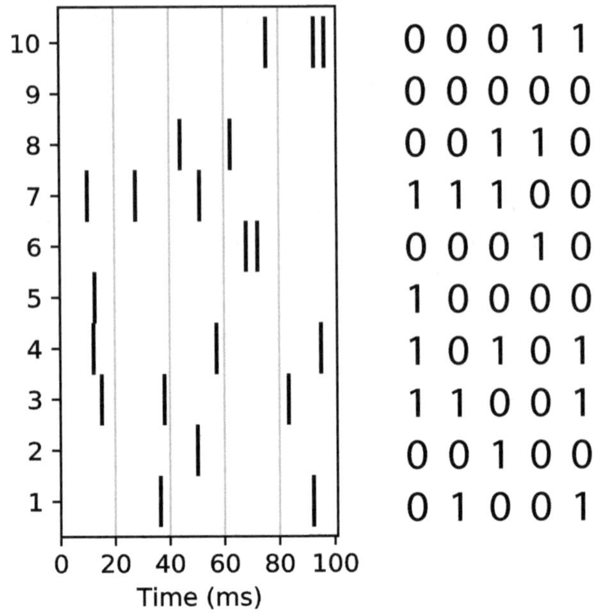

Fig. 1 Digitisation of spike trains from 10 neurons into a boolean matrix with bin size $\delta t = 20$ ms. Each row corresponds to the spike train of an individual neuron, with spikes represented as bars. Vertical gray lines indicate the boundaries of each bin. On the right, the resulting boolean matrix

correlations on scales larger than δt: this approach is suited to describe short-time correlations across space or properties of the encoding. Needless to say, the choice of δt can have important consequences on the results: in the limit of very large bin size, the pattern where all neurons fire simultaneously will be the only one to be observed; in the opposite limit of small δt, the *silent* pattern will be the most common, patterns with a single active neuron arbitrarily rare, and multi-neuron patterns absent. The results we discuss hold for bin sizes of the order of 5–20 ms, i.e. of the same order of magnitude as the typical correlation length between neurons; this value is commonly adopted in the literature [41].

To understand why Zipf laws are considered a signature of criticality, we will now illustrate the relationship, exposed by recent literature, between them and the critical points of models that have been used to describe neural activity, and are well known in physics.

2.2 Statistical Modelling

The activity patterns of individual neurons and neural networks invariably display stochastic characteristics. A common approach, which we can call *top-down*, of modelling the nature of this activity is to make (simplifying) assumptions on the actual workings of neurons, synapses, and networks, in order to set up a computational model the results of which can then be compared with experimental observations.

A large part, perhaps the largest, of computational neuroscience is based on this paradigm, predominantly by simulations of spiking neural networks.

Here, on the contrary, we are concerned with what we call *bottom-up* modelling, which seeks to infer properties of neural activity in an entirely data-driven way. Understanding the correlation structure, the distribution of firing rates, or the repetition of identical patterns from experimental data are examples of this approach. In other words, the data is described in terms of probabilities and other statistical descriptors, instead of parameters directly implied by the biological or physical theory.

In the bottom-up approach, a very broad family of models is available. We will restrict ourselves to *energy-based statistical models*, a number of models developed in the last decade which adopt a log-linear relation between probability and state variables. In an *energy-based* model probabilities are expressed in terms of an energy function E, in analogy with statistical physics:

$$P(\boldsymbol{\sigma}) = \frac{1}{Z} e^{-E(\boldsymbol{\sigma})},$$

where Z is the relevant normalisation factor. Many energy-based models used in neuroscience adopt the aforementioned *digital* description of spike trains in terms of binary variables: we will focus on these. In this case, for N neurons, $\boldsymbol{\sigma}$ can take 2^N different values, and determining the full population probability distribution requires specifying 2^N probabilities, which is an unrealistic task even for modest population sizes. Assumptions on the analytical form of the distribution are therefore required in order to infer a complete distribution from a relatively small number of samples.

2.3 *Maximum Entropy Models*

The first, and perhaps more elegant, strategy developed to this end is to adopt a *maximum entropy* approach [21], in which one first selects what features of the data should be exactly reproduced, and determines then the highest-entropy probability distribution consistent with those constraints. Schneidman et al. [41] and Shlens et al. [44] first applied this approach to neural data, using a Pairwise Maximum Entropy (PME) model, which exactly fits all $\langle \sigma_i \rangle$ and $\langle \sigma_i \sigma_j \rangle$, i.e. firing rates and pairwise correlations. Indeed, the question behind that research was primarily related to the importance of correlations in the vertebrate retina, including the study of higher-order interactions.

The PME probability distribution over all codewords has the following form:

$$P(\boldsymbol{\sigma}) = \frac{1}{Z(h, J)} \exp\left(\sum_{i=1}^{N} h_i \sigma_i + \sum_{i \neq j} J_{ij} \sigma_i \sigma_j \right). \tag{2}$$

The expression above is mathematically identical, in statistical physics, to that of the canonical ensemble for the Ising model with arbitrary couplings, a generalisation of the model originally used to describe ferromagnetism in solids [20].

By definition, a successful fit of a PME model correctly reproduces all firing rates and pairwise correlations present in the data from the considered neural population. Fitting based solely on second-order statistics does not imply that third-order correlations and other statistical measures are correctly reproduced. Reports that higher-order correlations are largely irrelevant were thus very surprising [41, 44, 46], although these observation may be restricted to low activity and high pairwise correlations [51]. Assessing whether a maximum entropy model can capture additional statistics of the data provides a source of interpretability: If, say, a PME model can account for third-order correlations, then the latter are not constrained further by the data. If, conversely, third-order correlations diverge from the PME prediction, we learn that the neural activity uses higher-order statistics to encode information. Whether this is the case depends on the system and on the distance between the neurons considered [37].

Several attempts have been made at improving the quality of the fit of statistical models, using different features as known statistics. The generalisation of Eq. (2)

$$P(\boldsymbol{\sigma}) = \frac{1}{Z(h, J)} \exp\left(\sum_{i=1}^{N} h_i \sigma_i + \sum_{i,j} J_{ij}^{(2)} \sigma_i \sigma_j + \sum_{i,j,k} J_{ijk}^{(3)} \sigma_i \sigma_j \sigma_k + \dots \right)$$

can describe any probability distribution of binary variables exactly. However, finding the values of $J^{(n)}$ is computationally expensive and the benefits typically do not outweigh the costs for $n \geq 3$.

As a different way to assess at least some aspects of the higher-order statistics, we can consider, for instance, the probability distribution of the number of neurons firing in a time bin, $p(K)$, where $K(t) = \sum_{i=1}^{N} \sigma_i$, was used as a target. This can be introduced as a further constraint in a maximum entropy model in combination with firing rates and pairwise correlations, leading to the *K-pairwise* model [31, 47]. It typically produces significantly better fits than a pure PME, and is much less computationally expensive than attempting to fit higher order cumulants. Another related approach, the *population tracking model*, fits $p(K)$ together with the conditional probabilities $P(\sigma_i = 1|K)$ of each neuron firing, given the current population firing rate, providing a lightweight and interpretable model [36].

An example of an energy-based model which does not rely on the maximum entropy principle, finally, is to use a restricted or semi-restricted Boltzmann machine (RBM/sRBM). Despite not directly aiming at fitting correlations, $p(K)$, and cumulants, as a maximum-entropy model would, RBMs were shown to perform at least comparably well in fitting all these aspects [24]. An advantage is that their complexity can be tuned, offering a choice of various degrees of accuracy and the corresponding computational costs. Additionally, *contrastive divergence*, the algorithm used for fitting, is an approximate but relatively fast and reliable algorithm, which lets one fit the simultaneous activity of a large number of units (up to several hundreds, whereas the exact learning algorithm for a PME model is not usable in practice over $N \approx 40$, although more efficient methods have been studied). Finally, RBMs can be interpretable models, specifically by studying the roles taken by hidden units.

Although a detailed discussion is beyond the scope of this chapter, we should at least mention the efforts to reproduce the time dynamics of the system, so that the statistical model fits both the distribution of single-time bin patterns and the conditional distribution of the pattern given the pattern in the previous time bin [11, 28, 32].

2.4 Phase Transitions in Models

Although this may initially seem less relevant from the point of view of research in neuroscience, we should remind ourselves that the Ising model is one of the earliest and most commonly studied paradigms of a phase transition. To understand its behaviour, let us make the temperature dependence of Eq. (2) explicit:

$$P_T(\boldsymbol{\sigma}) = \frac{1}{Z(h/T, J/T)} \exp\left(\frac{1}{T}\left[\sum_{i=1}^{N} h_i\sigma_i + \sum_{i\neq j} J_{ij}\sigma_i\sigma_j\right]\right) \tag{3}$$

Note that, in the high temperature limit, this converges to a uniform probability:

$$P_{T\to\infty}(\boldsymbol{\sigma}) = \frac{1}{2^N}, \quad \forall \boldsymbol{\sigma}.$$

Conversely, when $T \to 0$, only a small number of states with non-zero probability survive, the others becoming infinitely rare. If a system that obeys $P_{T\to 0}(\boldsymbol{\sigma})$ is perturbed in any way, it will eventually converge to this stable set, under any reasonable dynamics. In other words, the distribution becomes, in the zero temperature limit, a finite set of stable attractors, the same as the stationary distribution of a Hopfield network [19], where the attractors play the role of memory patterns.

Clearly, neither of the two limiting cases can be a realistic description of neural statistics, and the truth lays in between them, in a regime where the model is much more informative. The physics literature shows that there is sharp phase transition between a *disordered* phase and a *spin glass* phase, with the exact location of the critical point depending on the statistics of h and J [34]. It is then a natural question to ask whether the Ising model that results from a fit to neural activity is in one of the two phases, or poised near the critical point, and whether this relates to other concepts of criticality in neural systems.

The divergence of specific heat, also called heat capacity, in a macroscopic system is a classic signature of discontinuity in the properties of the system upon variation of a single parameter, typically temperature (generalisations of this idea will be discussed in the next section). The most classic example is the case of a change in the state of matter—solid to liquid, liquid to gas, etc.—where an infinitesimal change in temperature through the critical point requires a finite amount of energy (the *latent heat*). This is equally true for spin systems of the form we examined above. Tkačik et al. [48] fitted a model of the form (2) to binned spike trains from recordings of the salamander retina subjected to movies of naturalistic stimuli. They varied the temperature of the model around $T = 1$ (this value corresponding to the fit to neural data), and studied the specific heat as a function of T for an increasing number of neurons.

Their result clearly showed a peak in the specific heat of their models, with the peak temperature approaching $T = 1$ as N is increased. This is evidence that $T = 1$ coincides with the critical point, and therefore, the model is poised at criticality for parameter values exactly corresponding to those that fit the neural data. Similar observations were independently repeated, e.g. in [15, 31, 35], generating a debate on the nature of this observation and its biological interpretation, as will be discussed in later sections.

2.5 Model Criticality and Zipf's Law

Zipf laws can be related to statistical criticality in the sense of models, as shown in [48] (supplementary information), as follows. Call $p_1, ..., p_k, ..., p_{2^N}$ the probability of occurrence for each of the 2^N possible codewords. In statistical physics, microcanonical entropy can be defined as $S = \log \Omega$, where $\Omega = \Omega(E)$ is the number of states with energy lower than E. On the other hand, the energy level associated with a pattern is a function of its probability:

$$E_k = -\log p_k + \text{const.}$$

Now, Zipf's law states that, for every pattern, its rank $r_k \propto 1/p_k$. In the notation used above, note that $r_k = \Omega(E_k)$. Therefore, Zipf's law implies

$$\begin{aligned} \log p_k &= -\log r_k + \text{const.} \\ E_k &= S_k + \text{const.} \end{aligned} \tag{4}$$

If the above linear relation holds, then $d^2S/dE^2 = 0$. Since both specific heat and the variance of energy are inversely proportional to d^2S/dE^2, these thermodynamic quantities diverge. This is the classic signature of a second order phase transition.

The rank-probability relation defined by Zipf's law, therefore, is a model-independent way of showing criticality in this statistical sense. Its appearance guarantees the divergence of the specific heat of a PME model fit to the same data, but does not require complex and computationally expensive fitting procedures, and relies only on the statistical properties of the data.

3 Statistical and Dynamical Criticality

As we have mentioned, most energy-based models do not account for dynamics, as they are concerned only with fitting a single-time bin distribution. The formulation of the Ising model in physics describes a stationary distribution and does not include any dynamics. Transition probabilities from a state to another can be added through additional assumptions about the dynamics of the system. For instance, Glauber dynamics [13] generates a Markov chain whose stationary distribution coincides with the distribution (2). This is useful, for example, when sampling states from that probability distribution. Avalanches can be observed in high-dimensional Ising systems when they are driven out of stationarity by a change in temperature or applied magnetic field. Therefore, it is not expected that maximum entropy models reproduce any aspect related to avalanche dynamics.

It is not clear a priori, then, whether the observation of Zipf laws and diverging specific heats should be related to power-laws in the dynamics. In fact, finding a connection between the two concepts seems challenging. In the next sections, we will provide examples of how the two might be entirely distinct, which prompts questions on the nature, meaning, and relevance of statistical criticality.

3.1 The Eurich Model Is Dynamically, But Not Statistically Critical

For a discussion of the relationship between the two concepts of criticality, it is interesting to consider the Eurich model for neural avalanches as a "testbed" [9]. It is mathematically well-understood and can be conveniently tuned, because the parameter values for (quasi-)critical as well as sub- or super-critical behaviour are known analytically (even for finite systems). The very definition of the model is such that all neurons have identical properties, and the same for pairs, triplets, etc., of neurons. As a consequence, all N patterns with exactly one active neuron appear with the same frequency; all $N(N-1)/2$ patterns with exactly two active neurons appear with the same frequency, and so on, giving the rank-probability plot a step-like appearance, which cannot follow a Zipf law (Fig. 2).

It is tempting to consider the tail of the rank distribution. Although the number of states increases with the activity (for neurally plausible activity levels), their probability decreases strongly if the firing rate (per time bin) is low. Therefore, steps will disappear for higher ranks, which may or may not produce a power-law-like behaviour. However, in the statistical approach, typically small values of $N\delta t$ (see Eq. 1) are used, such that the potentially Zipf-like tail (Fig. 2, right) will be statistically irrelevant. It has also been claimed that the statistical approach is most likely restricted to low-activity patterns [51].

Note that the rank curve would be less step-like if some form of heterogeneity is introduced, as opposed to the complete symmetry between neurons that characterises

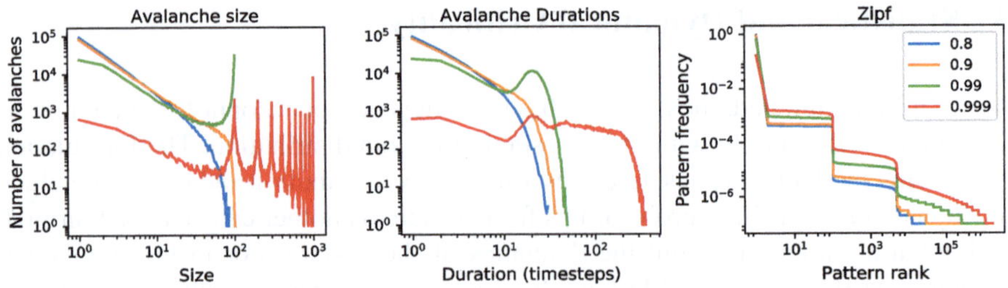

Fig. 2 Avalanche statistics compared to Zipf law for the Eurich model in different regimes [9]. Here $N = 100$, and the critical point is at $\alpha_c = 0.90$: in this case, the avalanche size and duration distributions most closely approach a power law with exponent $-3/2$, except for a cutoff due to the finite size. The subcritical case (blue), on the other hand, shows short-tailed distributions, and the supercritical cases (green and red) exhibit respectively one or many peaks at large values. In all cases, the Zipf plot does not show power-law dependence. Note that the smoothing of the step-like function is due to the finite sample size

the Eurich model. However, this would amount to an additional assumption that is, as the Eurich model shows, not necessary for criticality. In particular, we would need to assume that the patterns with a single active neuron already follow Zipf's law. This is a particular, but not unreasonable assumption, as this can be expected, for example, in a scale-free neural network. In fact, we cannot rule out that a Zipf profile, or at least an approximation, could be found just by tuning the distribution of firing rates, even in the absence of correlations. Such a finding would entirely rule out any relation to dynamical criticality, which appears exclusively as a consequence of emergent phenomena deriving from complex interactions. However, it would still require a specific distribution of firing rates among the neurons, an assumption that in itself would prompt questions about its functional reasons. Firing rate distributions have been studied extensively [30], and found to be highly skewed, with a small fraction of neurons responsible for the majority of emitted spikes. A clear theory on why this is an advantage for the encoding is still missing. In Sect. 3.3, we will consider a case in which the Zipf relation holds even when correlations are destroyed, which suggests the long-tailed firing rate distribution is sufficient for it to hold.

3.2 *Fitting Energy-Based Models to Critical Activity*

A natural way of checking if dynamical and statistical criticality are related could involve fitting a statistical model to neural models that exhibit various kinds of dynamics, and can be tuned to a supercritical (noisy), subcritical or critical regime. This was one of the goals of in Ref. [15]. The authors identified five different dynamical states of the cat and monkey cortex, studied their avalanche statistics, and evaluated the temperatures corresponding to peak specific heats of Ising models fitted to each dataset. The results did show a small but significant relationship between avalanche

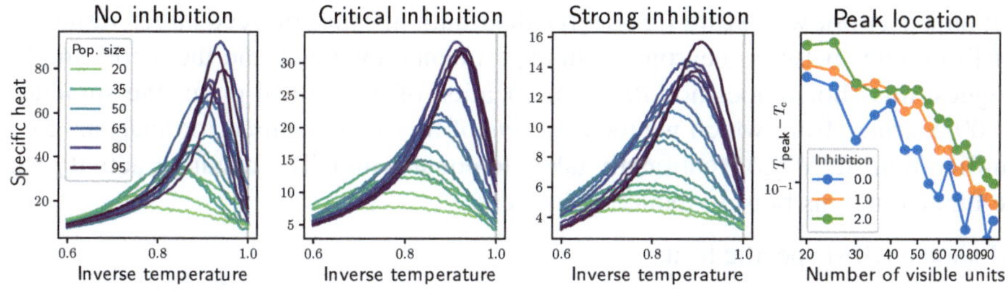

Fig. 3 RBM specific heats peak when fitted to a variety of datasets, with the peak approaching the temperature of the fit as the model size increases. RBMs were fitted to simulated data with different avalanche statistics: supercritical (left), critical (centre) and subcritical (right)

dynamics and specific heat peak location. However, it should be noted that some of the results in this work were obtained with small datasets of six neurons only, which may not offer insight on what happens in the thermodynamic limit.

We attempted a similar task fitting Bernoulli RBMs to the activity generated by a tunable model of a neural network, similar to the binary, non-leaky, integrate and fire model used by [12]. In this model, the strength of inhibition can be tuned, leading to a network with low, random-looking activity and a short-tailed avalanche distribution (high inhibition); or a network generating activity in large bursts (low or no inhibition). The critical regime lies in between the two. Details about the implementation of the model are given below.

We found that although the absolute value of the peak does depend on the correlations of the data, its location is always at a temperature near $T = 1$ (which is the value corresponding to the original fit), and further approaches this temperature as the number of units increases, i.e. in the thermodynamic limit (Fig. 3). These results are compatible with what was shown by [35], using a different dataset, for the K-pairwise model.

It seems, then, not only that the statistical model that we fitted does not accurately detect criticality in the dynamical sense, but it also exhibits statistical criticality no matter the dataset it was fitted to. This implies, on the one hand, that the dynamical criticality of a dataset and the statistical criticality of a model fitted to it are unrelated, and, on the other hand, that a model fitted to datasets of very different nature all tend to exhibit statistical criticality. This is compatible with an argument that was put forward by theoreticians [29], as will be discussed in Sect. 4.2.

Methods: network model

The binary neuron model used for the simulations is similar to the one presented by [12]. In this model, each neuron has a probability of firing given by a weighted sum of its inputs, divided by a factor dependent on its own firing history. A fifth of all neurons are inhibitory, while the rest are excitatory; the value of inhibition was tuned to 0.0, 1.0, or 2.0, to enforce different regimes, corresponding to different correlations and avalanche statistics. While in the original work the connectivity was all-to-all, with weights drawn from a uniform distribution, we modified it to identical couplings, but set on a network with scale-free degree distribution. This enforced larger variability of firing rates between neurons; both experimental evidence and the theory in [25] suggest this choice does not affect the location of the critical point. We simulated 1000 neurons, from which we took subsets of the sizes required for analysis, for 1 million time steps (conventionally taken to equal 1 ms). The resulting activity was re-binned in 5 ms bins, to reduce sparseness.

Methods: RBM specific heats

As we will argue in Sect. 4.2, the direction that best indicates the critical point coincides with the fist eigenvector (the one corresponding to the largest eigenvalue) of the Fisher information tensor. However, in practice, this is never orthogonal to the direction of increasing/decreasing temperature: thus, varying temperature is an acceptable way to look for a phase transition.

In statistical physics, the general expression for the probability of a pattern in an energy-based model at temperature T is

$$P_T(x) = \frac{1}{Z(T)} e^{-E(x)/T}.$$

The expression for the energy in the case of RBMs is

$$E(\mathbf{v}, \mathbf{h}) = -\mathbf{a}^\intercal \mathbf{v} - \mathbf{b}^\intercal \mathbf{h} - \mathbf{v}^\intercal J \mathbf{h}.$$

where $\mathbf{v}$ and $\mathbf{h}$ are vectors of visible and hidden binary variables respectively. Since this expression is linear in the parameters a_i, b_j and J_{ij} for all i, j, changing the temperature of a model coincides with rescaling these parameters by a linear factor $\beta = 1/T$. In the following, we have adopted the standard strategy of fitting an RBM to neural data, obtaining values for its parameters, and then rescaling them—this means $T = 1$ (no rescaling) coincides with the parameters as they were fitted, the values corresponding to a model that correctly reproduces the given data. Fits were obtained by 1-step persistent contrastive divergence.

We can then compute the specific heats at different temperatures. The marginal probability of $\mathbf{v}$ is

$$P_T(\mathbf{v}) = \frac{1}{Z} \sum_{h_{1\ldots N}=0,1} e^{\frac{1}{T}(\mathbf{a}^\intercal \mathbf{v} + \mathbf{b}^\intercal \mathbf{h} + \mathbf{v}^\intercal J \mathbf{h})} = \frac{e^{\mathbf{a}^\intercal \mathbf{v}/T}}{Z} \prod_{j=1}^{N_h} \left(1 + e^{b_j/T + (\mathbf{v}^\intercal J)_j/T}\right)$$

Disregarding an additive constant, the energy of a visible pattern can be expressed as the logarithm:

$$E_T(\mathbf{v}) = \frac{\mathbf{a}^\intercal \mathbf{v}}{T} + \sum_{j=1}^{N_h} \log\left(1 + e^{b_j/T + (\mathbf{v}^\intercal J)_j/T}\right)$$

In accordance with statistical physics, we can define

$$c(T) = \frac{\text{Var}(E_T)}{N_v T^2}.$$

This quantity can be computed from a sample. For each temperature value, in each dataset, we extracted 20 chains of 2000 samples, taken every 10 steps in order to reduce spurious correlations, and computed $c(T)$ by the expression above.

3.3 *Retinal Activity Obeys Zipf's Law, But Is Not Dynamically Critical*

The mammalian retina, a system that is often chosen when studying the statistics of neural activity, and whose encoding and dynamical properties are well known, is an example of the opposite case: It was the first system in which statistical criticality was observed, but it does not exhibit dynamical criticality.

Avalanches arise in the mammalian retina only during the period of development: for mice, in the first few days after birth, before eye opening, when the retina does not respond to light and the network activates spontaneously. During this stage, the activity of the retina consists of the so-called *retinal waves*, which are effectively power-law distributed avalanches. Direct comparison with a computational model showed that these are indeed the signature of a critical state between locally and globally connected activity [16]. However, these disappear in a functional retina: Fig. 4 shows the statistics of a 20-min recording of an untreated, adult mouse retina under an uncorrelated black-and-white checkerboard stimulation. It is evident that the avalanche statistics is short-tailed, and, at the same time, the probability-rank plot of pattern frequencies is well compatible with a Zipf law. Note that correlations between the activities of retinal ganglion cells change significantly with the statistics

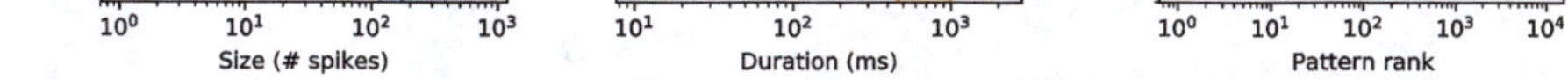

Fig. 4 Avalanche statistics compared to Zipf law in the neural activity of a healthy, adult (postnatal day 91) mouse retina stimulated by projection of a white noise checkerboard pattern. The detection of avalanches and the pattern count were repeated over 30 sets of 100 neighbouring neurons, each of which was recorded for 20 min. The sets may overlap. The solid lines are medians over these sets; the shaded area is delimited by the first and third quartiles. The grey line in the rightmost plot is for comparison with Zipf's law. The data were made available by G. Hilgen and E. Sernagor, University of Newcastle. We refer to [18] for experimental and data analysis methods

of the stimulus, and the avalanche statistics will consequently appear different. The example of adult retinas is complementary to Sect. 3.1 in the sense that, here, a system that does not show dynamical criticality can well obey Zipf's law.

It is worth mentioning that the observation of Zipf law in retinas is very robust to a number of external factors. We found no significant differences in the rank-frequency plots of patterns observed when the retina was treated with bicuculline (a GABA blocker) compared to a control; analogously for retinas under stimuli characterised by very different level of spatial correlations.

If Zipf's laws have a functional role, there is no expectation this phenomenon would survive in a non functioning neural system, such as a retina that has been pharmacologically treated in a way that breaks its normal operative mode. Here, we took data from the same mouse retina, before and after treatment with a 20 μM solution of bicuculline, which is a GABA_A antagonist. The results are shown in Fig. 6 (left): as it is evident, there is no clear difference between the two rank-probability plots. Of course, the only strong argument against the functional role of Zipf laws would be finding a functional retina in which this law is broken, which is not the case here. However, we can notice that even an intervention that significantly disrupts the retina's activity, by blocking inhibitory interactions, doesn't prevent this phenomenon to arise. This is despite the large change in the correlation between neurons induced by bicuculline.

Likewise, one may expect a dependence of pattern frequency-rank statistics on stimulus statistics. The retina, after all, is a neural system design to encode a stimulus—and the correlation structure of its neurons' activity strongly depends on the correlations in the stimulus. However, we found no significant difference in Zipf laws under different stimulations. Figure 6 (centre) shows a single group of 100 neurons selected in a retina that was stimulated with light patterns of different kinds. All stimulus presentations consisted of black-and-white random checkerboards, which are binarised versions of random noise of given frequency spectrum f^{-a} with $a = 0.5, 1.0, 1.5, 2.0$ in space and time: from near-white noise to the statistics of natural images (Fig. 5).

The independence from correlations is evident in the right panel of Fig. 6: here, the "control" curve is the same as in the left panel, and is compared with the

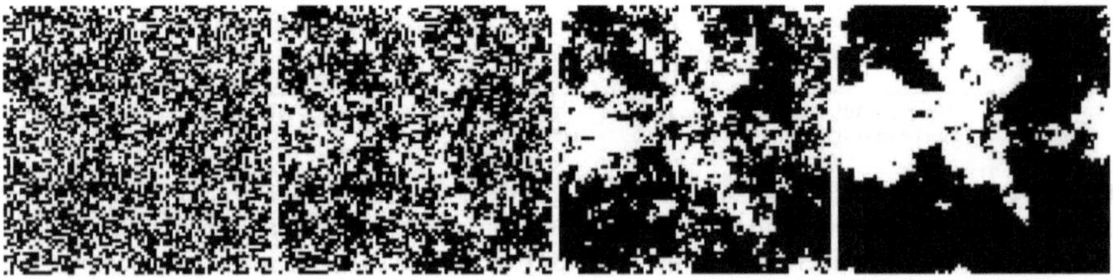

Fig. 5 Examples of stimulation frames. Correlations increase from left to right (dataset ID 0 to 3): the frequency spectra follow f^{-a} with $a = 0.5, 1.0, 1.5, 2.0$, i.e. from noise to the statistics of natural images. The correlation statistics extend to time

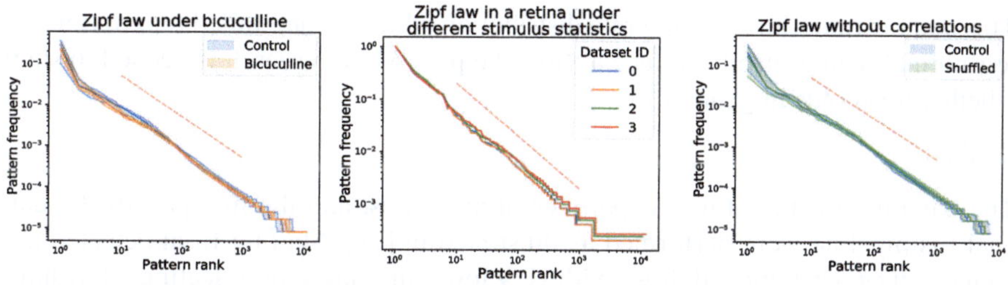

Fig. 6 Left: Zipf plots, before and after treatment with bicuculline. 30 groups of 100 neurons, selected as explained in the Methods paragraph. Centre: Zipf plots for a unique group of 100 neurons under stimuli of different statistics; the difference between datasets 0–3 consist in the different spatial frequency—from near-white noise to natural stimulus statistics. Right: the same data as in the left panel (control), and its shuffled version, where correlations have been destroyed, while keeping the same firing rates. The red dashed lines correspond to $1/x$ laws

rank-probability plot for a "shuffled" version of the same data, where the firing rates were kept the same, but spikes were moved in time in order to cancel neuron-neuron correlations. The difference between the two curves is clearly not significant.

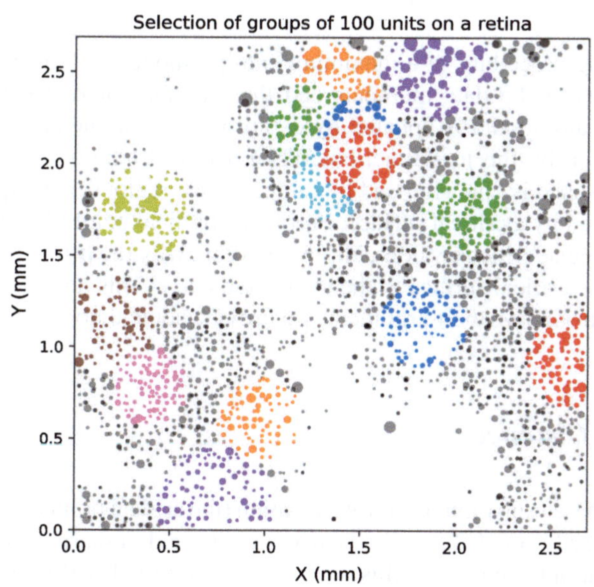

Fig. 7 All spike clusters in a dataset (P91 mouse retina under white noise checkerboard stimulation), arranged spatially. For Zipf analysis, a random cluster was selected, and the 100 nearest ones picked along it (coloured patches on the figure are examples) to form a 100-neuron group. The process was repeated 30 times to study error intervals. The size of the dot scales with the number of spikes in the cluster. Even if this image only represents detected spikes, the optic disc is noticeable at the bottom end; other inactive areas corresponds to cuts in the retina, unavoidable when placing it on a flat surface

This demonstrates how a firing rate distribution which is long-tailed (approximately log-normal) can in itself produce a Zipf-like plot. More research is needed to show whether this holds in general.

Methods

The handling of the retinas, experimental apparatus, and the first part of the data analysis pipeline were performed as illustrated in [18]. Starting from detected and sorted spikes, we removed those with very low amplitudes, by selecting a threshold corresponding roughly to the lowest 10%. This was to ensure only good-quality events were left. Then, we selected, for each Zipf plot, $N = 100$ clusters all pertaining to the same area of the retina (Fig. 7).

At this stage, spikes were binarised into a $N \times T$ matrix S of boolean variables, with $S(n, t) = 1$ if neuron n spiked between times t and $t + \delta t$ and $S(n, t) = 0$ otherwise. When multiple spikes from the same neuron occurred in a single time bin, the extra spikes were disregarded. For recordings shown in this chapter, $T =$ 120,000 or more, and $\delta t = 10$ ms, implying at least 20 min of neural activity were recorded.

4 Parametric Sensitivity

The basic fitting procedure of a maximum entropy model minimises the quadratic difference between the data moments and the moments predicted by the model. During fitting, any model is updated by exploring the parameter space, following a direction given by the loss function. When a model admits a phase transition, the parameter space is characterised by (at least) two regions, corresponding to the phases, separated by a critical surface. From a theoretical point of view, asking why a model is poised at criticality coincides with asking why the fitting process tends to lead towards the critical surface in the parameter space. This has been discussed by [29]; before introducing their argument, we provide some theoretical background.

4.1 Model Distance

Intuitively, a phase transition occurs at a location (the critical point or critical surface) where an arbitrarily small change in the parameters yields a sharp, qualitative change in the behaviour of the model. In this section, we will formalise this idea, and link the notion of model distance to the statistical physics framework that we have introduced above.

A common measure of the distance (in model space) of a probability distribution p from a given one q, both defined on a set S, is the Kullback-Leibler divergence

$$D_{KL}(p; q) = \int_S p(x) \log \frac{p(x)}{q(x)} dx.$$

It measures the amount of information that is lost when approximating p by q. The name *divergence* stresses that this quantity does not have the mathematical properties of a distance, namely not being symmetric. If the model space is parametrised by θ, and p and q are close to each other in this space, so that $q = P_\theta$ and $p = P_{\theta+\delta\theta}$, at second order in $\delta\theta$, the divergence can be approximated as

$$D_{KL}(P_{\theta+\delta\theta}; P_\theta) \approx \frac{1}{2}\delta\theta^T F(\theta)\delta\theta,$$

where F is called Fisher information tensor (FIT) which is given by

$$F_{ij}(\theta) = -\int P_\theta(x)\frac{\partial \log P_\theta(x)}{\partial\theta_i}\frac{\partial \log P_\theta(x)}{\partial\theta_j}dx.$$

Fisher information is here expressed as a statistical quantity, but it has an important relation to the physics of statistical models. Consider a Hamiltonian model, where the probability distribution is given by

$$P_\theta(x) = \frac{e^{-H_\theta(x)}}{Z_\theta}, \qquad H_\theta(x) = \sum_{k=1}^{n} \theta_k f_k(x), \tag{5}$$

which is an obvious generalisation of maximum entropy models. Calculating the Fisher information for this form of P (5), we retrieve the direct connection between the covariance (with respect to P_θ) of the physical quantities f and the FIT that is characteristic for probability distributions of the exponential type:

$$F_{ij}(\theta) = \mathrm{Cov}[f_i(x), f_j(x)].$$

This means that the FIT characterises the variances and correlations of the functions f which are now considered as stochastic variables and depend on the state x of the system.

Additionally, note that changing the temperature in the traditional canonical ensemble corresponds to scaling the Hamiltonian by a factor $\beta = 1/T$, similarly to Eq. (3). In the formulation above (5), this is equivalent to scaling all the θ_i by β. Given a point θ on the parameter manifold, the direction $\partial/\partial\beta$ can be expressed as

$$\frac{\partial}{\partial\beta} = \frac{1}{n}\sum_{k=1}^{n}\frac{\partial}{\partial\theta_k}$$

which is just a linear combination. The specific heat is given by

$$c(\beta) = \frac{\beta^2}{N}\mathrm{Var}[E].$$

Thus, we can arrange for the specific heat to be one of the entries of the Fisher information matrix, with a change of basis, which includes the β direction as a base vector together with other $n-1$ linearly independent ones. Analogous considerations can be made for the magnetic field and magnetic susceptibility. In this sense, the Fisher information tensor is a generalisation of specific heats and susceptibilities.

4.2 Fisher Information and Criticality

We can now look at the relationship between statistical criticality and the model's parameter space. Suppose any *generalised susceptibility* (i.e. a component of the Fisher tensor) diverges at a point θ_0. Then an eigenvalue of the Fisher information, say λ_k, diverges at θ_0. Call v_k the corresponding normalised eigenvector. For small α,

$$D_{KL}(P_{\theta_0+\alpha v_k}; P_{\theta_0}) \approx \frac{\alpha^2}{2} v_k^T F(\theta_0) v_k = \frac{\alpha^2}{2}\lambda_k,$$

and the r.h.s. diverges. This means that, moving from θ_0 in the v_k direction by an arbitrarily small step yields a model $P_{\theta_0+\alpha v_k}$ that is completely different from P_{θ_0}, as indicated by an infinite KL divergence.

We introduced this description in terms of Fisher information in order to give an interpretation of criticality from the point of view of modelling. A model is at a critical point whenever there is a direction in parameter space that leads to an infinitely different model by a finite change in parameters. This, incidentally, shows that the best way of measuring the distance from a critical point is not to vary temperature, but to use the first eigenvalue of the FIT and move in the direction of the corresponding eigenvector. Temperature is not always the most relevant control parameter.

Mastromatteo and Marsili [29] have argued that, because of this special property critical points have in the parameter space, they are particularly favoured by model fitting. In particular, they show that *distinguishable* models accumulate near critical points, whereas models farther away from these are largely indistinguishable. Their argument, in brief, goes a follows. Two models are considered indistinguishable if their Kullback-Leibler divergence is less than a given value ε. For small ε, D_{KL} is approximated by Fisher information, and the volume of parameter space occupied by models indistinguishable from θ_0 turns out to be proportional to $(\det F(\theta_0))^{-\frac{1}{2}}$. This quantity diverges at critical points due to the first eigenvalue diverging as explained above. Thus, *most models* actually are poised near a critical point, according to this metric. They conclude that criticality may be a feature induced by the inference process, rather than one intrinsic to the real system being studied by the model. This may be the reason why statistical models seem to be poised at a critical point, for a variety of training datasets, as we showed in Sect. 3.2. However, it does not affect Zipf laws, which are directly observed in the data.

4.3 Criticality and Parameter 'Sloppiness'

It is well known that the parameter spaces of many models often show only a small number of directions (linear combinations of parameter changes) along which the overall properties of the model strongly change ("stiff"), and a large number of directions which have little influence on the model ("sloppy"). This phenomenon, termed "model sloppiness", has been observed in a wide number of cases in systems science [14, 27].

For the specific case of neuronal networks, in Ref. [38], although for small numbers of neurons, "stiff" dimensions corresponding to large FIT eigenvalues were identified. The remaining "sloppy" dimensions, on the other hand, can change without much effect on the goodness of fit of the model. A further development of this approach has been reported in Ref. [17], where it was shown that about half of the dimensions in the data manifold are irrelevant for the modelling. As shown in paragraph 4.2, near a critical point, the direction pointing towards the critical surface has a diverging FIT eigenvalue, while the others are smaller. This hints there may be a connection between sloppiness and criticality, which, at the moment, we can only leave at the level of speculation.

Additionally, however, sloppiness indicates that a fitting algorithm for the data may be improved if different dimensions are differently weighted during the optimisation process. We can then ask whether using a natural gradient in the fitting procedure would lead to a different result while evaluating model criticality. In natural gradient optimisation, the components of the gradient are compensated by the inverse Fisher information, i.e. the divergence near a critical point of the model would disappear, at least theoretically when the Fisher information is exactly known. As a result, the fitting procedure is not homogeneous with respect to the set of the parameters, but with respect to the space of the parameters, taking into account its geometrical properties, and parameters can be identified equally well in all regions. In this way, the problem discussed in Ref. [29] may disappear—more research will be needed in order to verify this.

5 Discussion

Neuronal avalanches are an experimentally well-studied phenomenon, that can be explained as a consequence of the optimisation of information processing in the brain. It should be noted that an understanding of how the potential functional benefits of this "dynamical" criticality are realised is missing [42]—however, it has been shown that the maximisation of the dynamical range happens at criticality [23].

Statistical criticality is an equally complex phenomenon to explain theoretically. Like dynamical criticality, it can be taken to indicate the complexity of the neural data and the relevance of higher-order correlations or latent variables, but its functional implications are less clear. In this chapter, we have reviewed the concept, both in the context of fitted statistical models, and as a direct observation of Zipf laws in neural population data. Through experiments on restricted Boltzmann machines, we suggested that the divergence of model specific heat is not a reliable way to infer properties of the data. We mentioned how Fisher information provides the correct description of the parameter space and the critical surfaces, and reviewed a possible explanation of why statistical models tend to poise themselves at a critical point. Then, we tried to describe the connection between statistical and dynamical criticality, and argued there is no clear connection, by showing examples where one of the two was present without the other. Further insight on this matter might come from models that are capable of both, provided they can reproduce not only the equilibrium distribution of the data, but also the dynamics. A multi-time maximum entropy model might provide a starting point for this work.

Of course, it may well be that the observation of Zipf laws is simply a consequence of problems related to how we describe the data—these include the typically small sets of observables, the choice of binning size, failure to account for the real dynamics, and biases introduced by sampling. However, the ubiquity of Zipf laws in complex systems means its emergence in biological neural networks should not surprise us, and it could be explained in terms of mechanisms such as the one described by [1], or perhaps with preferential attachment. Conversely, an important open problem is an explanation on whether statistical criticality is something that is actively sought by the system because of some functional relevance. On this matter, we tried to analyse the Zipf profile of retinal activity under various conditions (various stimulus statistics, pharmacological treatment), but we found no significant differences in the cases examined. Interestingly, it seems to be possible to generate a Zipf profile simply by enforcing a long-tailed firing rate distribution, despite the absence of correlations. Even if this observation were confirmed, the question would simply shift towards finding a reason for such a skewed distribution of firing rates, which has not yet found a justification in terms of function.

Notably, recent research has started showing how Zipf laws appear in different kinds of parametric models, including “deep” ones, as soon as learning occurs. It has been shown that the Zipf property arises to different degrees in different layers of a deep network, and is maximal in the layers that attain an optimal trade-off between resolution and accuracy in generating samples [45]. This is a starting point in linking statistical criticality to function. It is not known whether similar principles are relevant in the case of biological neural networks, and finding such a link could be an interesting direction of future research.

References

1. Aitchison, L., Corradi, N., Latham, P.E.: Zipf’s law arises naturally when there are underlying, unobserved variables. PLOS Comput. Biol. **12**(12), 1–32 (2016). https://doi.org/10.1371/journal.pcbi.1005110

2. Athreya, K.B., Jagers, P.: Classical and Modern Branching Processes. IMA, vol. 84. Springer (1997)
3. Auerbach, F.: Das Gesetz der Bevölkerungskonzentration. Petermanns Geographische Mitteilungen **59**, 74–76 (1913) (Quote translated by J.M.H.)
4. Barabási, A.L., Albert, R.: Emergence of scaling in random networks. Science **286**, 509–512 (1999)
5. Beggs, J.M.: The criticality hypothesis: how local cortical networks might optimize information processing. Philos. Trans. R. Soc. Lond. A Math. Phys. Eng. Sci. **366**(1864), 329–343 (2008)
6. Beggs, J.M., Plenz, D.: Neuronal avalanches in neocortical circuits. J. Neurosci. **23**(35), 11167–11177 (2003)
7. Beggs, J.M., Timme, N.: Being critical of criticality in the brain. Front. Physiol. **3**, 163 (2012)
8. Cristelli, M., Batty, M., Pietronero, L.: There is more than power law in Zipf. Sci. Rep. **2**, 812(7) (2012)
9. Eurich, C.W., Herrmann, J.M., Ernst, U.A.: Finite-size effects of avalanche dynamics. Phys. Rev. E **66**(6), 066,137 (2002)
10. Gabaix, X.: Zipf's law and the growth of cities. Am. Econ. Rev. **89**(2), 129–132 (1999)
11. Gardella, C., Marre, O., Mora, T.: Blindfold learning of an accurate neural metric. In: Proceedings of the National Academy of Sciences, p. 201718710 (2018)
12. Gautam, S.H., Hoang, T.T., McClanahan, K., Grady, S.K., Shew, W.L.: Maximizing sensory dynamic range by tuning the cortical state to criticality. PLoS Comput. Biol. **11**(12), e1004,576 (2015)
13. Glauber, R.J.: Time-dependent statistics of the Ising model. J. Math. Phys. **4**(2), 294–307 (1963)
14. Gutenkunst, R.N., Waterfall, J.J., Casey, F.P., Brown, K.S., Myers, C.R., Sethna, J.P.: Universally sloppy parameter sensitivities in systems biology models. PLoS Comput. Biol. **3**(10), e189 (2007)
15. Hahn, G., Ponce-Alvarez, A., Monier, C., Benvenuti, G., Kumar, A., Chavane, F., Deco, G., Frgnac, Y.: Spontaneous cortical activity is transiently poised close to criticality. PLOS Comput. Biol. **13**(5), 1–29 (2017). https://doi.org/10.1371/journal.pcbi.1005543
16. Hennig, M.H., Adams, C., Willshaw, D., Sernagor, E.: Early-stage waves in the retinal network emerge close to a critical state transition between local and global functional connectivity. J. Neurosci. **29**(4), 1077–1086 (2009)
17. Herzog, R., Escobar, M.J., Cofre, R., Palacios, A.G., Cessac, B.: Dimensionality reduction on spatio-temporal maximum entropy models on spiking networks. Preprint bioRxiv:278606 (2018)
18. Hilgen, G., Sorbaro, M., Pirmoradian, S., Muthmann, J.O., Kepiro, I.E., Ullo, S., Ramirez, C.J., Encinas, A.P., Maccione, A., Berdondini, L., et al.: Unsupervised spike sorting for large-scale, high-density multielectrode arrays. Cell Rep. **18**(10), 2521–2532 (2017)
19. Hopfield, J.J.: Neural networks and physical systems with emergent collective computational abilities. Proc. Natl. Acad. Sci. **79**(8), 2554–2558 (1982)
20. Ising, E.: Beitrag zur Theorie des Ferromagnetismus. Z. für Phys. **31**(1), 253–258 (1925)
21. Jaynes, E.T.: Information theory and statistical mechanics. Phys. Rev. **106**(4), 620 (1957)
22. Jiang, B., Jia, T.: Zipf's law for all the natural cities in the United States: a geospatial perspective. Int. J. Geogr. Inf. Sci. **25**(8), 1269–1281 (2011)
23. Kinouchi, O., Copelli, M.: Optimal dynamical range of excitable networks at criticality. Nat. Phys. **2**, 348–352 (2006)
24. Köster, U., Sohl-Dickstein, J., Gray, C.M., Olshausen, B.A.: Modeling higher-order correlations within cortical microcolumns. PLoS Comput. Biol. **10**(7), e1003,684 (2014)
25. Larremore, D.B., Shew, W.L., Restrepo, J.G.: Predicting criticality and dynamic range in complex networks: effects of topology. Phys. Rev. Lett. **106**(5), 058,101 (2011)
26. Li, W.: Random texts exhibit Zipf's-law-like word frequency distribution. IEEE Trans. Inf. Theory **38**(6), 1842–1845 (1992)
27. Machta, B.B., Chachra, R., Transtrum, M.K., Sethna, J.P.: Parameter space compression underlies emergent theories and predictive models. Science **342**(6158), 604–607 (2013)

28. Marre, O., El Boustani, S., Frégnac, Y., Destexhe, A.: Prediction of spatiotemporal patterns of neural activity from pairwise correlations. Phys. Rev. Lett. **102**(13), 138,101 (2009)
29. Mastromatteo, I., Marsili, M.: On the criticality of inferred models. J. Stat. Mech. Theory Exp. **2011**(10), P10,012 (2011)
30. Mizuseki, K., Buzsáki, G.: Preconfigured, skewed distribution of firing rates in the hippocampus and entorhinal cortex. Cell Rep. **4**(5), 1010–1021 (2013)
31. Mora, T., Deny, S., Marre, O.: Dynamical criticality in the collective activity of a population of retinal neurons. Phys. Rev. Lett. **114**(7), 078,105 (2015)
32. Nasser, H., Marre, O., Cessac, B.: Spatio-temporal spike train analysis for large scale networks using the maximum entropy principle and Monte Carlo method. J. Stat. Mech. Theory Exp. **2013**(03), P03,006 (2013)
33. Newman, M.E.: Power laws, Pareto distributions and Zipf's law. Contemp. Phys. **46**(5), 323–351 (2005)
34. Nishimori, H.: Statistical Physics of Spin Glasses and Information Processing: An Introduction, vol. 111. Clarendon Press (2001)
35. Nonnenmacher, M., Behrens, C., Berens, P., Bethge, M., Macke, J.H.: Signatures of criticality arise from random subsampling in simple population models. PLoS Comput. Biol. **13**(10), e1005,718 (2017)
36. O'Donnell, C., Gonçalves, J.T., Whiteley, N., Portera-Cailliau, C., Sejnowski, T.J.: The population tracking model: a simple, scalable statistical model for neural population data. Neural Comput. (2016)
37. Ohiorhenuan, I.E., Mechler, F., Purpura, K.P., Schmid, A.M., Hu, Q., Victor, J.D.: Sparse coding and high-order correlations in fine-scale cortical networks. Nature **466**(7306), 617–621 (2010)
38. Panas, D., Amin, H., Maccione, A., Muthmann, O., van Rossum, M., Berdondini, L., Hennig, M.H.: Sloppiness in spontaneously active neuronal networks. J. Neurosci. **35**(22), 8480–8492 (2015)
39. Priesemann, V., Valderrama, M., Wibral, M., Le Van Quyen, M.: Neuronal avalanches differ from wakefulness to deep sleep–evidence from intracranial depth recordings in humans. PLoS Comput. Biol. **9**(3), e1002,985 (2013)
40. Redner, S.: How popular is your paper? An empirical study of the citation distribution. Eur. Phys. J. B Condens. Matter Complex Syst. **4**(2), 131–134 (1998)
41. Schneidman, E., Berry, M.J., Segev, R., Bialek, W.: Weak pairwise correlations imply strongly correlated network states in a neural population. Nature **440**(7087), 1007–1012 (2006)
42. Shew, W.L., Plenz, D.: The functional benefits of criticality in the cortex. The Neuroscientist **19**(1), 88–100 (2013)
43. Shew, W.L., Yang, H., Petermann, T., Roy, R., Plenz, D.: Neuronal avalanches imply maximum dynamic range in cortical networks at criticality. J. Neurosci. **29**(49), 15595–15600 (2009)
44. Shlens, J., Field, G.D., Gauthier, J.L., Grivich, M.I., Petrusca, D., Sher, A., Litke, A.M., Chichilnisky, E.: The structure of multi-neuron firing patterns in primate retina. J. Neurosci. **26**(32), 8254–8266 (2006)
45. Song, J., Marsili, M., Jo, J.: Emergence and relevance of criticality in deep learning (2017). arXiv preprint arXiv:1710.11324
46. Tang, A., Jackson, D., Hobbs, J., Chen, W., Smith, J.L., Patel, H., Prieto, A., Petrusca, D., Grivich, M.I., Sher, A., Hottowy, P., Dabrowski, W., Litke, A.M., Beggs, J.M.: A maximum entropy model applied to spatial and temporal correlations from cortical networks in vitro. J. Neurosci. **28**, 505518 (2008)
47. Tkačik, G., Marre, O., Amodei, D., Schneidman, E., Bialek, W., Berry II, M.J.: Searching for collective behavior in a large network of sensory neurons. PLoS Comput. Biol. **10**(1), e1003,408 (2014)
48. Tkačik, G., Mora, T., Marre, O., Amodei, D., Palmer, S.E., Berry, M.J., Bialek, W.: Thermodynamics and signatures of criticality in a network of neurons. Proc. Natl. Acad. Sci. **112**(37), 11508–11513 (2015)

49. Vázquez-Rodríguez, B., Avena-Koenigsberger, A., Sporns, O., Griffa, A., Hagmann, P., Larralde, H.: Stochastic resonance at criticality in a network model of the human cortex. Sci. Rep. **7**(1), 13,020 (2017)
50. Vitanov, N.K., Ausloos, M.: Test of two hypotheses explaining the size of populations in a system of cities. J. Appl. Stat. **42**(12), 2686–2693 (2015)
51. Yu, S., Yang, H., Nakahara, H., Santos, G.S., Nikolić, D., Plenz, D.: Higher-order interactions characterized in cortical activity. J. Neurosci. **30**(48), 17514–17526 (2011)
52. Zipf, G.K.: Human Behavior and the Principle of Least Effort. Addison-Wesley, Cambridge (1949)

Martino Sorbaro studied Physics at the University of Pavia, Italy, and has just completed his doctoral thesis in neuroinformatics at the University of Edinburgh and KTH, Stockholm. His Ph.D. research focused on statistical modelling of neural activity in large-scale recordings, from the spike sorting end, up to high-level modelling and function.

Matthias Hennig studied physics at Bochum University and received his Ph.D. in computational neuroscience at the University of Stirling in 2005. He is currently a Reader at the School of Informatics, University of Edinburgh. He employs statistical and computational modelling to investigate computations, stability and development in neuronal circuits. His group also develops tools for analysis of recordings from large neuronal populations enabled by novel technologies such as large scale, high density microelectrode arrays.

MIX
Papier aus verantwortungsvollen Quellen
Paper from responsible sources
FSC® C105338

If you have any concerns about our products,
you can contact us on
ProductSafety@springernature.com

In case Publisher is established outside the EU,
the EU authorized representative is:
Springer Nature Customer Service Center GmbH
Europaplatz 3, 69115 Heidelberg, Germany

Printed by Libri Plureos GmbH
in Hamburg, Germany